Algal Biorefinery

I0888250

This book enables readers to understand the theoretical aspects, key steps and scientific techniques with a detailed mechanism to produce biofuels from algae.

Each chapter provides the latest developments and recent advancements starting from algal cultivation techniques to the production of value-added green fuels, chemicals and products with wide applications. The volume brings together a broad range of international and interdisciplinary experts, including chemical and biological engineers, biotechnologists, process engineers, environmentalists, pharmacists and nutritionists, to one platform to explore the beneficial aspects and challenges for an algal-based biorefinery. Chapters address cutting-edge issues surrounding algal cultivation, including genetic modification of algal strains, design and optimization of photobioreactors and open-pond systems, algal oil extraction techniques and algal-derived fuel products (biodiesel, bio-gasoline, jet fuels and bio-oil). Finally, the book considers the potential environmental impacts for establishing a sustainable algal biorefinery through lifecycle analysis, techno-economic assessment and supply chain management.

This book will be an important resource for students, academics and professionals interested in algal cultivation, biofuels and agricultural engineering, and renewable energy and sustainable development more broadly.

Ajay K. Dalai is a Distinguished Professor and Canada Research Chair in the Department of Chemical and Biological Engineering at the University of Saskatchewan, Canada.

Vaibhav V. Goud is a Professor in the Department of Chemical Engineering at the Indian Institute of Technology, India.

Sonil Nanda is a Research Associate in the Department of Chemical and Biological Engineering at the University of Saskatchewan, Canada.

Venu Babu Borugadda is a Postdoctoral Research Fellow in the Department of Chemical and Biological Engineering at the University of Saskatchewan, Canada.

Routledge Studies in Bioenergy

Handbook of Bioenergy Crops
A Complete Reference to Species, Development and Applications
N. El Bassam

Biofuels, Food Security, and Developing Economies
Nazia Mintz-Habib

Bioenergy Crops for Ecosystem Health and Sustainability
Alex Baumber

Biofuels and Rural Poverty
Joy Clancy

Sugarcane Bioenergy for Sustainable Development
Expanding Production in Latin America and Africa
Edited by Luis A. B. Cortez, Manoel Regis L. V. Leal and Luiz A. Horta Nogueira

Miscanthus for Bioenergy Production
Crop Production, Utilization and Climate Change Mitigation
Michael B. Jones

Algal Biorefinery
Developments, Challenges and Opportunities
Edited by Ajay K. Dalai, Vaibhav V. Goud, Sonil Nanda and Venu Babu Borugadda

For further information about this series, please visit www.routledge.com/Routledge-Studies-in-Bioenergy/book-series/RSBIOEN

Algal Biorefinery

Developments, Challenges and Opportunities

Edited by
Ajay K. Dalai,
Vaibhav V. Goud,
Sonil Nanda and
Venu Babu Borugadda

Routledge
Taylor & Francis Group
LONDON AND NEW YORK

First published 2022
by Routledge
2 Park Square, Milton Park, Abingdon, Oxon OX14 4RN

and by Routledge
605 Third Avenue, New York, NY 10158

Routledge is an imprint of the Taylor & Francis Group, an informa business

British Library Cataloguing-in-Publication Data
A catalog record for this book is available from the British Library

Library of Congress Cataloging-in-Publication Data
A catalog record has been requested for this book

ISBN: 978-0-367-57005-7 (hbk)
ISBN: 978-0-367-57006-4 (pbk)
ISBN: 978-1-003-10031-7 (ebk)

Typeset in Bembo
by codeMantra

Contents

Figures

Tables

About the editors

Dr. Ajay K. Dalai

Dr. Ajay K. Dalai is a Distinguished Professor of Chemical Engineering and Canada Research Chair in Bioenergy and Environmentally Friendly Chemical Processing in the Department of Chemical and Biological Engineering at the University of Saskatchewan. He received his Ph.D. in Chemical Engineering from the University of Saskatchewan (1990); M.Tech. in Chemical Engineering from the Indian Institute of Technology Kanpur, India (1984); B.Sc. Tech. in Petro-Chemical Engineering from Laxminarayan Institute of Technology, Nagpur, India (1982); and B.Sc. in Chemistry from S.C.S. College, Utkal University, India (1979). After obtaining his experiences as a Postdoctoral Fellow at Texas A&M University in Texas and as a Research Associate at the University of Calgary in Canada, he joined the University of Saskatchewan as an Assistant Professor in 1996. He was promoted to Associate Professor in 1998 and Full Professor in 2002. He was promoted to Distinguished Professor at the University of Saskatchewan in 2020. Dr. Dalai's research focus is the novel catalyst development for gas-to-liquid technologies, biodiesel production, hydrogen/syngas production, hydroprocessing, value-added products from biomass, and pollution control. He has supervised over 150 M.Sc., Ph.D. and postdoctoral fellows and has received over $35 million for his research funding over the past 25 years. Dr. Dalai has published over 550 research papers mostly in heterogeneous catalysis and catalytic processes in international journals and conference proceedings. His groundbreaking research has led to over 28,000 citations of his work and an H-index of 78. He serves on the Editorial Board of several prestigious journals. Dr. Dalai has received fellowships from different societies such as the Royal Society of Canada, Royal Society of Chemistry in the UK, International Association of Advanced Materials, American Institute of Chemical Engineering, Canadian Academy of Engineering, Chemical Institute of Canada, Engineering Institute of Canada and Indian Institute of Chemical Engineers. Dr. Dalai is also a Fulbright Fellow from the USA, DAAD Fellow from Germany and JSPS Fellow from Japan. He has received several national and international awards. Recently, he received the Indian Institute of Chemical Engineers DRC Life Time Achievement Award, Canadian Society of Chemical Engineering R.S. Jane Memorial Award for outstanding contributions to the Chemical Engineering Profession, Royal Society of Canada's Miroslaw Romanowski Medal for Outstanding Contribution to Environmental Science, and University of Saskatchewan's George Ivany Internationalization Award.

Dr. Vaibhav V. Goud

Dr. Vaibhav V. Goud is a Professor in the Department of Chemical Engineering at the Indian Institute of Technology Guwahati, Assam, India. He received his Ph.D. degree from the Indian Institute of Technology Kharagpur in 2006. He worked as Lecturer at the Birla Institute of Technology and Science Pilani, Goa Campus, from 2006 to 2007, and as a postdoctoral fellow at the University of Saskatchewan, Canada, from 2008 to 2009. Dr. Goud's principal research interests are in the fields of processing of non-edible oilseeds, biofuels, bio-lubricants and oleo-chemicals. He also works on the extraction of natural plant products such as essential oils, oleo-resins, food colors, biopesticides, nutraceuticals and cosmeceuticals by the application of subcritical and supercritical fluids. He has published more than 120 papers in international peer-reviewed journals and 12 book chapters and presented his research at several national and international conferences. Dr. Goud was awarded the Rapid Grants for Young Investigators (RGYI) Scheme from the Department of Biotechnology, Government of India. He also received the Visiting Scientist Award from the Universidade Federal do Rio de Janeiro, Brazil, in 2018 for a Department of Biotechnology-sponsored Indo-Brazil project. He has active research collaborations with the University of Saskatchewan (Canada), Gifu University (Japan), Federal Universidade Federal do Rio de Janeiro (Brazil) and National Sun Yat-sen University (Taiwan). Dr. Goud is a life member of the Indian Institute of Chemical Engineers, Seabuckthorn Association of India and the International Society of Food Engineering.

Dr. Sonil Nanda

Dr. Sonil Nanda is a Research Associate at the University of Saskatchewan in Saskatoon, Saskatchewan, Canada. He received his Ph.D. degree in Biology from York University, Canada; M.Sc. degree in Applied Microbiology from Vellore Institute of Technology (VIT University), India; and B.Sc. degree in Microbiology from Orissa University of Agriculture and Technology, India. Dr. Nanda's research areas are related to the production of advanced biofuels and biochemical through thermochemical and biochemical conversion technologies such as gasification, pyrolysis, carbonization, torrefaction and fermentation. He has gained expertise in hydrothermal gasification of various organic wastes and biomass including agricultural and forestry residues, industrial effluents, municipal solid wastes, cattle manure, sewage sludge, food wastes, waste tires and petroleum residues to produce hydrogen fuel. His parallel interests are also in the generation of hydrothermal flames for the treatment of hazardous wastes, agronomic applications of biochar, phytoremediation of heavy metal contaminated soils, and carbon capture and sequestration. Dr. Nanda has published over 100 peer-reviewed journal articles and more than 50 book chapters and has presented at many international conferences. He is the editor of books entitled '*New Dimensions in Production and Utilization of Hydrogen*' (Elsevier), '*Recent Advancements in Biofuels and Bioenergy Utilization*' (Springer Nature), '*Biorefinery of Alternative Resources: Targeting Green Fuels and Platform Chemicals*' (Springer Nature), '*Fuel Processing and Energy Utilization*' (CRC Press), '*Bioprocessing of Biofuels*' (CRC Press) and '*Biotechnology for Sustainable Energy and Products*' (I.K. International Publishing House Pvt. Ltd.). Dr. Nanda serves as a Fellow Member of the *Society for Applied Biotechnology* in India and as a life member of the *Indian Institute of Chemical Engineers, Association of Microbiologists of India, Indian Science Congress Association* and the *Biotech Research Society of India*. He is also an active member of several chemical engineering societies across North America such as the *American Institute of Chemical Engineers*, the *Chemical Institute of Canada*, the *Combustion Institute – Canadian Section* and *Engineers Without Borders Canada*. Dr. Nanda is an Assistant Subject Editor of the *International Journal of Hydrogen Energy* (Elsevier) and an Associate Editor of *Environmental Chemistry Letters* (Springer Nature) and *Applied Nanoscience* (Springer Nature). He has also edited several special issues in renowned journals such as the *International Journal of Hydrogen Energy* (Elsevier), *Chemical Engineering Science* (Elsevier), *Biomass Conversion and Biorefinery* (Springer Nature), *Waste and Biomass Valorization* (Springer Nature), *Topics in Catalysis* (Springer Nature), *SN Applied Sciences* (Springer Nature) and *Chemical Engineering & Technology* (Wiley).

Dr. Venu Babu Borugadda

Dr. Venu Babu Borugadda is a MITACS Postdoctoral Fellow in the Department of Chemical and Biological Engineering at the University of Saskatchewan, Canada. Before joining the University of Saskatchewan, he was an Assistant Professor in the Department of Chemical Engineering at the National Institute of Technology Calicut, Kerala. He received his Ph.D. in Chemical Engineering from the Indian Institute of Technology Guwahati, India, in 2016. He received his Master of Technology degree with Petroleum Refinery Engineering specialization from the Indian Institute of Technology Guwahati in 2011. He has 10 years of research experience in biomaterial processing, bio-lubricants and the production of biofuels and biochemicals. Currently, the focus of his research is on the production of advanced biofuels through hydrothermal liquefaction, catalysis and chemical reaction engineering, as well as techno-economic and lifecycle analyses. He has published more than 25 research articles in peer-reviewed journals, 8 book chapters and more than 30 conference proceedings. He has been serving as a reviewer for many international journals such as *Fuel* (Elsevier), *Journal of Cleaner Production* (Elsevier), *International Journal of Hydrogen Energy* (Elsevier), *Industrial Crops and Products* (Elsevier), *Chemical Engineering Communications* (Taylor & Francis Group) and *International Journal of Industrial Chemistry* (Springer Nature).

Contributors

Hanieh Bamdad
Department of Chemical and Biochemical Engineering
Western University
London, Ontario, Canada

Rabindra Kangsha Banik
Centre for Energy
Indian Institute of Technology Guwahati
Guwahati, Assam, India

Franco Berruti
Department of Chemical and Biochemical Engineering
Western University
London, Ontario, Canada

Doljit Borah
Department of Energy
Tezpur University
Tezpur, Assam, India

Venu Babu Borugadda
Department of Chemical and Biological Engineering
University of Saskatchewan
Saskatoon, Saskatchewan, Canada

Ajay K. Dalai
Department of Chemical and Biological Engineering
University of Saskatchewan
Saskatoon, Saskatchewan, Canada

Velentina Das
Department of Energy
Tezpur University
Tezpur, Assam, India

Dhanapati Deka
Department of Energy
Tezpur University
Tezpur, Assam, India

Nongmaithem Debeni Devi
Centre for Energy
Indian Institute of Technology Guwahati
Guwahati, Assam, India

Khalifa S.H. Eldiehy
Department of Molecular Biology and Biotechnology
Tezpur University
Tezpur, Assam, India

Anjani Ravi Kiran Gollakota
Department of Safety, Health, and Environmental Engineering
National Yunlin University of Science and Technology
Douliou, Taiwan, China

Vaibhav V. Goud
Department of Chemical Engineering
Indian Institute of Technology Guwahati
Guwahati, Assam, India

Nidhi Hans
Centre for Rural Development and Technology
Indian Institute of Technology New Delhi
New Delhi, India

Bo Hu
Department of Bioproducts and Biosystems Engineering
University of Minnesota
Minneapolis, Minnesota, United States

Yulin Hu
Department of Chemical and Biochemical Engineering
Western University
London, Ontario, Canada

Pankaj Kalita
Centre for Energy
Indian Institute of Technology Guwahati
Guwahati, Assam, India

Kang Kang
Department of Chemical and Biochemical Engineering
Western University
London, Ontario, Canada

Naga Chaitanya Kavuri
Department of Civil Engineering
Koneru Lakshmaiah Education Foundation
Guntur, Andhra Pradesh, India

Dipesh Kumar
Centre for Energy
Indian Institute of Technology Guwahati
Guwahati, Assam, India

Vivek Kumar
Centre for Rural Development and Technology
Indian Institute of Technology New Delhi
New Delhi, India

Anushree Malik
Centre for Rural Development and Technology
Indian Institute of Technology New Delhi
New Delhi, India

Manabendra Mandal
Department of Energy
Tezpur University
Tezpur, Assam, India

Devendra Kumar Maravi
Center for Energy
National Institute of Technology Arunachal Pradesh
Yupia, Itanagar, Arunachal Pradesh

Shima Masoumi
Department of Chemical and Biological Engineering
University of Saskatchewan
Saskatoon, Saskatchewan, Canada

Satyanarayan Naik
Centre for Rural Development and Technology
Indian Institute of Technology New Delhi
New Delhi, India

Sonil Nanda
Department of Chemical and Biological Engineering
University of Saskatchewan
Saskatoon, Saskatchewan, Canada

Sharmila Narayanan
Department of Oncology
University of Saskatchewan
Saskatoon, Saskatchewan, Canada

Sadegh Papari
Department of Chemical and Biochemical Engineering
Western University
London, Ontario, Canada

Falguni Pattnaik
Centre for Rural Development and Technology
Indian Institute of Technology New Delhi
New Delhi, India

Santhi Raju Pilli
Department of Chemical Engineering Technology
Jazan University
Jazan, Saudi Arabia

Ali Shemsedin Reshad
Department of Chemical Engineering
Addis Ababa Science and Technology University
Addis Ababa, Ethiopia

Chi-Min Shu
Department of Safety, Health, and Environmental Engineering
National Yunlin University of Science and Technology
Douliou, Taiwan, China

Ajeet Singh
Government Adarsh Girls College
Sheopur, Madhya Pradesh, India

Garima Srivastava
Department of Chemical Engineering
Indian Institute of Technology Guwahati
Guwahati, Assam, India

Vikranth Volli
Department of Safety, Health, and Environmental Engineering
National Yunlin University of Science and Technology
Douliou, Taiwan, China

Chunbao Charles Xu
Department of Chemical and Biochemical Engineering
Western University
London, Ontario, Canada

Abbreviations

MPTP	1-Methyl-4-phenyl-1,2,3,6-tetrahydropyridine
5-FU	5-Fluorouracil
6-OHDA	6-Hydroxydopamine
ACh	Acetylcholine
AChE	Acetylcholinesterase
ARG9	Acetylornithine aminotransferase
AD	Alzheimer's disease
aadA	Aminoglycoside 3'-adenyltransferase
aphA	Aminoglycoside phosphotransferase gene aphVIII
AD	Anaerobic digestion
Abs	Antibodies
Anti-IAV	Anti-influenza A virus
AC	Ash Content
AMV	Avian myeloblastosis virus
Bcl-2	B-cell lymphoma 2
BOD	Biochemical oxygen demand
LFB	Bovine lactoferricin
BChE	Butyrylcholinesterase
CO_2	Carbon dioxide
CHG	Catalytic hydrothermal gasification
CaMV35S	Cauliflower mosaic virus
C/EBPα	CCAAT/enhancer-binding protein alpha
CNS	Central nervous system
CF-PLS	*Champia feldmannii* sulfated polysaccharides
CTB	Cholera toxin B subunit
CHP	Combined heat and power
CSP	Concentrated solar power
CST	Concentrated solar thermal
CF-PBR	Cyclic-flow photobioreactor
COX-2	Cyclooxygenase-2
DENV	Dengue virus
DNA	Deoxyribonucleic acid
DHA	Docosahexaenoic acid

DPA	Docosapentaenoic acid
EPA	Eicosapentaenoic acid
ESA	*Eucheuma serra* agglutinin
FT	Fischer–Tropsch
FC	Fixed Carbon
FPA-PBR	Flat-panel airlift photobioreactor
FDA	Food and Drug Administration
GGE	Gasoline gallon equivalent
GAD65	Glutamic acid decarboxylase-65
GBSS	Granule-bound starch synthase
GHG	Greenhouse Gas
HBcAg	Hepatitis B core antigen
HBsAg	Hepatitis B virus surface antigen
HBV	Hepatitis B virus
HSV	Herpes simplex virus
HDL	High-density lipoprotein
HHV	Higher Heating Value
HCMV	Human cytomegalovirus
HIV-1	Human immunodeficiency virus
HRV	Human rhinovirus
HDO	Hydrodeoxygenation
HTL	Hydrothermal liquefaction
iNOS	Inducible nitric oxide synthase
IFN-γ	Interferon-gamma
IRR	Internal rate of return
LCOF	Levelized cost of fuel
LPS	Lipopolysaccharide
LDL	Low-density lipoprotein
MCR	Macroalgae cellulosic residue
MAP	Microwave-assisted pyrolysis
MFSP	Minimum fuel selling price
MAPK	Mitogen-activated protein kinase
MC	Moisture Content
mAb	Monoclonal antibody
MCP-1	Monocyte chemotactic protein-1
NP-1	Neutrophil peptide-1
NIT1	Nitrate reductase
NO	Nitric oxide
NRP	Non-ribosomal peptide
NF-κB	Nuclear factor-κB
DHTL	One-stage direct hydrothermal liquefaction
ORP	Open raceway pond
PD	Parkinson's disease
PPARγ	Peroxisome proliferator-activated receptor gamma
PBR	Photobioreactor

AppA	Phytase
PDS	Phytoene desaturase
PbAMA1	*Plasmodium berghei* apical major antigen
PbMSP1	*Plasmodium berghei* major surface protein
PfAMA1	*Plasmodium falciparum* apical major antigen
PfMSP1	*Plasmodium falciparum* major surface protein
PUFAs	Polyunsaturated fatty acids
PGE2	Prostaglandin E2
RP	Raceway pond
RMLV	Rauscher murine leukemia virus
RNS	Reactive nitrogen species
ROS	Reactive oxygen species
rbcL	Ribulose bisphosphate carboxylase/oxygenase large subunit
RABR	Rotating algal biofilm reactor
RbcS2	Rubisco small subunit
SCWG	Supercritical water gasification
SNG	Synthetic natural gas
TEA	Techno-economic analysis
TMV	Tobacco mosaic virus
TB	Tuberculosis
TNF-α	Tumor necrosis factor-α
SEQHTL	Two-stage sequential hydrothermal liquefaction
Ubi1	Ubiquitin 1
UV	Ultraviolet
VSV-G	Vesicular stomatitis virus glycoprotein
VP28	Viral protein 28
VM	Volatile Matter
Aβ	β-Amyloid

Preface

The most pragmatic socio-environmental concerns of the modern world are pollution, global warming, climate change, energy demands, food quality and public health. Since the time of the industrial revolution, the most attractive source of energy for transportation as well as heat and power generation has been fossil fuels, especially gasoline, diesel, coal and natural gas. It is well recognized that the increased usage of fossil fuel results in massive emissions of greenhouse gases leading to global warming and climate change. Rapid industrialization and urbanization are also the chief causes for the pollution of air, water and soil. The deterioration of the environment has posed risks and threats to public health and the shifting of natural habitats for animals and plants. To address such alarming consequences, it is essential to seek alternative energy sources and natural agents that can remediate environmental pollution in an innocuous manner. Algae are of great importance as a biological material with numerous applications in such directions.

Algae are a group of photosynthetic eukaryotic organisms capable of scrubbing environmental pollutants including CO_2, produce lipids for conversion to biofuels and synthesize metabolites as precursors for a broad range of bioproducts. Algae can remediate wastewater and polluted air while fixing CO_2 from the atmosphere through photosynthesis. This makes algae a biological agent for carbon capture and sequestration. The lipids in algae can be extracted through mechanical, physicochemical, thermochemical and hydrothermal technologies like biofuels and biochemicals. The algal metabolites can be used as value-added food supplements, nutraceuticals, pharmaceuticals, therapeutic drugs and other bioactive products. Considering the versatility of algae, it is highly imperative to shed light on the aspects of metabolism, cultivation, biorefining, industrial applications, techno-economics and lifecycle studies. These are the important and timely scientific works that this book broadly covers.

Chapter 1 by Kang et al. reviews the aspects of algal metabolism, its physicochemical characteristics and research and development on high-lipid-containing algal strains. An introductory summary is made on the conversion of algae through physicochemical, thermochemical and biological technologies. The applications of algae in carbon sequestration and wastewater remediation are discussed.

Chapter 2 by Devi et al. provides insights into the efficient and large-scale production of microalgae and their economic assessment. The chapter highlights the key characteristics, commercialization, advantages and disadvantages of different microalgal cultivation systems including photobioreactors.

Chapter 3 by Eldiehy et al. provides a comprehensive discussion of different classical and modern approaches to induce lipid production in microalgae with enhancing recovery. The chapters shed light on enhancing microalgal lipid production through manipulating nutrient regime, culture conditions, controlling stress factors, additives, nanoparticles, biocatalytic enzymes and metabolic engineering strategies.

Chapter 4 by Srivastava et al. discusses the prospects of biodiesel production from algae through catalyst-assisted transesterification reactions. The chapter also comprehensively describes the factors affecting transesterification of algal oil such as polar lipids, moisture, free fatty acids, reaction temperature, reaction time, alcohol/oil molar ratio, co-solvents, storage conditions and catalysts.

Chapter 5 by Reshad and Borugadda describes the thermochemical conversion technologies of algal biomass for bio-oil production. This chapter reviews the literature on the organic and inorganic compositions of various macroalgal and microalgal species followed by their thermochemical conversion to produce bio-oil.

Chapter 6 by Hu and Xu summarizes the most recent advances in fast pyrolysis and hydrothermal liquefaction of algae to produce high-value bio-oil and bio-crude oil. The recently developed methods for lipid extraction from algae including microwave-, ultrasound- and supercritical fluid-based extraction approaches have been thoroughly discussed.

Chapter 7 by Kalita and Banik discusses the hydrothermal conversion techniques such as liquefaction, carbonization and gasification of algae to produce process-specific biofuel products such as bio-crude oil, hydrochar and syngas. The importance of operating parameters and catalysts and their impact on the biofuel yield are discussed.

Chapter 8 by Volli et al. describes different routes for the production of algal biofuels. The production of algal-delivered bio-crude oil, its properties, extraction techniques and limitations are discussed in detail. Besides, the upgrading of algal bio-crude oil via catalytic cracking, the chapter also discusses on hydroprocessing, supercritical fluids and bio-jet fuels.

Chapter 9 by Hans et al. provides a detailed overview of numerous nutraceuticals of microalgal and macroalgal origin such as dietary supplements, vitamins, carotenoids, fatty acids and other novel products. The chapter describes the biochemical properties of the algal nutraceuticals along with their health benefits and challenges such as bioavailability and biomagnification of toxic molecules through the food chain associated.

Chapter 10 by Narayanan and Pilli provides an overview of several pharmaceuticals of algal origin such as antimicrobial, antiviral, antifungal and neuroprotective products, therapeutic proteins and drugs. The chapter covers the biological properties of algal pharmaceuticals along with their therapeutic effects, challenges and applications.

Chapter 11 by Masoumi et al. discusses the techno-economic analysis for algal-derived biofuels, biochemicals, antioxidants, fatty acids, flavonoids, carbohydrates, fertilizers, wastewater phycoremediation, fish and animal feed applications and bioplastics. Moreover, the challenges in the commercialization of algal bioproducts are identified to improve their processing technologies and marketability.

Chapter 12 by Kumar and Goud highlights the attractiveness of different processing technologies for the conversion of algae to biofuels. The chapter covers the identification of hotspots in the algal biomass supply chain, technological aspects for algal conversion to clean fuels and chemicals, and lifecycle assessment of conversion technologies.

This book provides a comprehensive synopsis of the abovementioned topics for applications of algae in several cross-disciplinary areas of biotechnology, bioprocess engineering, chemical engineering, environmental sciences, food science and pharmacy. We are grateful to all the authors for timely contributing their high-quality manuscripts to develop this book. Moreover, we sincerely hope that this book will immensely benefit the students and researchers working on sustainable energy and bioproduct production and applications. Finally, we express our sincere thanks to Mr. John Baddeley from Routledge for his enthusiastic assistance and support while developing this book.

Dr. Ajay K. Dalai
Distinguished Professor and Canada Research Chair
Department of Chemical and Biological Engineering
University of Saskatchewan
Saskatoon, Saskatchewan Canada
ORCID: https://orcid.org/0000-0003-3907-6232

Dr. Vaibhav V. Goud
Professor
Department of Chemical Engineering
Indian Institute of Technology Guwahati
Guwahati, Assam, India
ORCID: https://orcid.org/0000-0001-7755-6451

Dr. Sonil Nanda
Research Associate
Department of Chemical and Biological Engineering
University of Saskatchewan
Saskatoon, Saskatchewan Canada
ORCID: https://orcid.org/0000-0001-6047-0846

Dr. Venu Babu Borugadda
MITACS Postdoctoral Fellow
Department of Chemical and Biological Engineering
University of Saskatchewan
Saskatoon, Saskatchewan Canada
ORCID: https://orcid.org/0000-0002-8887-3572

1 Algae as a bioresource for clean fuels, carbon fixation and wastewater reclamation

Kang Kang, Sadegh Papari, Hanieh Bamdad, Sonil Nanda, Ajay K. Dalai, and Franco Berruti

Introduction

Energy is necessary for the progression of human society. The high consumption of fossil energy sources, e.g., petroleum, coal and natural gas, has created serious issues regarding the future energy supply and environmental sustainability and safety. Although fossil energy is still the primary contributor to the global energy portfolio with nearly 80% of all energy sources, tremendous innovations achieved in alternative energy and increasingly serious environmental concerns have stimulated considerable interest in the exploration and exploitation of renewable energy sources (Zou et al., 2016). Biomass as an alternative energy source is playing a critical role in mitigating global warming (Jin et al., 2019). Due to its abundance and carbon neutrality, biomass shows great potential as an energy resource in social, economic and environmental terms.

After years of development, the first generation of biofuels produced using edible feedstocks, such as grains, seeds, plant oils, sugar and starch, had been demonstrated to have limitations in sustainability and efficiency (Mohr and Raman, 2013). The second-generation biofuels are based on more efficient renewable feedstocks, such as inedible lignocellulosic biomass, including agroforestry residues, sawdust and municipal solid wastes, which overcomes the drawbacks of the first generation (Nanda and Berruti, 2021a, 2021b; Okolie et al., 2021). However, more research efforts are required to achieve a competitive cost (Alalwan et al., 2019). Land-intensive bioenergy is limited by the constraint of land area and often leads to high carbon emissions due to land-use change, biomass production, harvesting and transportation (Reid et al., 2020). Therefore, land-intensive bioenergy functions reasonably well as a transitional element of the global energy matrix, playing an important role over the next few decades, and then might fade in the long run. While optimizing feedstock cultivation, conversion technology and utilization options of the second-generation biofuels, research interests in third-generation biofuels are now booming (Figure 1.1).

There are several benefits of considering algae as the better biomass feedstock. Firstly, algae are fast-growing organisms that can transform

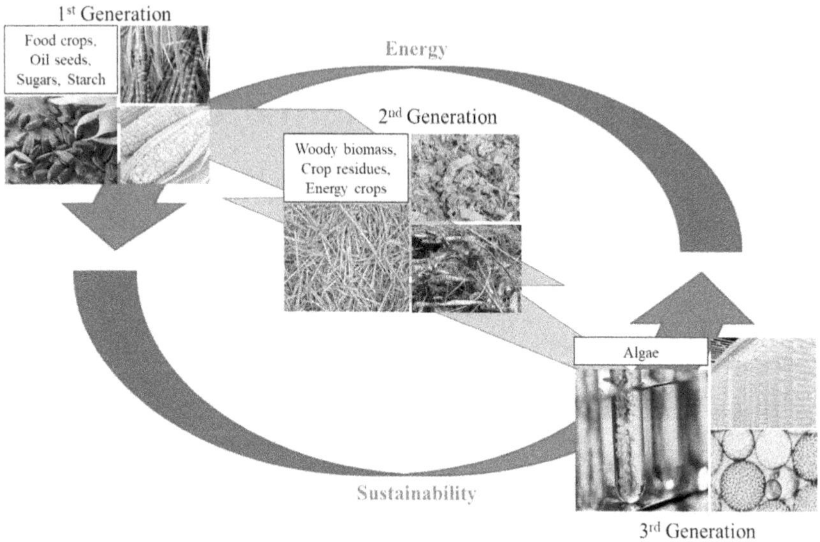

Figure 1.1 Feedstock for different generations of biofuels.

one-tenth of the solar energy into biomass, and the theoretical yield can reach approximately 77 g of biomass/m^2/day under the conditions of average sunlight irradiance (Khan et al., 2018). It was reported that nearly 13 kg dry weight/m^2 of brown algae could be produced within 7 months under the planting conditions that could produce sugarcane at the rate of 6.1–9.5 kg fresh weight m^2/year (Kraan, 2013). Secondly, as photosynthetic organisms, algae accumulate energy via the photosynthesis process, so the utilization of this energy under ideal conditions is not expected to increase the gross level of CO_2 in the atmosphere. In addition, compared to lignocellulosic biomass, algae could be cultivated within freshwater or saltwater bodies and can grow on infertile or marginal lands. Therefore, they are significantly less land-intensive. Since approximately 71% of the earth's surface is water-covered, algae win a tremendous edge in terms of location for cultivation. Moreover, the production of algae does not compete with land for cultivating food crops, and algae could efficiently utilize the water and fertilizers and are capable of producing high oil and biomass yields (Jones and Mayfield, 2012).

Algal biomass is a highly versatile feedstock and can be converted into various types of biofuels, such as bioethanol, biodiesel, bio-oil, biogas and biochar, via a range of biological and thermochemical conversion techniques (Rastogi et al., 2018). Other strategies, e.g., catalytic treatments, are also being investigated to convert algal biomass. To produce the algal biomass, both commercial and research activities are globally booming toward commodity algal-based products. Algae have attracted significant interest recently for their potential commercial utilization in the production of biofuels and nutritional supplements (Chang et al., 2011).

Currently, the most feasible way of the utilization of algal biomass is the conversion into liquid fuels with comparable fuel quality to conventional diesel. In combination with innovative techniques of using different types of solid wastes as fertilizers, the productivity and cost-effectiveness of algal-based liquid fuels will continue to improve (Campbell, 2008). Currently, the advantages of algal-based fuels have not been fully proven to justify large-scale algal cultivations, since the production costs are still high (Vassilev and Vassileva, 2016). Besides, the high water content causes low energy density of the un-processed algal biomass and increases the capital and operating costs for dewatering and transportation of the raw algae. In light of this, metabolic engineering might likely become a key solution in strain development to optimize the solar conversion efficiency of algae and form the basis for a fourth-generation biofuel production (Lü et al., 2011).

To comprehensively discuss the utilization of algae as a renewable energy and materials source, this chapter introduces the metabolism and physicochemical characteristics of algae. Furthermore, the progress made in converting algae into clean fuels using physicochemical, thermochemical and biological technologies is discussed. Finally, the advancements of using algae in environmental applications, such as biological carbon fixation and wastewater treatment, are reviewed.

Algal metabolism and high-lipid-containing algal strains

Algal metabolism

Algae are eukaryotic organisms, which belong to the Protista kingdom. Their cells consist of a nucleus and other organelles encapsulated in membranes. Algae can be categorized as macroalgae and microalgae. Macroalgae are multicellular and relatively larger in size compared to the unicellular microscopic microalgae. Based on their pigmentation, macroalgae could be classified into three broad groups, including red seaweed (*Rhodophyceae*), brown seaweed (*Phaeophyceae*) and green seaweed (*Chlorophyceae*) (Demirbas, 2010). Microalgae are a diverse group of aquatic organisms that include more than 50,000 photosynthetic types of microorganisms that can perform photosynthesis, reproduce quickly and survive in difficult growing environments (Adenle et al., 2013). The unicellular green algae *Chlamydomonas* have a simple lifecycle and provide easy isolation of mutants and a growing array of technologies for molecular genetic research (Harris, 2001).

During the growth phase, algae can adapt in both freshwater and saltwater. In addition, excellent flexibility allows them to live under different cultivation conditions. Microalgae cultivation with sunlight as an energy source can be carried out in ponds or photobioreactors (PBRs) designed according to tubular, flat plate or other configurations (Demirbas, 2010). Nutrient supply can also be achieved through the circulation of water from neighboring lands or by feeding the water from local wastewater treatment plants. This enables certain types of algae to be cultivated in contaminated water followed by their use as a feedstock for lipid extraction or conversion to biofuels. As proposed by De Bhowmick et al. (2019), a 'zero waste discharge' can be achieved with the simultaneous production of biofuel, biochar and other high-value products while keeping the greenhouse gases' impacts at very low levels.

Algae produce their nourishment via the photosynthesis process, during which the solar energy and atmospheric CO_2 are transformed into carbohydrates and oxygen. For their cultivation, CO_2 sources could be effluent from power plants or chemicals in different forms, e.g., soluble carbonates, including Na_2CO_3 and $NaHCO_3$ (Brennan and Owende, 2010). A deeper understanding of the light-absorbing mechanism and its effect on cellular systems facilitates the basic scientific research and commercialization of the photosynthetic organisms.

High-lipid-containing algal strains

The suitability of algae as fuel precursors is determined by many critical factors. The evaluation can be performed using parameters including growth rate, photosynthetic yield, biomass productivity and carbohydrate content (Sirajunnisa and Surendhiran, 2016). Due to the huge diversity, the screening of algae is one of the most important tasks to assess potential fuel production. Some of the processes used for lipid extraction from algae include mechanical methods, ultrasound, microwave, pulsed-electric field, surfactants, organic solvents, supercritical fluids (SCF), ionic liquids and deep eutectic solvents. The recovery and purification of lipids from algal cells via a few promising technologies could be a barrier due to substantial cost and energy requirement.

Lipid extraction from wet algae involves the disruption of the algal cell walls in the cultivation solution, and such process can be grouped into organic solvent-based and solvent-free approaches (Naghdi et al., 2016). During the solvent-based lipid extraction procedure, water acts as a barrier between the intracellular lipids and nonpolar organic solvents; therefore, the efficiency could be enhanced by increasing the polarity of the solvent. To deal with potential safety and environmental concerns, organic solvent-free methods have been developed and those that can be used for various algal strains with reasonable energy consumption and infrastructure are highly desirable. Some of the recently published results are given in Table 1.1.

Algae for biofuels

Biochemical and thermochemical techniques are typically used to convert microalgal biomass into various fuels and chemicals. In the biochemical method, large biomass molecules break down to shorter molecules using microorganisms or enzymes, while, in the thermochemical methods, biomass is degraded using heat (Tekin et al., 2014). Thermochemical techniques offer the advantages of being fast and do not require a high level of pretreatment.

Table 1.1 Summary of lipid yields from different algal strains

Strain	*Cultivation conditions*	*Lipid content (% of dry cell weight)*	*References*
Chlamydomonas reinhardtii (CC-124)	TAP medium, light from fluorescent lamps of 100 μmol photons/m^2/s at 26°C, cells exposed to 1 mM of colcemid	1.8 ± 0.1	Kwak et al. (2017)
Chlorella protothecoides (IOAC038F)	Cultivation using solid medium containing 1.3% agar plus L1 medium for marine samples, or MCM medium for freshwater samples. The solid plates were incubated at 25 ± 1°C using a 14:10 light:dark cycle at 20 μmol photon m^2/s.	20 ± 1.5	Li et al. (2015)
Scenedesmus sp. (IOAC744S)		20 ± 1.4	
Nannochloropsis sp. (IOAC759S)		22 ± 1.5	
Isochrysis galbana (IOAC724S)		42 ± 0.8	
Phaeodactylum tricornutum (IOAC742S)		21.2 ± 0.6	
Chlorella sp. (LS11)	Ten days of enrichment followed by streaked cells with a loop onto 1.5% agar BG-11 medium. The purified strains were cultured in conical flasks in a PBR at 80 μmol/m^2/s at 120 rpm for 10 days.	21.5 ± 2.1	Hena et al. (2015)
Botryococcus braunii (LS45)		42 ± 2.7	
Scenedesmus sp. (LS40)		29.5 ± 1.8	
Amphora sp. (Persian Gulf)	Samples enriched using 0.5 M Kent Marine salt at ratios of 1:10–1:100. They were then incubated in an adhesive 25-mL culture flask at 20°C under constant 3-klux photon flux of white and red LED lights.	24 ± 2.3	Talebi et al. (2013)
Dunaliella salina (UTEX)		24 ± 1.3	
Dunaliella sp. (Persian Gulf)		22 ± 2	
Chlorella pyrenoidosa (FACHB-9)	The microalgae were grown on the selenite enrichment medium, the pH was adjusted to 7.0–7.2, and mixed anaerobically digested and secondary treated starch wastewater was used	28 ± 1.7	Tan et al. (2019)

Mechanical pressing

Recently, researchers' attention has been drawn to microalgae oil due to the countless applications of this valuable product. Solvent extraction, enzymatic extraction and supercritical fluid extraction (SFE), along with mechanical milling, are typical technologies used to separate the microalgal oil from biomass. This value-added oily product has multiple applications in pharmaceuticals, nutraceuticals, cosmeceuticals and bioenergy (Mercer and Armenta, 2011; Olaizola, 2003). To find the 'best' oil extraction process for the commercial-scale production of oil products, economic and environmental aspects should be considered. Of the aforementioned technologies, mechanical milling produces fewer by-products, minimizes contamination and maintains the original chemical structure, making this process attractive (Greenwell et al., 2010). Pressing is the introduction of high pressure to the microalgal biomass, which ruptures biomass cells and releases the oil. In the homogenization step, biomass passes through an orifice, resulting in a sudden pressure change. Bead milling is used to recover oil more efficiently and wisely (Doucha and Lívanský, 2008). It has been reported that microalgal pretreatments, including acid wash and enzymatic process, significantly enhance the cell breakage during the mechanical milling process (Olaizola, 2003).

Hydrothermal liquefaction

Hydrothermal liquefaction (HTL) is a thermochemical process designed to convert a wide variety of biomass materials (e.g., wood, grass and algae) and wastes (e.g., manure, sewage sludge, bio-solid and digestate) at operating conditions at or below the critical point of water (374°C and 22.1 MPa). The process is composed of three main reactions, namely biomass depolymerization, monomer degradation and polymerization. The monomer degradation involves reactions such as dehydration, decarboxylation, bond cleavage and deamination. HTL produces bio-crudes, i.e., a mixture of aldehydes, phenols, alkenes, fatty acids, esters, aromatics and ketones, gaseous products composed predominantly of CO_2, H_2, CH_4 and C_2–C_3 hydrocarbons, and solid-phase biochar, which contains insoluble carbonates and ash (Raikova et al., 2019).

Operating temperatures (340–400°C), retention time (45–120 min), initial moisture content and type of catalyst are the significant variables during the HTL process. Microalgae are an excellent candidate for HTL due to their high ash and moisture content, which can cause fouling or involve excessive heat consumption during other processes like pyrolysis/combustion. Elliott et al. (2013) have demonstrated an HTL process for the conversion of macroalgal feedstock which yielded 19.2 wt.% bio-crude oil predominantly of phenolic compounds.

Zhou et al. (2010) found that a 30 min reaction time is ideal for the production of bio-crude. Typically, a longer reaction time favors polymerization and condensation reactions, resulting in higher solid product yield. Like the pyrolysis process, high heating rates could improve the bio-crude yields because short reaction times and high heating rates minimize secondary reactions, such as recombination and secondary cracking. A study on HTL of brown macroalga *Sargassum patens* resulted in yields of 32.1% of bio-crude (Li et al., 2012). Catalysts, such as sulfuric and acetic acids, and alkali catalysts, such as Na_2CO_3, KOH and K_2CO_3, are used to improve both yield and quality of products (Li et al., 2012). Duan and Savage (2011) investigated the influence of numerous catalysts on the liquefaction of *Nannochloropsis* sp. Their results showed that the presence of Pd/C led to the highest bio-crude yields. In addition to the above parameters, particle size and solvent type (i.e., toluene, n-heptane and anisole) are the other important HTL factors affecting the HTL performance (Gu et al., 2020; Kumar et al., 2020).

Hydrothermal gasification

Gasification is a versatile process that occurs in the presence of a limited supply of oxygen or air at temperatures ranging between 600°C and 900°C. Hydrothermal gasification (HTG) takes place in the presence of subcritical water or supercritical water (i.e., below or above 374°C and 22.1 MPa). Although HTG is an energy-intensive process, it produces nitrogen-free fuel, allowing the use of high-protein microalgae. This thermochemical process breaks down the microalgal molecules into syngas, i.e., a mixture of H_2 and CO as well as a fraction of CO_2, CH_4 and light hydrocarbons (C_2–C_3) (Li et al., 2012). Syngas obtained through the gasification process can be converted into H_2 via the water–gas shift reaction or transformed into liquid fuels via the Fischer–Tropsch synthesis process (Okolie et al., 2019). A study reported that the gasification of microalgal biomass (*Nannochloropsis oculata*) at 700°C in the presence of Ru/TiO_2 produced 40–53 wt.% H_2, 30–40 wt.% CO_2, 10 wt.% CH_4 and 6 wt.% CO (Li et al., 2012). A fundamental study of HTG of algae carried out by Minowa and Sawayama (1999) indicated that the gasification of *C. vulgaris* at 350°C in the presence of nickel catalyst resulted in 70 wt.% gas yield with a 50 wt.% CH_4. They found lower catalyst loadings were favoring H_2 production and CH_4 reduction (Biller and Ross, 2012; Latif et al., 2019; Patel et al., 2016).

Hydrothermal carbonization

Hydrothermal carbonization (HTC), or 'artificial coalification,' is a thermochemical process performed at temperatures around 180–220°C and at 2 MPa pressure for several hours to convert algal biomass to hydrochar as the main product and an aqueous phase as a by-product (Reza et al., 2014b). The conversion of algal biomass to hydrochar occurs in an environment saturated by water under elevated pressures (Titirici et al., 2012). Dehydration and decarboxylation are the most important chemical reactions occurring during the HTC process, resulting in an increase in heating value by enhancing the carbon content (Funke and Ziegler, 2010). Water plays a significant multifunctional role, including reactant and solvent (Reza et al., 2014a). This process offers the advantages of concentrating the biomass carbon without drying within a shorter time with less energy requirement than conventional carbonization (Tekin et al., 2014).

HTC biochar has specific properties, such as uniform particles, good porosity, specific functional groups (e.g., –OH and –COOH) and electronic charges (Titirici et al., 2012). Among the significant operating parameters, including reaction time, operating pressure and temperature, the temperature is the most important variable of the HTC process (Nizamuddin et al., 2017). Although lower temperature produces a higher yield of carbon-rich solid, the properties of biochar are significantly different than those produced at higher temperatures. A study reported 45.7 wt.% yields with 62 wt.% carbon recovered in algal char during an HTC process conducted at 200°C, 25 wt.%

Table 1.2 Results on hydrochar production via hydrothermal carbonization of algal biomass

Biomass	*Temperature (°C)*	*Reaction time*	*Solid yield (wt.%)*
Dunaliella salina	190–210	30–120 min	25.3–45.7
Arthrospira platensis	190–210	2–4 h	21.6–36.7
Nannochloropsis	180–220	15–30 min	30–47
Nannochloropsis oculata	180–215	15–45 min	41–51
Spirulina	175–215	30 min	23.3–49.3
Lipid-extracted *Spirulina*	175	30 min	44.6
Sargassum horneri	180–210	2–16 h	32.7–52.3
Laminaria digitata	200–250	1 h	18.4–21.8
Laminaria hyperborean	200–250	1 h	23.6–39.0
Alaria esculenta	200–250	1 h	23.7–30.0

Note: Reproduced with permission from Yu et al. (2017).

solid loadings and 0.5 h residence time (Heilmann et al., 2010). The results of algal hydrochar production in the HTC process are summarized in Table 1.2.

Transesterification

Biodiesel is produced through transesterification of microalgae via several techniques, including conventional, dry *in situ*, wet *in situ* and microwave-assisted transesterification. The traditional method suffers from long processing times and energy-intensiveness, as a drying process is needed before the biomass conversion stage, making it less favorable (Bagchi et al., 2015; Chen and Lee, 2018). In dry *in situ* transesterification, the lipid extraction part, conventionally carried out individually, is combined with the main stage, i.e., transesterification, making the whole process more cost-effective (Faried et al., 2017). Wet *in situ* transesterification takes advantage of avoiding the need for drying and lipid extraction (Wahidin et al., 2014); thus, the operating cost significantly decreases. There is still a technical challenge during the wet transesterification process as the produced biodiesel is hydrolyzed into free fatty acid and methanol (Huang et al., 2015).

In the microwave-assisted transesterification process, the required energy for rupturing biomass cell, via conduction and convection (Bundhoo, 2018), is significantly reduced with respect to conventional processes, even though it has not been commercialized yet (Patil et al., 2011). For example, Li et al. (2016) have demonstrated that the energy consumption of the microwave-assisted transesterification process is approximately two-thirds less than that of the conventional process. This process prefers wet biomass as the moisture accelerates the process of microwave irradiation (Cui and Liang, 2014), rapidly releasing the lipids. The faster process and lower energy consumption, along with higher biodiesel yields, make the microwave process interesting. Limitation of irradiation penetration through the biomass (i.e., less than 20 mm), hotspots in the heated biomass and the scalability of the reactor for large-scale biodiesel production are the main challenges of microwave technology (Chisti, 2013). Further research is required to produce biodiesel through the esterification of wet microalgae at the commercial scale (Goh et al., 2019).

Supercritical fluid extraction

Despite many advantages, the lipid and oil extraction stage for biodiesel production is energy- and cost-intensive, making the process non-feasible on a commercial scale (Chen et al., 2011). The mechanical approaches of lipid extraction from microalgae include oil expeller, press and microwave- or ultrasonic-assisted extraction (Mubarak et al., 2015). The chemical methods are reported as conventional solvent extraction, supercritical fluid extraction (SFE) and novel extraction (e.g., ionic liquid) (Li et al., 2019). Of the lipid extraction techniques, SFE is capable of meeting the safety, economics and environmental requirements and ultimately produces solvent-free lipids (Santana et al., 2012).

The preferred fluid used in the SFE for lipid extraction is supercritical CO_2 due to its low critical temperature of 31.1°C and moderate critical pressure of 7.4 MPa (Yang et al., 2019). Supercritical CO_2 is lipophilic (nonpolar) and can serve as a potential agent to extract polar lipids (e.g., chlorophylls, waxes and pigments) if combined with polar co-solvents (e.g., ethanol, methanol and toluene) (Mouahid et al., 2013).

A study by Tommasi et al. (2017) showed that using deep eutectic solvents and microwave heating, followed by supercritical CO_2, with dimethyl carbonate as the co-solvent, result in highly purified lipid extracts from *Phaeodactylum tricornutum*. In the SFE process using CO_2, the extraction yield of lipids can be influenced by temperature and pressure. The highest lipid yield (up to 92%) for *Scenedesmus obliquus* and *Scenedesmus obtusiusculus* was obtained at optimum conditions of 12 MPa and 20°C and the CO_2/biomass ratio of 100 (Lorenzen et al., 2017).

Anaerobic digestion

Anaerobic digestion (AD) is the natural biological process that transforms crop residues, manures and any other organic wastes into renewable biogas (50–70% CH_4) by the action of several types of microorganisms. Algae have increasingly captured attention as a source of biomass for the AD process, in contrast to combustion, pyrolysis and gasification in which the feedstock requires high energy inputs for moisture removal (Milledge et al., 2015). Additionally, the digestion of micro-/macroalgae residues can remediate nutrients (N and P) and produce CO_2, which are beneficial for microalgae cultivation (Uggetti et al., 2014). Lipid-extracted *Chlorella sorokiniana* residues from biodiesel production were found as suitable biomass for AD since the biomethane reached 38%, i.e., very similar to the same whole-cell type (41%)

(Ayala-Parra et al., 2017). Bucholc et al. (2014) used macroalgae (macrophyte) for producing biogas and minimizing eutrophication of the Baltic Sea.

There are still several technical/economic challenges associated with algal cultivation and integration with AD. The low rate of methane production and the risk of ammonia inhibition are the two major drawbacks of the AD treatment of micro-/macroalgae (Solé-Bundó et al., 2019). The effects of pretreatment techniques on AD performance and CH_4 yield, including thermal, mechanical, chemical, microwave, ultrasonic and biological, have been investigated (Saratale et al., 2018). For instance, adding alkaline reagents to microalgal cells leads to cleavage of resistant cell walls of biomass, thus improving methane conversion (Banu et al., 2018). The co-digestion of different materials, including sewage sludge, animal manure and waste cooking oil along with algae, is a practical solution to address the low biodegradability of algae and the risk of ammonia inhibition (Milledge et al., 2019). Zhong et al. (2012) reported that AD of blue algae with corn straw as a co-substrate improved the methane yield by 61.7% at 325 mL/g volatile solid.

Algae for biological carbon fixation

Rapid population growth and industrialization have inevitably caused intensive greenhouse gas emissions and, consequently, climate change. Changes in atmospheric temperature lead to direct/indirect detrimental impacts on the biological carbon cycles and environmental ecosystems, e.g., harsh weather conditions, melting polar ice and rising sea levels (Farrelly et al., 2013). To address these environmental issues, several methods have been proposed for sequestering carbon released from anthropogenic sources such as flue/fuel gases. The three main pathways for uptaking CO_2 are (i) chemical absorption, (ii) physical/chemical adsorption and (iii) biological fixation (Nanda et al., 2016).

The chemical absorption method can be applied for the removal of CO_2 utilizing solvents such as monoethanolamine (MEA) and diethylamine (DEA). This process shows relatively high capture efficiency. However, it is cost-intensive due to space and high energy requirements for the regeneration of the solvent (Yu et al., 2012). Porous solid adsorbents can capture CO_2 through physical/chemical adsorption. Solid adsorbents are categorized into carbon-based (e.g., biochar), silica-derived (e.g., MCM-41 silica) and metal oxide-derived (e.g., MOFs) (Bamdad et al., 2018, 2019). Although certain types of adsorbents exhibit considerable potential, major challenges are still needed to be overcome, such as the low production yields. As an alternative, the biological approach has been recognized as a long-term means for CO_2 fixation. Reforestation and ocean fertilization, using nutrients to increase photosynthesis by phytoplankton, are the two biological techniques for carbon storage (Singh and Dhar, 2019). Microalgae cultivation has been considered as a promising method for atmospheric and post-combustion CO_2

fixation and storage (Araújo et al., 2015). However, a more permanent solution to sequester carbon for longer time scales is the use of algal-derived biochar or hydrochar in soil amendment.

Biological CO_2 fixation is a process to convert inorganic carbon or CO_2 to organic carbon via photosynthesis (Ji et al., 2016). All autotrophic organisms, such as green plants, algae or seaweed, and specific types of bacteria can biologically fix carbon and incorporate them into the cells through photosynthesis (Petela, 2008). Photosynthesis is a complex reaction to produce glucose and O_2 using CO_2 and water under light (Eq. 1.1). This process occurs in two stages, according to both light-dependent and light-independent reactions (Calvin, 1989). The carbon fixation takes place in the second stage of photosynthesis by forming precursors of carbohydrates from CO_2 (Zhao and Su, 2014). Ultimately, the product (glucose) is converted into starch, cellulose, sucrose, amino acids, proteins and other organic compounds (Stepan et al., 2002).

$$6H_2O + 6CO_2 \rightarrow C_6H_{12}O_6 + 6O_2 \tag{1.1}$$

Microalgae and macroalgae are aquatic plants that can grow very rapidly compared to terrestrial plants, and both are regarded as a suitable option for CO_2 fixation. However, microalgae have attracted increased attention due to higher CO_2 fixation and adaptability to extreme conditions (Ho et al., 2011). Ramos-Ibarra et al. (2019) investigated the CO_2 tolerance of native microalgal strains (*Chlorella* sp. and *Scenedesmus* sp.) and found that they could endure the stressful conditions of industrial effluents (25% CO_2 and 75% CH_4). The CO_2 concentration profoundly affects the microalgal growth, and the optimal value of CO_2 depends on the microalgal species. A small number of microalgal strains could tolerate high concentrations of CO_2, even 100% (Aslam et al., 2017). For instance, green algae *Scenedesmus* sp. were able to grow up to 80% of CO_2. However, the optimal CO_2 concentration was observed within 10–30% (Hanagata et al., 1992). Culture medium temperature, light and pH are the three key factors influencing microalgal growth, CO_2 fixation and biomass production (Zhao and Su, 2014).

Ranglová et al. (2019) proposed a rapid 1-day screening test to estimate an optimal temperature according to photosynthetic activity for different macroalgal strains. They found an optimum temperature around 30°C for most of the tested strains (i.e., *Chlorella, Nostoc, Synechocystis, Scenedesmus* and *Cylindrospermum*). Some previous studies (Carneiro et al., 2020; Edmundson and Huesemann, 2015) have investigated the influence of light intensity on algal biomass production and reported up to 50% reduction of biomass yield compared to the daylight due to photosynthetic restriction. The lower pH of the growth medium, specifically below 6, impedes the growth of most microalgae due to the acidic environment (Xin et al., 2010), while higher pH improves the CO_2 adsorption rate (Yin et al., 2019). The impact of pH

(7.5–9.5) on *Spirulina* cultivation was evaluated, and pH 8.5 was identified as the best condition for higher biomass productivity and CO_2 bio-fixation rate (Mehar et al., 2019). Flue gases contain hazardous gases, such as SO_x and NO_x, that can significantly influence microalgal growth. The inhibitory role of SO_x in flue gases can be diminished by introducing neutralizing NaOH into the system (Yen et al., 2015). Moreover, NO_x impurity can be used as a nutrient in order to enhance the microalgae performance (Yoshihara et al., 1996). Thus, the application of microalgae for CO_2 fixation can potentially eliminate simultaneously both SO_x and NO_x in flue gases.

Two common microalgal CO_2 fixation systems are presented in the literature: open ponds and closed PBRs. The open-pond approach is simple and cost-effective and involves cultivation in circular ponds or raceway ponds (SundarRajan et al., 2019). Raceway ponds are superior due to the higher efficiency of CO_2 fixation and, subsequently, preferable for commercial-scale open-air cultivation (Smith et al., 2018). The PBR types are categorized into flat panel, vertical/horizontal tube, plastic bag and hybrid. PBRs display several advantages compared to open systems, such as less susceptibility to contamination, less space required and lower water/CO_2 losses (Pires et al., 2012). An attractive opportunity could be achieved by designing an integrated system for microalgae cultivation in which wastewater and emitted industrial flue gases are purified at the same time.

Algae for wastewater bioremediation

Algal cultivation systems

For the industrial application of algae resources, the enrichment of the biomass under artificially controlled cultivation conditions is essential. As mentioned earlier, there are two main categories of algae cultivation systems, such as simple open-pond systems and closed PBRs. The open-pond system is simple and easy to maintain, whereas the advanced closed PBRs are normally designed for the cultivation of specific algal strains to produce value-added chemicals (Suh and Lee, 2003). Although the growth rate and yield of phototrophic microorganisms vary by species, the performance could be improved by growing them under favorable environmental conditions. For the factors, light from the sun or an artificial source and CO_2 supply are critical participants in photosynthesis, which have significant impacts on cell growth (Chang et al., 2017). Other important factors, including pH, oxygen concentration, temperature and agitation, also influence cell growth by changing the cellular functions, including their behavior in the photosynthesis process.

Although the open systems may all have similar basic structures, they could be further optimized to enhance energy utilization efficiency. For example, computational fluid dynamics (CFD) codes were applied for the modeling of different raceway bend configurations for the raceway pond (Liffman et al., 2013). The simulations generated an improved bend design which reduced the energy consumption by approximately 87% with respect to the conventional bend. For the closed systems, the tubular-type structure PBR design is the most popular, consisting of an enclosed vessel with illumination, designed for biomass production under controlled conditions. Despite the higher cost, PBRs have many advantages over the open systems (Singh and Sharma, 2012):

i minimized contamination
ii allowance of axenic cultivation of mono-strains of algae
iii precise control of conditions, e.g., pH, temperature, light and CO_2 concentration
iv reduced CO_2 loss and decreased water evaporation
v higher cell concentrations
vi production of complex biopharmaceuticals.

In addition, the scaling flexibility of the tubular PBR could be tremendous, especially when following the modular design. This makes the tubular PBR an excellent choice for both laboratory research purposes and intensified industrial production.

Algal cultivation in wastewater

Water pollution is affecting different regions of the world and harming the health of the entire ecosystem. Algae are important producers in the aquatic ecosystem and can act as valuable indicators of ecosystem health, as they respond rapidly in both species composition and densities to the variations in the water chemistry. For instance, changes in the concentrations of acid-forming chemicals and industry-released heavy metals will change the composition of genera that can tolerate these chemicals (Bruun, 2012). Microalgae offer an attractive alternative for the treatment of wastewater from urban, industrial and agricultural sources as a tertiary biotreatment (de la Noüe et al., 1992).

Since part of the nutrients for their growth is dissolved in the water, the algae can be cultivated in wastewater for simultaneous purification and biomass accumulation. Algae help the removal of nutrients through different mechanisms, including precipitation, volatilization, biosorption, nitrification and denitrification. In addition, using highly efficient feeding sources, like municipal wastewater, would simultaneously reduce the treatment price and diminish biofuel prices by \$0.55–0.59/L (Nhat et al., 2018). Some of the recently published results obtained using algae in wastewater treatment are listed in Table 1.3.

Higgins et al. (2018) studied the synergy between algal and bacterial species contributing to the efficiency of winery wastewater treatment. It was observed that the mixed cultivation of algae with the local wastewater microbial species inhibited the production of algal photosynthate and promoted reductions of chemical oxygen demand (COD) and nutrients, and both *Auxenochlorella*

Table 1.3 Selected results on algal cultivation using wastewater

Strain	*Wastewater type*	*Nitrogen removal (%)*	*Phosphorus removal (%)*	*Carbon removal (%)*	*Residence Time (days)*	*References*
Scenedesmus quadricauda SDEC-8 (KF999643)	Domestic	> 90	~100	~80 COD	10	Yang et al. (2018)
Auxenochlorella protothecoides (UTEX 2341) and *Chlorella sorokiniana* (UTEX 2805)	Winery	> 90	> 50	30–60 COD	5	Higgins et al. (2018)
Leptolyngbya	Synthetic	99.9	68.1	94.7 DOC	24	Cai et al. (2019)
Chlorella vulgaris	Piggery	48–72	81–99	85–94 TOC	26	García et al. (2017)
Chlorella vulgaris	Synthetic	88.79	97.2	97.0 COD	5	Zhang et al. (2020)
Diplosphaera sp.	Synthetic	90.38	88.9	86.3 COD	5	Zhang et al. (2020)
Scenedesmus obliquus	Synthetic	82.97	70.9	95.1 COD	5	Zhang et al. (2020)
Scenedesmus obliquus and *Chlorella vulgaris*	Meat processing	61.2	59.8	89 COD	9	Hu et al. (2019)
Chlorella protothecoides + *Scenedesmus obliquus*	Meat processing	34.4	72.4	83 COD	9	Hu et al. (2019)

Note: Chemical oxygen demand (COD).

protothecoides (UTEX 2341) and *Chlorella sorokiniana* (UTEX 2805) boosted the growth of bacterial strains. The contribution to pathogen reduction of the algal wastewater system has been reported by Delanka-Pedige et al. (2019). The results showed that *Galdieria sulphuraria* reduced the total bacteria number and total coliform in the primary effluent by 1.7 logs and 7.4 logs, respectively, and the algal system reduced all the putative pathogenic bacteria genera in the existing system to non-detectable levels.

Zhang et al. (2020) observed high total nitrogen, NH_4^+-N, COD and total phosphorus removal to levels over 90% when using pine sawdust as an algal biofilm carrier. Based on their results, pine sawdust carriers not only supported the robust algal attachment but also acted efficiently as bio-sorbents to remove certain amounts of pollutants. Some toxicants, such as NH_4^+-N, PAHs, heavy metals and antibiotics, were partly removed by pine sawdust thanks to its bio-sorption capability. It also allowed algal cultivation in the raw or low-diluted wastewaters.

The process efficiencies of different algal wastewater treatment processes, namely photoautotrophic and mixotrophic modes, were compared by Nirmalakhandan et al. (2019). The comparison of results from both pilot-scale experiments and those reported in the literature showed significantly increased volumetric removal rates of biological oxygen demand (BOD), ammoniacal nitrogen and phosphates of the mixotrophic method utilizing *Galdieria sulphuraria*. During the treatment of diluted piggery wastewater, it was observed that the pollutant removal efficiency and the microalgae and bacterial community structure obtained under indoor conditions could not be directly used to extrapolate those of the outdoor PBRs (García et al., 2017). The combination of *Scenedesmus obliquus*, *Chlorella vulgaris* and *Chlorella sorokiniana* improved the adaptability of these microalgae significantly to the heavily polluted meat processing wastewater (Hu et al., 2019). In this research, the researchers also found that the microalgae coexisted with bacteria and promoted the digestion of organics, and enhanced nitrogen and phosphorus removal. Therefore, the synergistic effect among different algal species, as well as their interactions with bacterial during the treatment of wastewater from different sources, should be studied further. Other than the data presented in Table 1.3, the applications and benefits of algal-based technologies in the removal of pharmaceuticals and personal care products (PPCP) from wastewater have been discussed by Wang et al. (2017). It was pointed out that algae and algae–bacteria consortia are proven to have great capability in PPCP removal. However, for industrial-scale applications, the algal strain, operating conditions and reactor configurations should be further optimized.

Li et al. (2018) designed and demonstrated the potential of an integrated system for the conversion of municipal wastewater-grown algae into bio-oil, fuel gas and fertilizer. Their results suggest that 68% of the chemical energy contained in the feedstock was recovered, with 44% in the bio-oil and 23% in the fuel gas. Besides, 44% and 91% of N and P contained in the feedstock could be recovered as fertilizers in the forms of ammonium sulfate and struvite, respectively. Therefore, the combined municipal wastewater treatment and bio-based chemical/fuel production seems to be the most appealing direction for future research and development activities.

Conclusions

To realize the utilization of algae as a source of renewable energy and materials, we discussed its metabolism and physicochemical characteristics. In this chapter, we also presented the progress made in converting algae into fuels using physicochemical, thermochemical and biological technologies. Subsequently, we also reviewed the advancements of using algae in environmental applications, such as biological carbon fixation and wastewater treatment. Based on the review, it is evident that algae could play an increasingly important role in the energy and environmental sector. However, a deeper understanding of the light-absorbing mechanism and its effect on cellular systems should be further studied.

Compared to conventional bioresources, algae are unique in their growth patterns. Further efforts are still required to obtain high lipid yield and improve the energy density of the biomass. Moreover, the organic solvent-free methods must be optimized to tackle potential safety and environmental issues, and those that can be applied to different algal strains with reasonable energy consumption and infrastructure are highly desirable. Regarding the conversion routes, the hydrothermal processes seem to be more advantageous with the capability of avoiding an extra drying step. The conventional dry processes could be improved by further research efforts. The combined municipal wastewater treatment and chemical/fuel production is highly desirable and requires more research efforts.

References

Adenle, A.A., Haslam, G.E., Lee, L., 2013. Global assessment of research and development for algae biofuel production and its potential role for sustainable development in developing countries. *Energy Policy* 61, 182–195.

Alalwan, H.A., Alminshid, A.H., Aljaafari, H.A.S., 2019. Promising evolution of biofuel generations. Subject review. *Renewable Energy Focus* 28, 127–139.

Araújo, O.D.Q.F., de Medeiros, J.L., Yokoyama, L., Morgado, C.D.R.V., 2015. Metrics for sustainability analysis of post-combustion abatement of CO_2 emissions: Microalgae mediated routes and CCS (carbon capture and storage). *Energy* 92, 556–568.

Aslam, A., Thomas-Hall, S.R., Mughal, T.A., Schenk, P.M., 2017. Selection and adaptation of microalgae to growth in 100% unfiltered coal-fired flue gas. *Bioresource Technology* 233, 271–283.

Ayala-Parra, P., Liu, Y., Field, J.A., Sierra-Alvarez, R., 2017. Nutrient recovery and biogas generation from the anaerobic digestion of waste biomass from algal biofuel production. *Renewable Energy* 108, 410–416.

Bagchi, S.K., Rao, P.S., Mallick, N., 2015. Development of an oven drying protocol to improve biodiesel production for an indigenous chlorophycean microalga *Scenedesmus* sp. *Bioresource Technology* 180, 207–213.

Bamdad, H., Hawboldt, K., MacQuarrie, S., 2018. A review on common adsorbents for acid gases removal: Focus on biochar. *Renewable and Sustainable Energy Reviews* 81, 1705–1720.

Bamdad, H., Hawboldt, K., MacQuarrie, S., Papari, S., 2019. Application of biochar for acid gas removal: Experimental and statistical analysis using CO_2. *Environmental Science and Pollution Research* 26, 10902–10915.

Banu, J.R., Sugitha, S., Kannah, R.Y., Kavitha, S., Yeom, I.T., 2018. Marsilea spp.—A novel source of lignocellulosic biomass: Effect of solubilized lignin on anaerobic biodegradability and cost of energy products. *Bioresource Technology* 255, 220–228.

Biller, P., Ross, A.B., 2012. Hydrothermal processing of algal biomass for the production of biofuels and chemicals. *Biofuels* 3, 603–623.

Brennan, L., Owende, P., 2010. Biofuels from microalgae—A review of technologies for production, processing, and extractions of biofuels and co-products. *Renewable and Sustainable Energy Reviews* 14, 557–577.

Bruun, K., 2012. Algae can function as indicators of water pollution. Walpa org., Washington.

Bucholc, K., Szymczak-Żyła, M., Lubecki, L., Zamojska, A., Hapter, P., Tjernström, E., Kowalewska, G., 2014. Nutrient content in macrophyta collected from southern Baltic Sea beaches in relation to eutrophication and biogas production. *Science of the Total Environment* 473, 298–307.

Bundhoo, Z.M., 2018. Microwave-assisted conversion of biomass and waste materials to biofuels. *Renewable and Sustainable Energy Reviews* 82, 1149–1177.

Cai, W., Zhao, Z., Li, D., Lei, Z., Zhang, Z., Lee, D.-J., 2019. Algae granulation for nutrients uptake and algae harvesting during wastewater treatment. *Chemosphere* 214, 55–59.

Calvin, M., 1989. Forty years of photosynthesis and related activities. *Photosynthesis Research* 21, 3–16.

Campbell, M.N., 2008. Biodiesel: Algae as a renewable source for liquid fuel. *Guelph Engineering Journal* 1, 2–7.

Carneiro, M., Cicchi, B., Maia, I., Pereira, H., Zittelli, G.C., Varela, J., Malcata, F.X., Torzillo, G., 2020. Effect of temperature on growth, photosynthesis and biochemical composition of *Nannochloropsis oceanica*, grown outdoors in tubular photobioreactors. *Algal Research* 49, 101923.

Chang, J.-S., Pau Loke, S., Ling, T., Chen, C.-Y., Ho, S.-H., Tan, C.H., Nagarajan, D., Nee, W., 2017. "Photobioreactors" in *Current Developments in Biotechnology and Bioengineering: Bioprocesses, Bioreactors and Controls*, pp. 313–352. Elsevier.

Chang, R.L., Ghamsari, L., Manichaikul, A., Hom, E.F.Y., Balaji, S., Fu, W., Shen, Y., Hao, T., Palsson, B.Ø., Salehi-Ashtiani, K., Papin, J.A., 2011. Metabolic network reconstruction of *Chlamydomonas* offers insight into light-driven algal metabolism. *Molecular Systems Biology* 7, 518.

Chen, C.-Y., Yeh, K.-L., Aisyah, R., Lee, D.-J., Chang, J.-S., 2011. Cultivation, photobioreactor design and harvesting of microalgae for biodiesel production: A critical review. *Bioresource Technology* 102, 71–81.

Chen, J.-J., Lee, Y.-R., 2018. Optimization of the transesterification reaction of microalgal *Monoraphidium* sp. *Renewable Energy* 129, 717–723.

Chisti, Y., 2013. Constraints to commercialization of algal fuels. *Journal of Biotechnology* 167, 201–214.

Cui, Y., Liang, Y., 2014. Direct transesterification of wet *Cryptococcus curvatus* cells to biodiesel through use of microwave irradiation. *Applied Energy* 119, 438–444.

De Bhowmick, G., Sarmah, A.K., Sen, R., 2019. Zero-waste algal biorefinery for bioenergy and biochar: A green leap towards achieving energy and environmental sustainability. *Science of the Total Environment* 650, 2467–2482.

Delanka-Pedige, H.M.K., Munasinghe-Arachchige, S.P., Cornelius, J., Henkanatte-Gedera, S.M., Tchinda, D., Zhang, Y., Nirmalakhandan, N., 2019. Pathogen reduction in an algal-based wastewater treatment system employing *Galdieria sulphuraria*. *Algal Research* 39, 101423.

de la Noüe, J., Laliberté, G., Proulx, D., 1992. Algae and waste water. *Journal of Applied Phycology* 4, 247–254.

Demirbas, A., 2010. Use of algae as biofuel sources. *Energy Conversion and Management* 51, 2738–2749.

Doucha, J., Lívanský, K., 2008. Influence of processing parameters on disintegration of Chlorella cells in various types of homogenizers. *Applied Microbiology and Biotechnology* 81, 431.

Duan, P., Savage, P.E., 2011. Hydrothermal liquefaction of a microalga with heterogeneous catalysts. *Industrial & Engineering Chemistry Research* 50, 52–61.

Edmundson, S.J., Huesemann, M.H., 2015. The dark side of algae cultivation: Characterizing night biomass loss in three photosynthetic algae, *Chlorella sorokiniana*, *Nannochloropsis salina* and *Picochlorum* sp. *Algal Research* 12, 470–476.

Elliott, D.C., Hart, T.R., Schmidt, A.J., Neuenschwander, G.G., Rotness, L.J., Olarte, M.V., Zacher, A.H., Albrecht, K.O., Hallen, R.T., Holladay, J.E., 2013. Process development for hydrothermal liquefaction of algae feedstocks in a continuous-flow reactor. *Algal Research* 2, 445–454.

Faried, M., Samer, M., Abdelsalam, E., Yousef, R., Attia, Y., Ali, A., 2017. Biodiesel production from microalgae: Processes, technologies and recent advancements. *Renewable and Sustainable Energy Reviews* 79, 893–913.

Farrelly, D.J., Everard, C.D., Fagan, C.C., McDonnell, K.P., 2013. Carbon sequestration and the role of biological carbon mitigation: A review. *Renewable and Sustainable Energy Reviews* 21, 712–727.

Funke, A., Ziegler, F., 2010. Hydrothermal carbonization of biomass: A summary and discussion of chemical mechanisms for process engineering. *Biofuels, Bioproducts and Biorefining* 4, 160–177.

García, D., Posadas, E., Grajeda, C., Blanco, S., Martínez-Páramo, S., Acién, G., García-Encina, P., Bolado, S., Muñoz, R., 2017. Comparative evaluation of piggery wastewater treatment in algal-bacterial photobioreactors under indoor and outdoor conditions. *Bioresource Technology* 245, 483–490.

Goh, B.H.H., Ong, H.C., Cheah, M.Y., Chen, W.-H., Yu, K.L., Mahlia, T.M.I., 2019. Sustainability of direct biodiesel synthesis from microalgae biomass: A critical review. *Renewable and Sustainable Energy Reviews* 107, 59–74.

Greenwell, H.C., Laurens, L., Shields, R., Lovitt, R., Flynn, K., 2010. Placing microalgae on the biofuels priority list: A review of the technological challenges. *Journal of the Royal Society Interface* 7, 703–726.

Gu, X., Martinez-Fernandez, J., Pang, N., Fu, X., Chen, S., 2020. Recent development of hydrothermal liquefaction for algal biorefinery. *Renewable and Sustainable Energy Reviews* 121, 109707.

Hanagata, N., Takeuchi, T., Fukuju, Y., Barnes, D.J., Karube, I., 1992. Tolerance of microalgae to high CO_2 and high temperature. *Phytochemistry* 31, 3345–3348.

Harris, E.H., 2001. *Chlamydomonas* as a model organism. *Annual Review of Plant Biology* 52, 363–406.

Heilmann, S.M., Davis, H.T., Jader, L.R., Lefebvre, P.A., Sadowsky, M.J., Schendel, F.J., Von Keitz, M.G., Valentas, K.J., 2010. Hydrothermal carbonization of microalgae. *Biomass and Bioenergy* 34, 875–882.

Hena, S., Abida, N., Tabassum, S., 2015. Screening of facultative strains of high lipid producing microalgae for treating surfactant mediated municipal wastewater. *RSC Advances* 5, 98805–98813.

Higgins, B.T., Gennity, I., Fitzgerald, P.S., Ceballos, S.J., Fiehn, O., VanderGheynst, J.S., 2018. Algal–bacterial synergy in treatment of winery wastewater. *npj Clean Water* 1, 1–10.

Ho, S.-H., Chen, C.-Y., Lee, D.-J., Chang, J.-S., 2011. Perspectives on microalgal CO_2-emission mitigation systems—A review. *Biotechnology Advances* 29, 189–198.

Hu, X., Meneses, Y.E., Stratton, J., Wang, B., 2019. Acclimation of consortium of micro-algae help removal of organic pollutants from meat processing wastewater. *Journal of Cleaner Production* 214, 95–102.

Huang, R., Cheng, J., Qiu, Y., Li, T., Zhou, J., Cen, K., 2015. Using renewable ethanol and isopropanol for lipid transesterification in wet microalgae cells to produce biodiesel with low crystallization temperature. *Energy Conversion and Management* 105, 791–797.

Ji, Y., Xu, Z., Zou, D., Gao, K., 2016. Ecophysiological responses of marine macroalgae to climate change factors. *Journal of Applied Phycology* 28, 2953–2967.

Jin, V.L., Schmer, M.R., Stewart, C.E., Mitchell, R.B., Williams, C.O., Wienhold, B.J., Varvel, G.E., Follett, R.F., Kimble, J., Vogel, K.P., 2019. Management controls the net greenhouse gas outcomes of growing bioenergy feedstocks on marginally productive croplands. *Science Advances* 5, eaav9318.

Jones, C.S., Mayfield, S.P., 2012. Algae biofuels: Versatility for the future of bioenergy. *Current Opinion in Biotechnology* 23, 346–351.

Khan, M.I., Shin, J.H., Kim, J.D., 2018. The promising future of microalgae: Current status, challenges, and optimization of a sustainable and renewable industry for biofuels, feed, and other products. *Microbial Cell Factories* 17, 36–36.

Kraan, S., 2013. Mass-cultivation of carbohydrate rich macroalgae, a possible solution for sustainable biofuel production. *Mitigation and Adaptation Strategies for Global Change* 18, 27–46.

Kumar, M., Sun, Y., Rathour, R., Pandey, A., Thakur, I.S., Tsang, D.C., 2020. Algae as potential feedstock for the production of biofuels and value-added products: Opportunities and challenges. *Science of the Total Environment* 716, 137116.

Kwak, M., Park, W.-K., Shin, S.-E., Koh, H.-G., Lee, B., Jeong, B.-r., Chang, Y.K., 2017. Improvement of biomass and lipid yield under stress conditions by using diploid strains of *Chlamydomonas reinhardtii*. *Algal Research* 26, 180–189.

Latif, N.I.S.A., Ong, M.Y., Nomanbhay, S., 2019. Hydrothermal liquefaction of Malaysia's algal biomass for high-quality bio-oil production. *Engineering in Life Sciences* 19, 246–269.

Li, D., Chen, L., Xu, D., Zhang, X., Ye, N., Chen, F., Chen, S., 2012. Preparation and characteristics of bio-oil from the marine brown alga *Sargassum patens C. Agardh*. *Bioresource Technology* 104, 737–742.

Li, J., Dai, J., Liu, G., Zhang, H., Gao, Z., Fu, J., He, Y., Huang, Y., 2016. Biochar from microwave pyrolysis of biomass: A review. *Biomass and Bioenergy* 94, 228–244.

Li, L., Cui, J., Liu, Q., Ding, Y., Liu, J., 2015. Screening and phylogenetic analysis of lipid-rich microalgae. *Algal Research* 11, 381–386.

Li, P., Sakuragi, K., Makino, H., 2019. Extraction techniques in sustainable biofuel production: A concise review. *Fuel Processing Technology* 193, 295–303.

Li, Y., Tarpeh, W.A., Nelson, K.L., Strathmann, T.J., 2018. Quantitative evaluation of an integrated system for valorization of wastewater algae as bio-oil, fuel gas, and fertilizer products. *Environmental Science & Technology* 52, 12717–12727.

Liffman, K., Paterson, D.A., Liovic, P., Bandopadhayay, P., 2013. Comparing the energy efficiency of different high rate algal raceway pond designs using computational fluid dynamics. *Chemical Engineering Research and Design* 91, 221–226.

Lorenzen, J., Igl, N., Tippelt, M., Stege, A., Qoura, F., Sohling, U., Brück, T., 2017. Extraction of microalgae derived lipids with supercritical carbon dioxide in an industrial relevant pilot plant. *Bioprocess and Biosystems Engineering* 40, 911–918.

Lü, J., Sheahan, C., Fu, P., 2011. Metabolic engineering of algae for fourth generation biofuels production. *Energy & Environmental Science* 4, 2451–2466.

Mehar, J., Shekh, A., Nethravathy, M., Sarada, R., Chauhan, V.S., Mudliar, S., 2019. Automation of pilot-scale open raceway pond: A case study of CO_2-fed pH control on Spirulina biomass, protein and phycocyanin production. *Journal of CO_2 Utilization* 33, 384–393.

Mercer, P., Armenta, R.E., 2011. Developments in oil extraction from microalgae. *European Journal of Lipid Science and Technology* 113, 539–547.

Milledge, J.J., Nielsen, B.V., Maneein, S., Harvey, P.J., 2019. A brief review of anaerobic digestion of algae for bioenergy. *Energies* 12, 1166.

Milledge, J.J., Staple, A., Harvey, P.J., 2015. Slow pyrolysis as a method for the destruction of Japanese wireweed, *Sargassum muticum*. *Environment and Natural Resources Research* 5, 28–37.

Minowa, T., Sawayama, S., 1999. A novel microalgal system for energy production with nitrogen cycling. *Fuel* 78, 1213–1215.

Mohr, A., Raman, S., 2013. Lessons from first generation biofuels and implications for the sustainability appraisal of second generation biofuels. *Energy Policy* 63, 114–122.

Mouahid, A., Crampon, C., Toudji, S.-A.A., Badens, E., 2013. Supercritical CO_2 extraction of neutral lipids from microalgae: Experiments and modelling. *The Journal of Supercritical Fluids* 77, 7–16.

Mubarak, M., Shaija, A., Suchithra, T., 2015. A review on the extraction of lipid from microalgae for biodiesel production. *Algal Research* 7, 117–123.

Naghdi, F.G., González, L.M., Chan, W., Schenk, P.M., 2016. Progress on lipid extraction from wet algal biomass for biodiesel production. *Microbial Biotechnology* 9, 718–726.

Nanda, S., Berruti, F., 2021b. Municipal solid waste management and landfilling technologies: A review. *Environmental Chemistry Letters* 19, 1433–1456.

Nanda, S., Berruti, F., 2021a. A technical review of bioenergy and resource recovery from municipal solid waste. *Journal of Hazardous Materials* 403, 123970.

Nanda, S., Reddy, S.N., Mitra, S.K., Kozinski, J.A., 2016. The progressive routes for carbon capture and sequestration. *Energy Science & Engineering* 4, 99–122.

Nhat, P.V.H., Ngo, H.H., Guo, W.S., Chang, S.W., Nguyen, D.D., Nguyen, P.D., Bui, X.T., Zhang, X.B., Guo, J.B., 2018. Can algae-based technologies be an affordable green process for biofuel production and wastewater remediation? *Bioresource Technology* 256, 491–501.

Nirmalakhandan, N., Selvaratnam, T., Henkanatte-Gedera, S.M., Tchinda, D., Abeysiriwardana-Arachchige, I.S.A., Delanka-Pedige, H.M.K., Munasinghe-Arachchige, S.P., Zhang, Y., Holguin, F.O., Lammers, P.J., 2019. Algal wastewater treatment: Photoautotrophic vs. mixotrophic processes. *Algal Research* 41, 101569.

Nizamuddin, S., Baloch, H.A., Griffin, G.J., Mubarak, N.M., Bhutto, A.W., Abro, R., Mazari, S.A., Ali, B.S., 2017. An overview of effect of process parameters on hydrothermal carbonization of biomass. *Renewable and Sustainable Energy Reviews* 73, 1289–1299.

Okolie, J.A., Nanda, S., Dalai, A.K., Kozinski, J.A., 2021. Chemistry and specialty industrial applications of lignocellulosic biomass. *Waste and Biomass* 12, 2145–2169.

Okolie, J.A., Rana, R., Nanda, S., Dalai, A.K., Kozinski, J.A., 2019. Supercritical water gasification of biomass: A state-of-the-art review of process parameters, reaction mechanisms and catalysis. *Sustainable Energy & Fuels* 3, 578–598.

Olaizola, M., 2003. Commercial development of microalgal biotechnology: From the test tube to the marketplace. *Biomolecular Engineering* 20, 459–466.

Patel, B., Guo, M., Izadpanah, A., Shah, N., Hellgardt, K., 2016. A review on hydrothermal pre-treatment technologies and environmental profiles of algal biomass processing. *Bioresource Technology* 199, 288–299.

Patil, P.D., Gude, V.G., Mannarswamy, A., Cooke, P., Munson-McGee, S., Nirmalakhandan, N., Lammers, P., Deng, S., 2011. Optimization of microwave-assisted transesterification of dry algal biomass using response surface methodology. *Bioresource Technology* 102, 1399–1405.

Petela, R., 2008. An approach to the exergy analysis of photosynthesis. *Solar Energy* 82, 311–328.

Pires, J., Alvim-Ferraz, M., Martins, F., Simões, M., 2012. Carbon dioxide capture from flue gases using microalgae: Engineering aspects and biorefinery concept. *Renewable and Sustainable Energy Reviews* 16, 3043–3053.

Raikova, S., Allen, M.J., Chuck, C.J., 2019. Hydrothermal liquefaction of macroalgae for the production of renewable biofuels. *Biofuels, Bioproducts and Biorefining* 13, 1483–1504.

Ramos-Ibarra, J.R., Snell-Castro, R., Neria-Casillas, J.A., Choix, F.J., 2019. Biotechnological potential of *Chlorella* sp. and *Scenedesmus* sp. microalgae to endure high CO_2 and methane concentrations from biogas. *Bioprocess and Biosystems Engineering* 42, 1603–1610.

Ranglová, K., Lakatos, G.E., Manoel, J.A.C., Grivalský, T., Masojídek, J., 2019. Rapid screening test to estimate temperature optima for microalgae growth using photosynthesis activity measurements. *Folia Microbiologica* 64, 615–625.

Rastogi, R.P., Pandey, A., Larroche, C., Madamwar, D., 2018. Algal Green Energy – R&D and technological perspectives for biodiesel production. *Renewable and Sustainable Energy Reviews* 82, 2946–2969.

Reid, W.V., Ali, M.K., Field, C.B., 2020. The future of bioenergy. *Global Change Biology* 26, 274–286.

Reza, M.T., Andert, J., Wirth, B., Busch, D., Pielert, J., Lynam, J.G., Mumme, J., 2014a. Hydrothermal carbonization of biomass for energy and crop production. *Applied Bioenergy* 1, 11–29.

Reza, M.T., Becker, W., Sachsenheimer, K., Mumme, J., 2014b. Hydrothermal carbonization (HTC): Near infrared spectroscopy and partial least-squares regression for determination of selective components in HTC solid and liquid products derived from maize silage. *Bioresource Technology* 161, 91–101.

Santana, A., Jesus, S., Larrayoz, M., 2012. Supercritical carbon dioxide extraction of algal lipids for the biodiesel production. *Procedia Engineering* 42, 1755–1761.

Saratale, R.G., Kumar, G., Banu, R., Xia, A., Periyasamy, S., Saratale, G.D., 2018. A critical review on anaerobic digestion of microalgae and macroalgae and co-digestion of biomass for enhanced methane generation. *Bioresource Technology* 262, 319–332.

Singh, J., Dhar, D.W., 2019. Overview of carbon capture technology: microalgal biorefinery concept and state-of-the-art. *Frontiers in Marine Science* 6, 29.

Singh, R.N., Sharma, S., 2012. Development of suitable photobioreactor for algae production – A review. *Renewable and Sustainable Energy Reviews* 16, 2347–2353.

Sirajunnisa, A.R., Surendhiran, D., 2016. Algae – A quintessential and positive resource of bioethanol production: A comprehensive review. *Renewable and Sustainable Energy Reviews* 66, 248–267.

Smith, R.G., Smith, I.J., Smith, B.D., 2018. A novel strategy for sequestering atmospheric CO_2: The use of sealed microalgal cultures located in the open-oceans. *Renewable and Sustainable Energy Reviews* 83, 85–89.

Solé-Bundó, M., Passos, F., Romero-Güiza, M.S., Ferrer, I., Astals, S., 2019. Co-digestion strategies to enhance microalgae anaerobic digestion: A review. *Renewable and Sustainable Energy Reviews* 112, 471–482.

Stepan, D.J., Shockey, R.E., Moe, T.A., Dorn, R., 2002. *Carbon dioxide sequestering using microalgal systems*. University of North Dakota.

Suh, I.S., Lee, C.-G., 2003. Photobioreactor engineering: Design and performance. *Biotechnology and Bioprocess Engineering* 8, 313.

SundarRajan, P., Gopinath, K.P., Greetham, D., Antonysamy, A.J., 2019. A review on cleaner production of biofuel feedstock from integrated CO_2 sequestration and wastewater treatment system. *Journal of Cleaner Production* 210, 445–458.

Talebi, A.F., Mohtashami, S.K., Tabatabaei, M., Tohidfar, M., Bagheri, A., Zeinalabedini, M., Hadavand Mirzaei, H., Mirzajanzadeh, M., Malekzadeh Shafaroudi, S., Bakhtiari, S., 2013. Fatty acids profiling: A selective criterion for screening microalgae strains for biodiesel production. *Algal Research* 2, 258–267.

Tan, X.-B., Zhao, X.-C., Yang, L.-B., 2019. Strategies for enhanced biomass and lipid production by *Chlorella pyrenoidosa* culture in starch processing wastewater. *Journal of Cleaner Production* 236, 117671.

Tekin, K., Karagöz, S., Bektaş, S., 2014. A review of hydrothermal biomass processing. *Renewable and Sustainable Energy Reviews* 40, 673–687.

Titirici, M.-M., White, R.J., Falco, C., Sevilla, M., 2012. Black perspectives for a green future: Hydrothermal carbons for environment protection and energy storage. *Energy & Environmental Science* 5, 6796–6822.

Tommasi, E., Cravotto, G., Galletti, P., Grillo, G., Mazzotti, M., Sacchetti, G., Samorì, C., Tabasso, S., Tacchini, M., Tagliavini, E., 2017. Enhanced and selective lipid extraction from the microalga *P. tricornutum* by dimethyl carbonate and supercritical CO_2 using deep eutectic solvents and microwaves as pretreatment. *ACS Sustainable Chemistry & Engineering* 5, 8316–8322.

Uggetti, E., Sialve, B., Latrille, E., Steyer, J.-P., 2014. Anaerobic digestate as substrate for microalgae culture: The role of ammonium concentration on the microalgae productivity. *Bioresource Technology* 152, 437–443.

Vassilev, S.V., Vassileva, C.G., 2016. Composition, properties and challenges of algae biomass for biofuel application: An overview. *Fuel* 181, 1–33.

Wahidin, S., Idris, A., Shaleh, S.R.M., 2014. Rapid biodiesel production using wet microalgae via microwave irradiation. *Energy Conversion and Management* 84, 227–233.

Wang, Y., Liu, J., Kang, D., Wu, C., Wu, Y., 2017. Removal of pharmaceuticals and personal care products from wastewater using algae-based technologies: A review. *Reviews in Environmental Science and Bio/Technology* 16, 717–735.

Xin, L., Hong-ying, H., Ke, G., Jia, Y., 2010. Growth and nutrient removal properties of a freshwater microalga *Scenedesmus* sp. LX1 under different kinds of nitrogen sources. *Ecological Engineering* 36, 379–381.

Yang, Y.-C., Wang, C.-S., Wei, M.-C., 2019. Kinetics and mass transfer considerations for an ultrasound-assisted supercritical CO_2 procedure to produce extracts enriched in flavonoids from Scutellaria barbata. *Journal of CO_2 Utilization* 32, 219–231.

Yang, Z., Pei, H., Hou, Q., Jiang, L., Zhang, L., Nie, C., 2018. Algal biofilm-assisted microbial fuel cell to enhance domestic wastewater treatment: Nutrient, organics removal and bioenergy production. *Chemical Engineering Journal* 332, 277–285.

Yen, H.W., Ho, S.H., Chen, C.Y., Chang, J.S., 2015. CO_2, NO_x and SO_x removal from flue gas via microalgae cultivation: A critical review. *Biotechnology Journal* 10, 829–839.

Yin, D., Wang, Z., Wen, X., Ding, Y., Hou, X., Geng, Y., Li, Y., 2019. Effects of carbon concentration, pH, and bubbling depth on carbon dioxide absorption ratio in microalgae medium. *Environmental Science and Pollution Research* 26, 32902–32910.

Yoshihara, K.-I., Nagase, H., Eguchi, K., Hirata, K., Miyamoto, K., 1996. Biological elimination of nitric oxide and carbon dioxide from flue gas by marine microalga NOA-113 cultivated in a long tubular photobioreactor. *Journal of Fermentation and Bioengineering* 82, 351–354.

Yu, C.-H., Huang, C.-H., Tan, C.-S., 2012. A review of CO_2 capture by absorption and adsorption. *Aerosol and Air Quality Research* 12, 745–769.

Yu, K.L., Lau, B.F., Show, P.L., Ong, H.C., Ling, T.C., Chen, W.-H., Ng, E.P., Chang, J.-S., 2017. Recent developments on algal biochar production and characterization. *Bioresource Technology* 246, 2–11.

Zhang, Q., Wang, L., Yu, Z., Zhou, T., Gu, Z., Huang, Q., Xiao, B., Zhou, W., Ruan, R., Liu, Y., 2020. Pine sawdust as algal biofilm biocarrier for wastewater treatment and algae-based byproducts production. *Journal of Cleaner Production* 256, 120449.

Zhao, B., Su, Y., 2014. Process effect of microalgal-carbon dioxide fixation and biomass production: A review. *Renewable and Sustainable Energy Reviews* 31, 121–132.

Zhong, W., Zhang, Z., Luo, Y., Qiao, W., Xiao, M., Zhang, M., 2012. Biogas productivity by co-digesting Taihu blue algae with corn straw as an external carbon source. *Bioresource Technology* 114, 281–286.

Zhou, D., Zhang, L., Zhang, S., Fu, H., Chen, J., 2010. Hydrothermal liquefaction of macroalgae *Enteromorpha prolifera* to bio-oil. *Energy & Fuels* 24, 4054–4061.

Zou, C., Zhao, Q., Zhang, G., Xiong, B., 2016. Energy revolution: From a fossil energy era to a new energy era. *Natural Gas Industry* B 3, 1–11.

2 Algal cultivation systems and photobioreactor designs

Nongmaithem Debeni Devi, Vaibhav V. Goud, and Bo Hu

Introduction

Microalgae belong to prokaryotic or eukaryotic unicellular microorganisms. They accumulate biomass through photosynthesis (Metting, 1996). The cells of microalgae are also known as mini-factory as they convert solar energy and carbon dioxide (CO_2) into biomolecules including carbohydrates, proteins, lipids, triglycerides, carotenoids, and other pigments, vitamins and steroids. These biochemical components possess potential microalgae-derived products for commercialization, ranging from biofuels (such as biodiesel, bioethanol and biohydrogen), human nutrition, aquaculture and animal feed, cosmetics, to biofertilizers. (Yen et al., 2013). Microalgae are exploited in various applications such as wastewater treatment plants, atmospheric CO_2 sequestration and reduction of greenhouse gases. Their ability of rapid growth, greater photosynthetic efficiency and adaptation to extreme environments make microalgae superior to other terrestrial oil crops in terms of biomass and biodiesel yield (Chisti, 2008).

Figure 2.1 demonstrates the benefits of microalgae in multiple aspects. Core genesis of microalgae technology lies within the type of cultivation system, photobioreactor (PBR) designs and evaluation of optimal conditions for target species for high performance of growth. In this chapter, the basic ideas and principles of emerging microalgae cultivation systems and PBR designs are presented. The advantages and disadvantages of the commonly used cultivation systems, as well as the types of PBR designs, are compared. Furthermore, some examples of the commercial microalgae cultivation designs and processes are overviewed to understand the updates on the commercialization and development of microalgae-based products.

Conditions influencing microalgal growth and lipid yield

Carbon supply

The most vital factor in the growth of microalgae is carbon. Generally, diversified carbon sources, i.e., either inorganic (CO_2, bicarbonates, etc.) or organic (methanol, acetate, glucose and so on) carbon, are used for

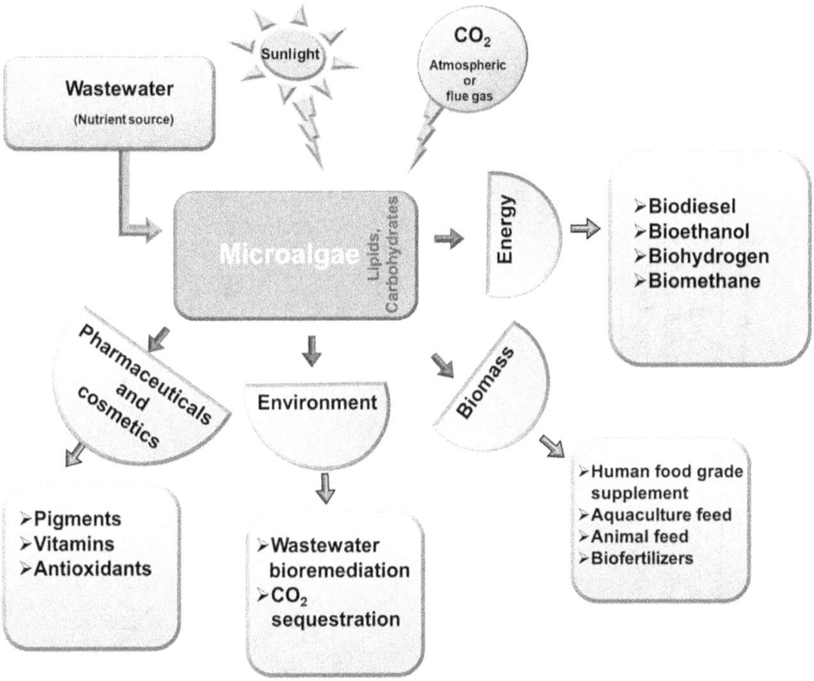

Figure 2.1 Schematic diagram representing the usefulness of microalgal biomass and its biochemical content in various aspects.

microalgal cultivations. Algae can be cultivated into three different modes, viz. phototrophic, heterotrophic and mixotrophic modes. Each mode of cultivation is described in the forthcoming sections.

Phototrophic mode

During the phototrophic mode, microalgal cells harvest light energy and assimilate inorganic carbon sources (CO_2 or bicarbonates) to generate chemical energy via photosynthesis (Brennan and Owende, 2010). As in Table 2.1, under phototrophic cultivation using various carbon sources, different microalgal species may diverge in their biomass and lipid productivity and lipid content. The biomass productivity and lipid content can reach up to 0.02–15.79 g/L/d and 8.1–63%, respectively, showing significant differences between the various microalgal species (Chen and Walker, 2011). Using 2.2% CO_2 under phototrophic conditions could enhance the lipid productivity of *Scenedesmus obliquus* up to 69.2 mg/L/d (El Baky et al., 2012). The major benefit of phototrophic cultivation is the utilization of CO_2 emitting from the flue gas of industrial activities. In a literature survey, De Morais and Costa

Table 2.1 Biomass, lipid productivity and lipid content of different microalgal species under different modes of cultivations

Microalgae	*Mode of cultivation*	*Carbon source*	*Biomass productivity (g/L/d)*	*Lipid (%)*	*Lipid productivity (mg/L/d)*	*References*
Botryococcus braunii SAG-30.81	Phototrophic	5% CO_2	0.61	33	68.42	Sydney et al. (2010)
Chaetoceros calcitrans CS 178	Phototrophic	3.3% CO_2	0.04	39.8	17.6	Rodolfi et al. (2009)
Chaetoceros muelleri F&M-M43	Phototrophic	3.3% CO_2	0.07	33.6	21.8	Rodolfi et al. (2009)
Chlorella ellipsoidea FACHB-4	Heterotrophic	18 g/L glucose	2.96	33.3	–	Han et al. (2012)
Chlorella emersonii CCAP 211/11N	Phototrophic	5% CO_2	0.03–0.05	29.0–63.0	8.1–49.9	Illman et al. (2000)
Chlorella minutissima MCC27	Heterotrophic	Crude glycerol	0.446	34.36	165.15	Katiyar et al. (2017)
Chlorella muelleri MMDL50116	Phototrophic	10% CO_2	0.272	43.3	51.07	Wang et al. (2014)
Chlorella protothecoides	Heterotrophic	15–60 g/L glucose	2.2–7.4	50.3–57.8	1,209.6–3,701.1	Xiong et al. (2008)
Chlorella protothecoides UTEX 256	Heterotrophic	Biodiesel-derived crude glycerol	5.5	–	2,900	Chen and Walker (2011)
Chlorella pyrenoidosa FACHB-9	Heterotrophic	18 g/L glucose	2.09	35.3	–	Han et al. (2012)
Chlorella pyrenoidosa SJTU-2	Phototrophic	10% CO_2	0.144	24.25	26.85	Tang et al. (2011)
Chlorella regularis FACHB-729	Heterotrophic	10 g/L glucose	0.72	42.30%	310	Fu et al. (2017)
Chlorella sorokiniana	Heterotrophic	Aquaculture wastewater	0.353	31.9	138.17	Guldhe et al. (2017)
Chlorella sp. HS2	Phototrophic	0.5 g/L $NaHCO_3$ and 1% CO_2	0.53	–	–	Nayak et al. (2018)
Chlorella vulgaris	Phototrophic	1 g/L $NaHCO_3$	0.99	–	–	Mokashi et al. (2016)
Chlorella vulgaris #259	Heterotrophic	Glucose and acetate	0.08–0.15	23.0–36.0	27.0–35.0	Liang et al. (2009)
Chlorella vulgaris #259	Heterotrophic	Glucose and acetate	0.08–0.15	23.0–36.0	27.0–35.0	Liang et al. (2009)

Chlorella vulgaris #259	Mixotrophic	Glucose and glycerol	0.09–0.25	21.0–34.0	22.0–54.0	Liang et al. (2009)
Chlorella vulgaris #259	Mixotrophic	Glucose and glycerol	0.09–0.25	21.0–34.0	22.0–54.0	Liang et al. (2009)
Chlorella vulgaris 31	Heterotrophic	2 and 10 g/L glucose	0.159	–	0.19	Kong et al. (2020)
Chlorella vulgaris 31	Heterotrophic	2 and 10 g/L maltose	0.144	–	0.162	Kong et al. (2020)
Chlorella vulgaris 31	Heterotrophic	2 and 10 g/L sodium acetate	0.071	–	0.104	Kong et al. (2020)
Chlorella vulgaris 31	Mixotrophic	2 and 10 g/L glucose	0.196	–	0.226	Kong et al. (2020)
Chlorella vulgaris 31	Mixotrophic	2 and 10 g/L maltose	0.152	–	0.195	Kong et al. (2020)
Chlorella vulgaris 31	Mixotrophic	2 and 10 g/L sodium acetate	0.113	–	0.170	Kong et al. (2020)
Chlorella vulgaris ESP-3	Mixotrophic	Glucose and CO_2	–	45–53	67–144	Yeh and Chang (2012)
Chlorella vulgaris FACHB-31	Heterotrophic	18 g/L glucose	2.77	30.1	–	Han et al. (2012)
Nannochloropsis oculata NCTU-3	Phototrophic	2% CO_2	0.497	30.7	15.1	Chiu et al. (2009)
Nannochloropsis sp.	Mixotrophic	2 g/L glucose	0.587	25.3	148.3	Cheirsilp and Torpee (2012)
Scenedesmus obliquus	Phototrophic	12% CO_2	0.02	33.14	69.23	El Baky et al. (2012)
Scenedesmus obliquus	Mixotrophic	Glucose	0.10–0.51	6.6–11.8	11.6–58.6	Mandal and Mallick (2009)
Scenedesmus obliquus CNW-N	Phototrophic	2.5% CO_2	0.44 ± 15.79	10.32	45.48	Ho et al. (2012)
Scenedesmus sp. CCNM 1077	Phototrophic	0.6 g/L $NaHCO_3$	–	34.44	–	Pancha et al. (2015a,b)
Schizochytrium limacinum SR21 (ATCC MYA-1381)	Mixotrophic	Biodiesel-derived crude glycerol	3.88	–	–	Ethier et al. (2011)
Schizochytrium limacinum SR21 (ATCCMYA-1381)	Heterotrophic	75–150 g/L crude glycerol	1.05–3.11	–	160–500	Chi et al. (2007)

(2007) reported the raceway pond (RWP) cultivation of *Spirulina* through fixation of CO_2 from flue gas of the coal-combustion thermoelectric power plant (Usina Termelétrica Presidente Médici, Brazil).

The utilization of CO_2 as a substrate reduces carbon footprint when the grown microalgal biomass can be converted into microalgae-derived biofuels and high-value-added products (Yen et al., 2015). Thus, phototrophic cultivation is considered to be ideal and rational from the perspective of commercial viability for scaling up in the outdoor open or RWP systems (Mata et al., 2010).

Heterotrophic mode

Under heterotrophic mode, the algal species can directly use organic carbon (such as acetate, glucose, glycerol and maltose) in either dark or illuminated conditions (Han et al., 2012). The photosynthetic products such as simple organic sugar (glucose or sucrose) are converted into polysaccharide, viz. starch in plants or microalgae. In dark conditions, starch is then hydrolyzed and utilized for microalgal cell metabolism. This phenomenon is known as dark respiration, and it is the key source of energy for maintaining the cellular metabolisms of photosynthetic plants or microorganisms in dark conditions (Hu et al., 2006). The important benefit of heterotrophic mode is to circumvent the problem associated with the light restraint. The restriction of light penetration in high microalgal cell density in PBRs during phototrophic cultivation can be overcome via the heterotrophic mode (Berberoglu et al., 2007). Hence, the lipid yield and productivity reported in the literature could reach up to 50.3–57.8% and 1,209.6–3,701.1 mg/L/d, respectively, by *Chlorella protothecoides* using 15–60 g/L glucose (Table 2.1) (Xiong et al., 2008). In the past few years, studies have been focused tremendously on the discovery of cheaper and effective quality organic carbon sources as well as an integrated waste treatment.

As illustrated in Table 2.1, *Chlorella sorokiniana* could effectively utilize aquaculture wastewater under heterotrophic condition, achieving the maximum biomass (0.353 g/L/d) and lipid (138.17 mg/L/d) productivities and significant removal of phosphate (PO_4^{-3}, 100%), ammonium (NH_4^+, 98.21%) and nitrate (NO_3^-, 75.76%) (Guldhe et al., 2017). Crude glycerol, which is a by-product of biodiesel, is an excellent organic carbon source for microalgae cultivation. As shown in Table 2.1, *Chlorella protothecoides* and *Schizochytrium limacinum* can utilize biodiesel-derived crude glycerol to yield substantial amounts of total lipid content and docosahexaenoic acid (DHA, C22:6 *n*−3) under heterotrophic cultivation (Chen and Walker, 2011; Chi et al., 2007).

Mixotrophic mode

Microalgae that are cultivated in mixotrophic mode find the advantages of both phototrophic and heterotrophic conditions. Under this condition, microalgae undergo photosynthesis by assimilating organic carbon and CO_2

as a carbon substrate. The important benefit of this condition is that during algal cell respiration, CO_2 is released and then the cells recapture and recycle CO_2 during the phototrophic mode. As indicated in Table 2.1, *Chlorella vulgaris* 31 shows higher lipid productivity 0.226, 0.195 and 0.170 g/L/d by utilizing glucose, maltose and sodium acetate, respectively, by altering the culture conditions from heterotrophic to mixotrophic (Kong et al., 2020). Like in heterotrophic conditions, the high biomass productivity of 3.88 g/L/d and DHA productivity of 0.52 g/L/d could be achieved by *Schizochytrium limacinum* SR21 (ATCC MYA-1381) under mixotrophic condition using biodiesel-derived crude glycerol.

Nutrient supply

Lipid productivity is the resultant of both biomass production and the total lipid yield of the algae. Hence, it is particularly important to be taken into account in large-scale microalgal lipid production (Rodolfi et al., 2009). Suitable media formulation or composition of nutrients at various concentrations is a crucial factor to increase the lipid production performance of microalgal species (Mandalam and Palsson, 1998). Sharma et al. (2012) reported that under nutrient stress conditions, several microalgal species exhibit a common trend of accumulating higher neutral lipids (predominantly triacylglycerols (TAG)) or carbohydrates.

Under nitrogen stress conditions, *Chlorella vulgaris* could considerably increase its lipid content by 40%. Hu et al. (2006) conducted a study on the nitrogen stress response, indicating algae, cyanobacteria and diatoms with 10-20% increased lipid yield when compared to normal condition. It has been suggested that under nitrogen stress, the algal protein content will be disintegrated, thereby converting to lipids as storage energy for the long term (Hu et al., 2006; Siaut et al., 2011). Besides, phosphorus limitation could significantly increase the lipid yield and fatty acid composition of *Monodus subterraneus, Pavlova lutheri, Chaetoceros* sp. and *Isochrysis galbana,* but the lipid content of *Tetraselmis* sp. and *Nannochloris atomus* was found to decrease (Khozin-Goldberg and Cohen, 2006; Reitan et al., 1994). The proportion of phospholipids of *Monodus subterraneus* decreased from 8.3% to 1.4% of total lipids, while the TAG content increased from 6.5% up to 39.3% of total lipid under phosphorus limitation (Khozin-Goldberg and Cohen, 2006).

Feng et al. (2012) suggested that nitrogen deficiency condition is the best and effective strategy to enhance TAG production, indicating the maximum lipid productivity (87.1 mg/L/d) by *Chlorella zofingiensis* under nitrogen starvation than that of phosphorus. In recent years, wastewaters, mostly agricultural and municipal wastewaters, are becoming a potential supply of phosphorus and nitrogen (Chiu et al., 2015; Guldhe et al., 2017). In a literature survey, Chiu et al. (2015) reported the feasibility of growing different microalgal species (*Chlorella vulgaris, Chlorella zofingiensis,* etc.) using various types of wastewater, containing total phosphorus (3–500 mg/L), total nitrogen (30–900 g/L) and ammonium (NH_3^+) (5–900 mg/L).

Salinity

Algae have received considerable attention as they adapt well in varying salinities. Their adaptions to extreme salinity involve osmotic adjustment by accumulating certain metabolites and balance the osmotic stress of the surroundings (Srivastava et al., 2017). Generally, microalgae and cyanobacteria can thrive in a varied salt concentration up to 1.7 M, while *Dunaliella* species can acclimate exceptionally up to 5.5 M salt concentrations (Alkayal et al., 2010). Takagi et al. (2006) reported that an increase in salt concentration from 0.5 to 1.0 M could enhance the intracellular lipid yield from 60% to 70% in *Dunaliella* cells.

In some studies, varying concentrations of sodium salt were reported to enhance lipid yield. *Desmodesmus abundans* could increase the lipid productivity up to 67.08 mg/L/d when 20 g/L NaCl salt was used (Xia et al., 2014). Pancha et al. (2015b) reported that under two-stage salinity stress cultivation with 400 mM NaCl, *Scenedesmus* sp. could accumulate 74.87% of neutral lipid and 24.77% of total lipid yield. *Chlorella vulgaris* has also been grown in a 5-L fermenter by applying the salinity stress strategies, resulting in a lipid yield of 3.81 g/L at 180 h, which was 30.1 higher than the control condition (Duan, 2012). Salinity stress is also suggested as another approach to lipid overproduction as being feasible.

Light

Photosynthesis and cell photo-acclimatization are both light-driven processes. Hence, types of light sources are a major constraint for microalgal growth. The response of photosynthetic cells to light intensity, absorption spectra and photoperiod may vary for different microalgal species. The light energy spectrum, ranging from 400 to 700 nm wavelength, is considered as the desirable range for the photosynthesis of microalgae and cyanobacteria (Berberoglu et al., 2007). The light intensity required for growing microalgae in phototrophic mode is divided into several light phases: limitation phase, saturation phase and inhibition phase (Ogbonna and Tanaka, 2000).

The light saturation phase varying from 50 to 200 $\mu mol/m^2/s$ is the maximum intensity that can be accommodated with the photosynthetic cells of different microalgae (Goldman, 1979). *Scenedesmus* cells could favorably grow under the light intensity ranging from 50 to 400 $\mu mol/m^2/s$. At 400 $\mu mol/m^2/s$, *Scenedesmus* could achieve 3.88 g/L of the maximum cell concentration, 41.1% of total lipid yields and 32.9% of neutral lipid (Liu et al., 2012). In PBRs, the high cell concentration can experience photoinhibition due to the self-shading effect between individual cells, causing a productivity decrease. Even distribution of light saturation phase in the whole microalgal cultivation can induce the maximum biomass productivity. To circumvent the effect of photo-shading in the high cell density, the light path should be reduced and proper agitation of culture will help to distribute the optimum light intensity (Goldman, 1979).

Temperature

Temperature is another element for the growth of algae as it highly influences to dissolve nutrient, nutrient assimilation and cellular protein structure. Change in temperature influences the accumulation of lipid in microalgae by diverting the pathway of metabolism of biosynthesis to the mechanism of lipid formation. At high temperatures, essential cellular enzymes and proteins get denatured causing an inhibitory effect on the physiology of the growing cells. Hence, to maintain the growth of microalgae in optimal temperature, a controller is required in the PBR cultivation system, while for large-scale microalgae cultivation in the outdoor system, temperature change is dependent on photo-duration and seasonal change.

Several studies have been reported for the growth and lipid content of different microalgae due to varying temperatures. *Monoraphidium* sp. achieved its maximum cell concentration (650 mg/L) and lipid production (29.2 mg/L/d) at 30°C. For the highest biomass growth of *Chlorella minutissima*, 25°C was found to be the optimal temperature (Cao et al., 2014). Another study reported that the increased cultivation temperature increased monounsaturated fatty acids (MUFAs) and saturated fatty acids (FAs), while the neutral lipid and polyunsaturated FAs decreased (Wei et al., 2014) With increased temperature, the 'cetane number' of fatty acid methyl esters (FAME) surged from 52.3 till 60.3 and 45.3 to 47.6 in *Nannochloropsis oculata* and *Tetraselmis subcordiformis*, respectively (Wei et al., 2014). Thus, suitable temperatures are required to enhance the growth of algae and FAME content on a large scale.

pH value

pH is influencing the rate of absorption of inorganic carbon in the culture medium. In general, pH 7–9 is considered ideal for microalgal growth. The cytosol of the majority of microalgae is within neutral-to-marginal basic pH, wherein the cytosolic enzymes become active at these pH ranges. Thus, the extreme pH undesirably affects the fixation of CO_2 and disrupts the cellular metabolism, thereby inhibiting microalgal growth (Kumar et al., 2011). Liu et al. (2007) reported that *Chattonella marina* could grow favorably within a range of pH 7.5–8.5, while pH beyond 9 reduced the growth significantly. However, in some studies, *Spirulina* sp. could thrive efficiently at pH greater than 10, thus minimizing the chances of contamination when cultivation was done in the RWP system (Kumar et al., 2011). Similarly, Belkin and Boussiba (1991) observed that the range of pH, i.e., pH 9.0 to pH 10.0, was favorable for the growth of *Spirulina platensis*. This indicates that microalgae and cyanobacterial species may require varying pH ranges for optimal growth.

Mixing

Appropriate mixing helps even transfer of gas and nutrient supplement to the entire culture. It also helps to avoid settling of cells and self-shading, and hence, the cells will receive the efficient light, thereby maximizing the biomass productivity and microalgal growth (Goldman, 1979). In open systems, mixing is generally achieved by rotating arms or paddle wheels, while in a closed PBR, mixing is provided in various forms such as air/CO_2 supply, impeller blades or baffle. However, vigorous mixing can severely damage the cell wall and even cause cell lysis. It was highly recommended to provide an appropriate velocity for mixing to reduce the shear stress and achieve higher biomass productivity (Roncallo et al., 2013).

Photobioreactors

Various physiological and technological approaches have been proposed in both the laboratory and pilot-scale systems for maximizing the productivity of microalgae. In the late 1960s, an open circular pond system was operated in Japan for the cultivation of *Chlorella* via connection of a series of pipes furnished with jets as inlets for fresh culture and media (Gupta et al., 2015). With progress in technology, the development of different PBRs with efficient performance has become a concern for mass-scale cultivation of algae.

A few fundamental principles required to efficiently operate the PBR cultivation system are: (i) increasing effective light penetration, (ii) use of appropriate light sources and (iii) maintaining proper microalgal cell concentration during a continuous run. The basic requirement to have effective light penetration is to increase the surface/volume ratio of the PBR design. The main design may be an outdoor open-pond system, horizontal tubular, flat-plate, etc., that are naturally illuminated with great surface areas for illumination (Ugwu et al., 2008).

The provision of irradiation with an appropriate source with spectral quality and intensity is also a key for efficient PBRs. Generally, laboratory-scale PBRs are illuminated with different types of light sources such as light-emitting diodes (LEDs), halogen lamps and fluorescent lamps (Katuwal, 2017). Photosynthesis and the growth of microalgae are directly affected by the levels of light intensities that result in high biomass and lipid productivity. In large-scale cultivation systems, microalgae are commonly cultivated in batch mode condition, wherein a part of microalgal cells is collected periodically during the cultivation. In this way, an equivalent volume of fresh nutrient broth is reloaded inside the culture growth. Besides, degassing O_2 during photosynthesis and efficient mixing are also keys to operate the innovative design of PBRs to achieve commercially viable production (Roncallo et al., 2013). The microalgae cultivation system is done either in open systems or closed PBRs. Descriptions that are more detailed are discussed below.

Open systems

Open systems are classified into two types: natural water bodies (lagoons, ponds, lakes, etc.) and artificial ponds or tanks. Mass cultivation of microalgae and cyanobacteria is done in open ponds of circular or raceway type under the illumination of sunlight (Rao et al., 2012). Among them, RWPs are the most used system. In RWPs, the microalgae cultures, nutrients and media are distributed continuously via the agitation of the rotating paddle wheels. The major advantages of RWPs are as follows: They are easy to build and operate, the systems are easy to mix, they have higher growth rate, and they have lower costs than closed systems. Mendoza et al. (2013) stated that nearly 95% of the total microalgal biomass around the globe is generated from the cultivation of the RWP system. Besides, an average total cost of production for biodiesel, including financial costs ($12.73/gallon), was approximately 2.5 times lower than that of PBR cultivation (Richardson et al., 2012).

Figure 2.2a shows the paddle wheel-driven RWP with a volumetric productivity of 0.03–0.08 g/L/d by *Nannochloropsis* sp. operated at the AlgaePARC pilot facilities, Wageningen UR (the Netherlands) (De Vree et al., 2015). Although the open-pond systems are economically feasible and have

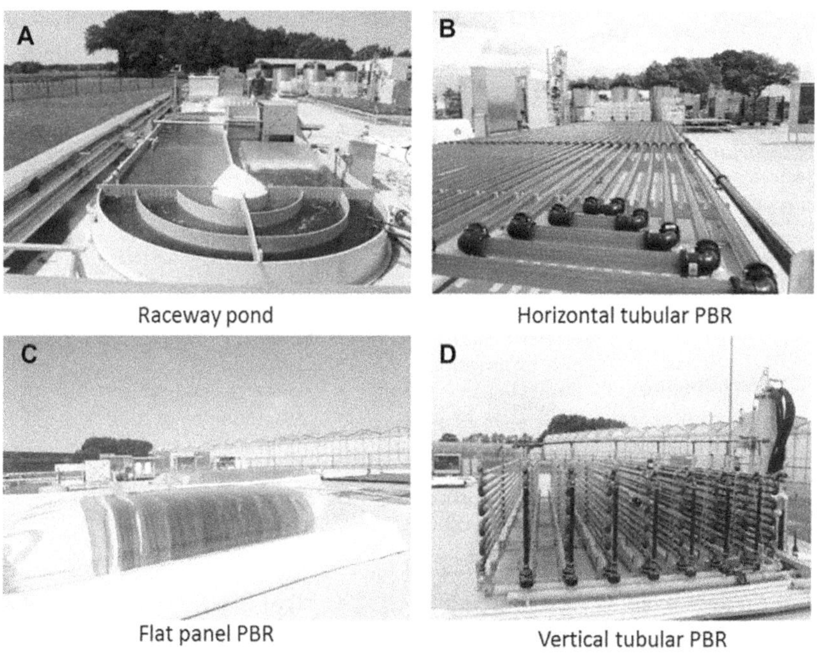

Figure 2.2 Outdoor pilot plant. (a) Open raceway pond system, (b) horizontal tubular PBR, (c) flat-panel PBR in bag style and (d) vertical bubble-tubular PBRs currently operated at AlgaePARC, Wageningen UR, the Netherlands (De Vree et al., 2015). Reproduced with permission from De Vree et al. (2015).

high productivity, there could be certain limitations involved with them. Some of the major problems are insufficient utilization of the light by the cells due to self-shading, loss of media due to evaporation, easy diffusion of CO_2 to the atmosphere, the necessity of large land, and easy contamination by other fast-growing unwanted microorganisms (Dunford, 2015). The advantages and disadvantages associated with the open cultivation system are summarized in Table 2.2. Nevertheless, the controllable and easily manageable factors make open systems an attractive option for the industrial cultivation of microalgae.

Table 2.2 Comparison of the variance of photobioreactors

Cultivation system	*Mixing*	*Advantages*	*Disadvantages*	*References*
Open pond	Paddle wheel	• Economical • Easy for scaling up • Easy and flexible configuration	• Less biomass productivity • Difficult to control high cell density and overgrowth cultures • Easy contamination • Need of extra cultivation area • Diffusion of CO_2 • Evaporation of water	Dunford (2015)
Horizontal tubular PBR	Airlift or bubble	• High surface/volume ratio • Easy to operate outdoors • High microalgal cell concentration • Economical • Lower photoinhibition	• Elevation of dissolved O_2 • Susceptible to photoinhibition and self-shading due to the growth of cells on the wall of tubes • Poor temperature regulation	Ugwu et al. (2008)
Flat-panel PBR	Recirculation through pump	• High surface/volume ratio • The requirement of less space • Good performance for photosynthesis • Inexpensive • Low dissolved oxygen	• Poor light penetration • Development of foul odor • Cleansing issues • Poor temperature control • Limited scale-up opportunities	Dunford (2015)
Vertical tubular PBR	Air bubble or pump	• Less photoinhibition • Increased mass transfer • Proper mixing with minimal stress	• High cost of building materials and manufacturing charges • Issues of self-shading • Design and configuration constraints	Das (2015) and Gupta et al. (2015)

Closed systems

The designs of closed PBRs have provisions to avoid the exchange of gases from the atmosphere and cross-contamination with unwanted microorganisms from the outside environment. Transparent glass, polyvinyl chloride (PVC), polycarbonate or plexiglass is commonly used to design PBRs. The closed-system PBRs offer several advantages over open systems. The important advantages of using closed systems are mentioned below (Kumar et al., 2011):

i PBRs have the provision of efficient light supply
ii They have a high surface/volume ratio
iii They prevent water evaporation
iv They have well-equipped controllers to maintain optimal pH and temperature of the culture system,
v They introduce CO_2 in a controlled way, thereby increasing mass transfer, and
vi They prevent the problems of contamination, and thus, culture modes such as phototrophic, heterotrophic and mixotrophic modes can be operated effectively.

There are various types of closed-system PBRs operating in a laboratory, pilot plant level and an industrial level, including horizontal tubular, flat-plate and vertical tubular PBRs. A brief description of the PBR configurations is given in the forthcoming sections.

Horizontal tubular PBR

Tubular PBR is a type of closed PBR, which is the most suitable system for mass cultivation. Tubular PBRs are typically made up of narrow internal diameters of transparent, polypropylene acrylic, plastic, glass and PVC pipes. Mixing or agitation of the cultures is done by airlift or pump systems. An illustration of horizontal tubular PBR configuration in a loop system that is connected to a bubble column to remove oxygen is shown in Figure 2.2b (De Vree et al., 2015).

In Table 2.2, the key merits and limitations of using horizontal tubular PBRs are summarized. High dissolved CO_2 is built up inside the tubular PBR due to increased air residence. The arrangements provide a large illumination surface/volume area along with resistance to contaminants. Tubular PBRs could be operated using artificial light as well as sunlight (Ugwu et al., 2008), whereas the prime hurdle using tubular PBRs lies in their inefficient mass transfer when scaled up. Another problem is light shading that arises due to microalgal growth along pipe the walls of pipes, hindering the infiltration of light. The rise in dissolved O_2 inside the tubes could inhibit photosynthesis and consequently decrease biomass productivity (Ugwu et al., 2008). Nevertheless, tubular PBRs are operated in the majority as compared to the other designs.

Flat-panel PBR

Flat-plate PBRs are recently in consideration for the growth of phototrophic microalgae owing to the high surface/volume ratio. The bubbled air is circulated inside the panel. The air is circulated inside the panel through the rise of bubbles inside the tubes. The simple configuration of flat-plate PBRs is easily constructed with provision for the desired light path. However, certain limitations involved with the flat-panel system are wall growth, requirements of suitable light intensity, difficulty in cleansing, and less efficiency for mass production per unit space. A rise in the hydrostatic pressure, along with an increase in the capacity of the reactor, becomes a constraint for outdoor applications (Dunford, 2015). To reduce the cost of constructing materials, a low-cost sustainable and novel design of flat-panel PBR that was made up of transparent plastic bag on a rigid frame has been proposed. Ten vertical panels were assembled to set up the flat-panel system, resulting in 0.9 g/L/d volumetric biomass yield (De Vree et al., 2015).

Vertical column PBR

Vertical-column PBRs are constructed with transparent tubes/vessels arranged vertically. Based on the liquid flow inside the reactor, vertical column PBR is further grouped into bubble-column and airlift PBR (Das, 2015).

In the bubble column, the air sparging provides optimum agitation and uniform air transfer. The system has a high surface/volume ratio, satisfactory heat and mass transfer and efficient release of O_2 and residual gas mixture. In tall bubble pilot-scale PBRs, 'perforated plates' are used as the spargers and the bubbles released through the perforation are recirculated inside the column to provide aeration (Das, 2015). Figure 2.2d represents the vertical tubular PBR consists of seven bubble columns used for oxygen removal, temperature control and nutrient and antifoam addition (De Vree et al., 2015).

In airlift PBRs, the reactor is physically separated by two interconnecting zones. They are the riser (up-flow) and the downcomer (downflow). The air is sparged through the riser resulting in upward flow, while the nutrient broth drifts down the bottom and recycles the movement within the riser and downcomer (Das, 2015; Gupta et al., 2015). The two commonly used airlift configurations are internal and external loop vessels. In the internal loop vessel, an internal baffle partitions the riser and the downcomer in the internal loop vessels, while in an external loop vessel, a small horizontal section connects the main vessels with the extended separate tube through the top and bottom. This type of configuration separates the riser and downcomer further apart and thus favors an effective circulation of the flow (Gupta et al., 2015)

Other systems

In recent years, intensive studies have been done to develop a new configuration of PBRs for enhanced microalgal growth. However, some limitations still hinder microalgal biofuel research and development (Rashid et al., 2019). The major limitation is the requirement of extra operational cost during harvesting of biomass from the suspension condition of open or closed PBRs. The conventional harvesting techniques such as centrifugation, filtration and chemical flocculation are energy-intensive, high-cost and time-consuming (Rashid et al., 2014). With the advent of technologies, the attention on developing biofilm-based cultivation as well as co-cultivation/consortium systems is increased (Christenson and Sims, 2012). A brief description of both the attached and co-cultivation systems is given below.

Attached cultivation

In the attached cultivation system, microalgal biomass develops as a biofilm by adhering to the supporting materials, which are initiated by acid–base interactions, hydrophobic interactions and much other chemical bonding between the surface and microalgae (Ista et al., 2004). The system is characterized by easier harvesting efficiency and enhanced biomass accumulation with less moisture content. In general, a nutrient reservoir equipped with a pump supplies the media to the biofilm growth through the gravitational flow. Thus, the biofilm receives nutrient broths by transfusion through the surface. Normally, light is illuminated from the other side of microalgae biofilm.

In the case of a transparent or porous surface (e.g., glass or membrane), the light can be illuminated from the inner membrane. In nylon mesh attached system, light penetrates deeper and stronger owing to achieving the maximum lipid productivity of 1.9 $g/m^2/d$ and increased aerial biomass productivity which is 2.8 times higher than that of the suspended pond system (Lee et al., 2014). Recently, various configurations of attached systems have been designed with different types of supporting materials. For instance, Bernstein et al. (2014) developed a large-scale rotating algal biofilm reactor (RABR) with microalgae attachment cotton cord coiled onto aluminum wheels and operated using municipal wastewater (Figure 2.3).

Gross and Wen (2014) designed RABR with microalgae adherence material made of cotton duct fabrics in a triangle shape and a modified vertical configuration that looks like a conveyor belt in a vertical orientation (Figure 2.4). The RABR achieved a 302% increase in biomass productivity compared to a standard RWP, with 18.9 $g/m^2/d$ of maximum biomass productivity. The biofilm of microalgae grows thicker in vertical design than the triangular RABR. RABR is generally used for wastewater treatment coupled with algal growth and dewatering while achieving high biomass concentration. Lipid content and production in biofilm reactors coupled can

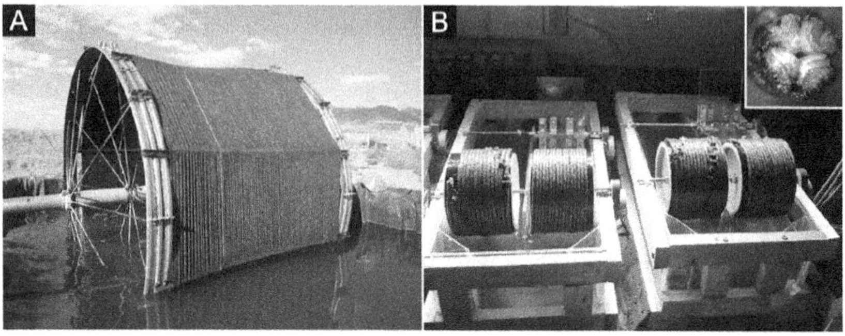

Figure 2.3 Microalgal biofilm growth over the cotton cord as a substratum in (a) outdoor pilot-scale RABR and (b) laboratory-scale RABR (inset represents the cross section of cotton cord with biofilm growth). Reproduced with permission from Elsevier (Bernstein et al., 2014).

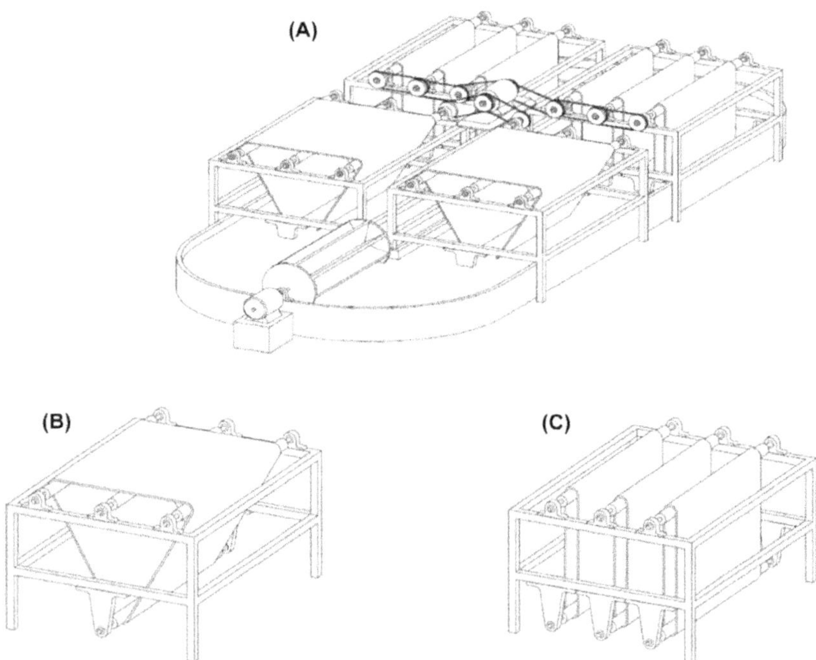

Figure 2.4 (a) Triangular vertical RABR system operated in an open raceway pond. Schematic diagrams of (b) triangular RABR system and (c) vertical RABR system. Reproduced with permission from Elsevier (Gross and Wen, 2014).

achieve comparable lipid content and biomass productivity as suspension cultures (Gross and Wen, 2014).

Co-cultivation system

Microalgae typically coexist with other microorganisms in a very complex relationship in natural habitat. The relationship can be either mutualistic, cooperative, or parasitic (Ashen and Goff, 2000). In the microalgae–bacteria consortium, bacteria can liberate CO_2 and valuable metabolites as a coproduct via respiration, which would subsequently assist the microalgal growth. The released O_2 during photosynthesis of algal cells can support bacterial growth (Shen et al., 2017). Likewise, in fungi–algae interactions, also known as lichens, fungi consume the photosynthetic carbon released from the algae as nutrients, while the growing cluster of fungal cells traps water, harboring mineral nutrients, which algae can take up readily (Rajendran and Hu, 2016). The coexistence of microalgae with other microalgae or cyanobacteria exhibits commensalism conditions. In this condition, the synergistic effect developed by the alliance of species allows each other to support and induce the production of high metabolites.

Extracellular polymeric substances (EPS), released from bacteria or cyanobacteria, contain mainly proteins and polysaccharides. These biochemical compositions of EPS perform a key part in the symbiosis of microalgae with other species, enhancing the aggregation of minute algal cells and aiding efficient biomass harvesting (Watanabe et al., 1999). The consortium systems are performed widespread, ranging from laboratory to pilot scale for many reasons such as the generation of aquaculture feed, waste treatment and biomass production (Rajendran and Hu, 2016).

Leong et al. (2018) cultivated *Chlorella* sp. with activated sludge in a ratio of 1:0.75, achieving the maximum lipid content of 130 mg/L, total nitrogen removal of 97–98% and flocculation efficiency of 42%. A consortium of the fungus *Rhodotorula glutinis* and microalga *Chlorella pyrenoidosa* with a ratio of 1:3 could attain a maximum microalgal cell concentration of 6.12 and 2.48 g/L of oil content, respectively. In addition, the consortium culture could enhance the total fatty acid (TFA) by twofold than the axenic culture, resulting in the maximum TFA productivity of 175.64 ± 2.32 mg/L/d (Liu et al., 2018). Rashid et al. (2019) reported the effective consortium of two microalgae (*Chlorella* and *Ettlia*) at the ratio of 1:08, achieving more biomass productivity of 0.74 ± 0.06 g/L/d as compared to 0.41 ± 0.06 g/L/d in the axenic culture of *Ettlia* in mixotrophic condition. Thus, the co-cultivation strategy would be an ideal strategy for enhanced biomass and lipid productivity of algae in cost-effective harvesting.

Techno-economic aspects of PBRs

The capital cost of the open-pond system has been estimated at the range of US$10,000–79,000/ha considering the costs of the construction and CO_2 and the energy cost for paddle wheel, while in closed PBRs, the total cost of wall PBR has been estimated at US$25,000/ha considering the manufacturer charge and the cost of building materials (Dogaris et al., 2015). Paddle wheel-driven raceway-shape culture in the USA, Israel, Japan, China and other countries could maintain approximately 0.5 g/L of culture by using fertilizer, achieving productivity of about 25 g/m^2/d. Thus, regardless of their low construction and operation cost, the cost required for the production of algae in the range between US$1 and 15/kg dry cell weight (dcw) (Lee, 2001).

Based on the economic analysis as reported by Banerjee and Ramaswamy (2019), the microalgae production cost for flat-panel PBR ranged between US$2,895 and 9,564/ton. Another common type of system, which is widely used in scaling up microalgae cultivation, is sterile plastic bag cultivation (nearly 1,000 L with 0.5 m of diameter). These systems have poor mixing performance and expensive labor charge that escalates the production price of microalgae. The production costs are nearly US$50/kg dcw, while the costs for a smaller system can reach up to US$300–600/kg. The estimated costs are more expensive and higher than the production of *Dunaliella*, *Spirulina* and *Chlorella*, which are in the range of US$9–25/kg dcw (Borowitzka, 1999).

Current status on industrial microalgae cultivation

In recent decades, many companies are engaged in the commercial cultivation of microalgae. Various methods and technologies incorporated designs of cultivation systems, and PBRs have been developed for efficient cultivation and high biomass productivity. Some of the leading companies include Algenol Biofuels, Sapphire Energy, Earthrise Nutritionals, Duke Energy, Algae Systems, East India Distilleries (EID) Parry Limited and The Chinese Microalgae Industry Alliance (CMIA). A description of each company is given below as well as in Table 2.3.

Algenol Biofuels designed a proprietary method, called 'Direct to Ethanol.' In this method, ethanol is produced from the biomass of microalgae or cyanobacteria without killing or harvesting the microorganisms (Carr, 2015). The cyanobacteria that are modified genetically for overexpressing enzymes involved in the fermentation pathway are cultivated in flexible plastic-film PBRs. The system could achieve a 20 times higher ethanol yield than the ethanol produced from corn (Koller, 2015). Recently, the company has a strategic collaboration with some other leading biofuel companies such as Indian Reliance Industries Limited and BioFields SAPI de C.V., Mexico, thus increasing the marketplace around the globe.

Sapphire Energy and Duke Energy are operated using CO_2 emitted in the flue gas of coal combustion. In Sapphire Energy, the marine algal strains are cultivated using non-potable brine in pond system with atmospheric CO_2 as carbon substrate and solar energy in the desert (Koller, 2015). The biomass is periodically harvested to produce 'green crude,' a renewable alternative for petroleum. They anticipated that the first commercial unit of the company

Table 2.3 Leading companies in commercial microalgae production

Company name	*Products*	*Cultivation conditions*	*Market*	*References*
Algae System (USA)	• Treated water • Biofuels • Biochar • Fertilizers	• Operates algae-immobilized floating membrane PBRs	• 40,000 gallons of wastewater is treated per day	Carr (2015)
Algenol Biofuels (USA)	• Ethanol • Gasoline • Jet fuel	• Cultivation of the genetically modified species in plastic-film PBRs	• Targets to generate 1 billion gallons of ethanol per annum with less than US$1/gallon by 2020.	Carr (2015) and Koller (2015)
Duke Energy (USA)	• Algal biomass • Biofuel • Aquaculture • Bioplastic	• Pilot-scale column airlift PBR with a capacity volume of 1,135–26,000 L	• PBR mitigates CO_2 in the coal flue gas to algal biomass	Carr (2015)
Earthrise Nutritionals (USA)	• *Spirulina* derivatives as food supplements, nutraceuticals, etc.	• Open raceway ponds of about 37 with roughly 1.3 acres each	• Produce 600 tons of *Spirulina*-based powder annually in the world • Extract of *Spirulina*-based food colorant, worth US$10 million.	Carr (2015)
East India Distilleries EID Parry Limited (India)	• *Chlorella* and *Spirulina* powder and tablets • Biofertilizer • β-Carotene from *Dunaliella salina*	• Open raceways ponds and multiple smaller tanks	• Produced 600 tons of *Spirulina* per year. • Algae products market targets to reach the US$3,318 million by 2022.	Das (2015)
Sapphire Energy (USA)	• Bio-crude oil	• Open raceway ponds. • More than 70 ponds with different sizes of 14–300 ft production capacity of 1 million liters are operated	• Anticipated to produce 5,000–10,000 barrels/d of bio-crude oil. • Aims to produce 1 million to 1 billion gallons per year of jet fuel and biodiesel by 2025.	Carr (2015) and Koller (2015)
The Chinese Microalgae Industry Alliance (China)	• Phycocyanin tablets and powder derived from *Spirulina* and *Chlorella*	• 750 ha of cultivation base in the 1-acre facility	• Produced 9,600 ton dry weight cell per year with an annual retail value of over US$650 million productivity.	Chen et al. (2016)

would produce green crude oil of nearly 5,000–10,000 barrels/d. The company has signed with the Linde Group (USA) with the interest of partnership on a commercial-scale system that could deliver CO_2 to open-pond algae. Besides, Sapphire also collaborated with Monsanto (USA) to develop genetically engineered algal strains that might apply to Monsanto's work in agriculture (Carr, 2015; Voegele, 2014). Duke Energy has carried out the pilot-scale column airlift PBR with a capacity volume of 1,135–26,000 L. In PBR, the microalgal cells consume CO_2 in the flue gas for the growth through photosynthesis, yielding high-value-added products, aquaculture feed and other derivatives like microalgae-based bioplastic (Carr, 2015).

Earthrise Nutritionals, EID Parry Limited and CMIA are famous for the world's largest *Spirulina* produce. All companies adopt open-pond systems (Carr, 2015; Chen et al., 2016; Das, 2015). Earthrise is operated on a 108-acre site comprising about 37 ponds and producing more than 600 tons of *Spirulina* powder each year. Earthrise supplies *Spirulina* biomass and its derivatives like nutraceuticals, cosmetics, animal feed, nutritional supplement and beverage in above 20 countries (Carr, 2015). In EID Parry Limited, Spirulina is grown in open RWPs, covering an area of 130 acres, as well as in multiple tanks (Costa and de Morais, 2014). The company is one of the great producers of carotenoids, and even *Dunaliella*-derived carotenoids (15% concentration of carotene) are available in the trade name 'carotenoids mix.' Parry Limited targets the algae market to reach the US$3,318.1 million by 2022 (Das, 2015). CMIA consists of an assemblage of 14 leading enterprises for microalgae producer. Some of the leading enterprises include King Dnarmsa Spirulina Co., Ltd (Hainan, China) and Yantai Hairong Biology Technology Co., Ltd (Shandong, China) (Chen et al., 2016). Under CMIA, 750 ha of land is cultivated with more than 60 *Spirulina* plants, resulting in 9,600 ton dcw/year with about US$650 million/year as an annual marketing price. In Yantai Hairong Biology Technology Co., Ltd, microalgal strain *Nannochloropsis* is cultivated in open RWPs with CO_2 flue gas as a carbon source. The company became the first to commercialize *Nannochloropsis* into the aquaculture market (Chen et al., 2016).

In Algae Systems, algae are cultivated in immobilized floating membrane PBRs utilizing municipal wastewater as nutrient supply and atmospheric CO_2 as a carbon substrate under solar irradiation. The algae consume nitrogen and phosphorus from the wastewater, thereby reducing the cost of treating wastewater and solving the problems of water–energy nexus. Algae Systems treats 40,000 gallons of wastewater per day in a 1-acre facility on Mobile Bay. The integrated system produces microalgal biomass and converts it to crude oil and biochar agricultural fertilizers as a by-product (Camargo, 2014).

Conclusions

In recent decades, great efforts have been made for the advancement of algal biotechnology and bioengineering of PBRs for large production. Some of the types of cultivation systems show a promising diverse application on a large scale such as bioenergy production and animal feed production. Nevertheless, certain challenges that limited the scaling-up of the PBRs in efficient ways are limitation of CO_2, photoinhibition, solar irradiation issue, modification in reactor sizes, plate thickness, the requirement of dedicated large land area, etc. Hence, the existing models of PBRs can be upgraded with the motive of high performance efficiency and maintenance of maximum biomass productivities.

To increase performance, the PBR technology should be assisted by proper mixing, efficient illumination, dewatering and processing. Enclosed PBR can be a promising system for the production of microalgae-derived products such as pharmaceutical and food supplements. The mode of culture system of specific microalgae or cyanobacteria cannot be inferred from other culture conditions. It is because the species may respond differently by altering the PBRs or culture conditions.

Future research must focus on algal biofuel coupling with other sustainable technologies for environmental security, such as wastewater bioremediation with the production of microalgae-based valuable coproducts, and CO_2 sequestration from flue gas. Newly developed algal technology and design should be analyzed for the total capital cost and lifecycle assessment to understand its impact on the environment as well as economic viability. Furthermore, potential microalgal strains and cultivation sites should be identified and evaluated as the outdoor cultivation is influenced by sunlight duration and varying temperature as season varies. In tropical developing countries, seasons are typically neither too hot nor cold that can favor the growth of microalgae for a relatively long period. Thus, in future research, the tropical regions may be chosen as prominent sites for the industrial cultivation of algae.

Acknowledgments

The authors would like to acknowledge the Indo-US Science and Technology Forum (IUSSTF) and Department of Biotechnology (DBT), Government of India, for financial assistantship under Bioenergy-Awards for Cutting Edge Research (B-ACER) program [Award Letter No. IUSSTF B-ACER 2019 Internship/5/Nongmaithem Debeni Devi, 03/01/2020].

References

Alkayal, F., Albion, R. L., Tillett, R. L., Hathwaik, L. T., Lemos, M. S., and Cushman, J. C. (2010). Expressed sequence tag (EST) profiling in hyper saline shocked *Dunaliella salina* reveals high expression of protein synthetic apparatus components. *Plant Science* 179(5):437–449.

Ashen, J. B., and Goff, L. J. (2000). Molecular and ecological evidence for species specificity and coevolution in a group of marine algal-bacterial symbioses. *Applied and Environmental Microbiology* 66(7):3024–3030.

Banerjee, S., and Ramaswamy, S. (2019). Dynamic process model and economic analysis of microalgae cultivation in flat panel photobioreactors. *Algal Research* 39:101445.

Belkin, S., and Boussiba, S. (1991). Resistance of *Spirulina platensis* to ammonia at high pH values. *Plant and Cell Physiology* 32(7):953–958.

Berberoglu, H., Yin, J., and Pilon, L. (2007). Light transfer in bubble sparged photobioreactors for H_2 production and CO_2 mitigation. *International Journal of Hydrogen Energy* 32(13):2273–2285.

Bernstein, H. C., Kesaano, M., Moll, K., Smith, T., Gerlach, R., Carlson, R. P., and Sims, R. C. (2014). Direct measurement and characterization of active photosynthesis zones inside wastewater remediating and biofuel producing microalgal biofilms. *Bioresource Technology* 156:206–215.

Borowitzka, M. A. (1999). Commercial production of microalgae: Ponds, tanks, and fermenters. *Progress in Industrial Microbiology* 35(C):313–321.

Brennan, L., and Owende, P. (2010). Biofuels from microalgae-A review of technologies for production, processing, and extractions of biofuels and co-products. *Renewable and Sustainable Energy Reviews* 14(2):557–577.

Camargo, O. (2014). Algae Systems Testing Algae Bio-Fuel Production. *Businesswire The Dilenschneider Group.* http://www.mfrtech.com/news/513283/algae_systems_testing_algae_biofuel_production.html

Cao, J., Yuan, H. L., Li, B. Z., and Yang, J. S. (2014). Significance evaluation of the effects of environmental factors on the lipid accumulation of *Chlorella minutissima* UTEX 2341 under low-nutrition heterotrophic condition. *Bioresource Technology* 152:177–184.

Carr, M. (2015). Algae Industry Project Book 2015. *Algae Biomass Organization.* www.algaebiomass.org

Cheirsilp, B., and Torpee, S. (2012). Enhanced growth and lipid production of microalgae under mixotrophic culture condition: Effect of light intensity, glucose concentration and fed-batch cultivation. *Bioresource Technology* 110:510–516.

Chen, J., Wang, Y., Benemann, J. R., Zhang, X., Hu, H., and Qin, S. (2016). Microalgal industry in China: Challenges and prospects. *Journal of Applied Phycology* 28(2):715–725.

Chen, Y. H., and Walker, T. H. (2011). Biomass and lipid production of heterotrophic microalgae *Chlorella protothecoides* by using biodiesel-derived crude glycerol. *Biotechnology Letters* 33(10):1973–1983.

Chi, Z., Pyle, D., Wen, Z., Frear, C., and Chen, S. (2007). A laboratory study of producing docosahexaenoic acid from biodiesel-waste glycerol by microalgal fermentation. *Process Biochemistry* 42(11):1537–1545.

Chisti, Y. (2008). Biodiesel from microalgae beats bioethanol. *Trends in Biotechnology* 26(3):126–131.

Chiu, S. Y., Kao, C. Y., Chen, T. Y., Chang, Y. B., Kuo, C. M., and Lin, C. S. (2015). Cultivation of microalgal *Chlorella* for biomass and lipid production using wastewater as nutrient resource. *Bioresource Technology* 184:179–189.

Chiu, S. Y., Kao, C. Y., Tsai, M. T., Ong, S. C., Chen, C. H., and Lin, C. S. (2009). Lipid accumulation and CO_2 utilization of *Nannochloropsis oculata* in response to CO_2 aeration. *Bioresource Technology* 100(2):833–838.

Christenson, L. B., and Sims, R. C. (2012). Rotating algal biofilm reactor and spool harvester for wastewater treatment with biofuels by-products. *Biotechnology and Bioengineering* 109(7):1674–1684.

Costa, J. A. V., and de Morais, M. G. (2014). An open pond system for microalgal cultivation. In Pandey, A., Chisti, Y., Lee, D.J. and Soccol, C.R. (eds.) *Biofuels from Algae.* Elsevier, pp. 1–22.

Das, D. (ed.) (2015). *Algal Biorefinery: An Integrated Approach*. Springer.

De Morais, M. G., and Costa, J. A. V. (2007). Carbon dioxide fixation by *Chlorella kessleri, C. vulgaris, Scenedesmus obliquus* and *Spirulina* sp. cultivated in flasks and vertical tubular photobioreactors. *Biotechnology Letters* 29(9):1349–1352.

De Vree, J. H., Bosma, R., Janssen, M., Barbosa, M. J., and Wijffels, R. H. (2015). Comparison of four outdoor pilot-scale photobioreactors. *Biotechnology for Biofuels* 8:1.

Dogaris, I., Welch, M., Meiser, A., Walmsley, L., and Philippidis, G. (2015). A novel horizontal photobioreactor for high-density cultivation of microalgae. *Bioresource Technology* 198:316–324.

Duan, X., Ren, G. Y., Liu, L. L., and Zhu, W. X. (2012). Salt-induced osmotic stress for lipid overproduction in batch culture of *Chlorella vulgaris*. *African Journal of Biotechnology* 11(27):7072–7078.

Dunford, N. (2015). Photobioreactor Design for Algal Biomass Production. *Food Technology Fact Sheet*, FAPC-192. http://pods.dasnr.okstate.edu/docushare/dsweb/Get/Document-9676/FAPC-191web.pdf

El Baky, H. H. A., El-Baroty, G. S., Bouaid, A., Martinez, M., and Aracil, J. (2012). Enhancement of lipid accumulation in *Scenedesmus obliquus* by Optimizing CO_2 and Fe^{3+} levels for biodiesel production. *Bioresource Technology* 119:429–432.

Ethier, S., Woisard, K., Vaughan, D., and Wen, Z. (2011). Continuous culture of the microalgae *Schizochytrium limacinum* on biodiesel-derived crude glycerol for producing docosahexaenoic acid. *Bioresource Technology* 102(1):88–93.

Feng, P., Deng, Z., Fan, L., and Hu, Z. (2012). Lipid accumulation and growth characteristics of *Chlorella zofingiensis* under different nitrate and phosphate concentrations. *Journal of Bioscience and Bioengineering* 114(4):405–410.

Fu, L., Cui, X., Li, Y., Xu, L., Zhang, C., Xiong, R., and Crittenden, J. C. (2017). Excessive phosphorus enhances *Chlorella regularis* lipid production under nitrogen starvation stress during glucose heterotrophic cultivation. *Chemical Engineering Journal* 330:566–572.

Goldman, J. C. (1979). Outdoor algal mass cultures-II. Photosynthetic yield limitations. *Water Research* 13(2):119–136.

Gross, M., and Wen, Z. (2014). Yearlong evaluation of performance and durability of a pilot-scale Revolving Algal Biofilm (RAB) cultivation system. *Bioresource Technology* 171:50–58.

Guldhe, A., Ansari, F. A., Singh, P., and Bux, F. (2017). Heterotrophic cultivation of microalgae using aquaculture wastewater: A biorefinery concept for biomass production and nutrient remediation. *Ecological Engineering* 99:47–53.

Gupta, P. L., Lee, S. M., and Choi, H. J. (2015). A mini review: Photobioreactors for large scale algal cultivation. *World Journal of Microbiology and Biotechnology* 31(9):1409–1417.

Han, F., Huang, J., Li, Y., Wang, W., Wang, J., Fan, J., and Shen, G. (2012). Enhancement of microalgal biomass and lipid productivities by a model of photoautotrophic culture with heterotrophic cells as seed. *Bioresource Technology* 118:431–437.

Ho, S. H., Chen, C. Y., and Chang, J. S. (2012). Effect of light intensity and nitrogen starvation on CO_2 fixation and lipid/carbohydrate production of an indigenous microalga *Scenedesmus obliquus* CNW-N. *Bioresource Technology* 113:244–252.

Hu, Q., Zhang, C., and Sommerfeld, M. (2006). Biodiesel from algae: Lessons learned over the past 60 years and future perspectives. *Journal of Phycology* 42(1):7–12.

Illman, A. M., Scragg, A. H., and Shales, S. W. (2000). Increase in *Chlorella* strains calorific values when grown in low nitrogen medium. *Enzyme and Microbial Technology* 27(8):631–635.

Ista, L. K., Callow, M. E., Finlay, J. A., Coleman, S. E., Nolasco, A. C., Simons, R. H., Callorw, J. A., and Lopez, G. P. (2004). Effect of substratum surface chemistry and surface energy on attachment of marine bacteria and algal spores. *Applied and Environmental Microbiology* 70(7):4151–4157.

Katiyar, R., Gurjar, B. R., Bharti, R. K., Kumar, A., Biswas, S., and Pruthi, V. (2017). Heterotrophic cultivation of microalgae in photobioreactor using low cost crude glycerol for enhanced biodiesel production. *Renewable Energy* 113:1359–1365.

Katuwal, S. (2017). Designing and development of a photobioreactor for optimizing the growth of micro algae and studying its growth parameters. *Electronic Theses and Dissertations- open prairie.* https://openprairie.sdstate.edu/etd/2161

Khozin-Goldberg, I., and Cohen, Z. (2006). The effect of phosphate starvation on the lipid and fatty acid composition of the fresh water eustigmatophyte *Monodus subterraneus. Phytochemistry* 67(7):696–701.

Koller, M. (2015). Design of closed photobioreactors for algal cultivation. In Koller, M. (ed.) *Algal Biorefineries: Volume 2: Products and Refinery Design.* Springer, pp. 133–186.

Kong, W., Yang, S., Wang, H., Huo, H., Guo, B., Liu, N., and Niu, S. (2020). Regulation of biomass, pigments, and lipid production by *Chlorella vulgaris* 31 through controlling trophic modes and carbon sources. *Journal of Applied Phycology* 32:1569–1579.

Kumar, K., Dasgupta, C. N., Nayak, B., Lindblad, P., and Das, D. (2011). Development of suitable photobioreactors for CO_2 sequestration addressing global warming using green algae and cyanobacteria. *Bioresource Technology* 102:4945–4953.

Lee, S. H., Oh, H. M., Jo, B. H., Lee, S. A., Shin, S. Y., Kim, H. S., and Ahn, C. Y. (2014). Higher biomass productivity of microalgae in an attached growth system, using wastewater. *Journal of Microbiology and Biotechnology* 24(11):1566–1573.

Lee, Y. K. (2001). Microalgal mass culture systems and methods: Their limitation and potential. *Journal of Applied Phycology* 13:307–315.

Leong, W. H., Lim, J. W., Lam, M. K., Uemura, Y., Ho, C. D., and Ho, Y. C. (2018). Co-cultivation of activated sludge and microalgae for the simultaneous enhancements of nitrogen-rich wastewater bioremediation and lipid production. *Journal of the Taiwan Institute of Chemical Engineers* 87:216–224.

Liang, Y., Sarkany, N., and Cui, Y. (2009). Biomass and lipid productivities of *Chlorella vulgaris* under autotrophic, heterotrophic and mixotrophic growth conditions. *Biotechnology Letters* 31(7):1043–1049.

Liu, J., Yuan, C., Hu, G., and Li, F. (2012). Effects of light intensity on the growth and lipid accumulation of microalga *Scenedesmus* sp. 11-1 under nitrogen limitation. *Applied Biochemistry and Biotechnology* 166(8):2127–2137.

Liu, L., Chen, J., Lim, P. E., and Wei, D. (2018). Dual-species cultivation of microalgae and yeast for enhanced biomass and microbial lipid production. *Journal of Applied Phycology* 30(6):2997–3007.

Liu, W., Au, D. W. T., Anderson, D. M., Lam, P. K. S., and Wu, R. S. S. (2007). Effects of nutrients, salinity, pH and light:dark cycle on the production of reactive oxygen species in the alga *Chattonella marina. Journal of Experimental Marine Biology and Ecology* 346(1–2):76–86.

Mandal, S., and Mallick, N. (2009). Microalga *Scenedesmus obliquus* as a potential source for biodiesel production. *Applied Microbiology and Biotechnology* 84(2):281–291.

Mandalam, R. K., and Palsson, B. (1998). Elemental balancing of biomass and medium composition enhances growth capacity in high-density *Chlorella vulgaris* cultures. *Biotechnology and Bioengineering* 59(5):605–611.

Mata, T. M., Martins, A. A., and Caetano, N. S. (2010). Microalgae for biodiesel production and other applications: A review. *Renewable and Sustainable Energy Reviews* 14(1):217–232.

Mendoza, J. L., Granados, M. R., de Godos, I., Acién, F. G., Molina, E., Heaven, S., and Banks, C. J. (2013). Oxygen transfer and evolution in microalgal culture in open raceways. *Bioresource Technology* 137:188–195.

Metting, F. B. (1996). Biodiversity and application of microalgae. *Journal of Industrial Microbiology and Biotechnology* 17(5–6):477–489.

Mokashi, K., Shetty, V., George, S. A., and Sibi, G. (2016). Sodium bicarbonate as inorganic carbon source for higher biomass and lipid production integrated carbon capture in *Chlorella vulgaris*. *Achievements in the Life Sciences* 10(1):111–117.

Nayak, M., Suh, W. I., Lee, B., and Chang, Y. K. (2018). Enhanced carbon utilization efficiency and FAME production of *Chlorella* sp. HS2 through combined supplementation of bicarbonate and carbon dioxide. *Energy Conversion and Management* 156:45–52.

Ogbonna, J. C., and Tanaka, H. (2000). Light requirement and photosynthetic cell cultivation - Development of processes for efficient light utilization in photobioreactors. *Journal of Applied Phycology* 12(3–5):207–218.

Pancha, I., Chokshi, K., Ghosh, T., Paliwal, C., Maurya, R., and Mishra, S. (2015a). Bicarbonate supplementation enhanced biofuel production potential as well as nutritional stress mitigation in the microalgae Scenedesmus sp. CCNM 1077. *Bioresource Technology* 193:315–323.

Pancha, I., Chokshi, K., Maurya, R., Trivedi, K., Patidar, S. K., Ghosh, A., and Mishra, S. (2015b). Salinity induced oxidative stress enhanced biofuel production potential of microalgae *Scenedesmus* sp. CCNM 1077. *Bioresource Technology* 189:341–348.

Rajendran, A., and Hu, B. (2016). Mycoalgae biofilm: Development of a novel platform technology using algae and fungal cultures. *Biotechnology for Biofuels* 9(1):112.

Rao, A. R., Ravishankar, G. A., and Sarada, R. (2012). Cultivation of green alga *Botryococcus braunii* in raceway, circular ponds under outdoor conditions and its growth, hydrocarbon production. *Bioresource Technology* 123:528–533.

Rashid, N., Ryu, A. J., Jeong, K. J., Lee, B., and Chang, Y. K. (2019). Co-cultivation of two freshwater microalgae species to improve biomass productivity and biodiesel production. *Energy Conversion and Management* 196:640–648.

Rashid, N., Ur Rehman, M. S., Sadiq, M., Mahmood, T., and Han, J. I. (2014). Current status, issues and developments in microalgae derived biodiesel production. *Renewable and Sustainable Energy Reviews* 40:760–778.

Reitan, K. I., Rainuzzo, J. R., and Olsen, Y. (1994). Effect of nutrient limitation on fatty acid and lipid content of marine microalgae. *Journal of Phycology* 30(6):972–979.

Richardson, J. W., Johnson, M. D., and Outlaw, J. L. (2012). Economic comparison of open pond raceways to photo bio-reactors for profitable production of algae for transportation fuels in the Southwest. *Algal Research* 1(1):93–100.

Rodolfi, L., Zittelli, G. C., Bassi, N., Padovani, G., Biondi, N., Bonini, G., and Tredici, M. R. (2009). Microalgae for oil: Strain selection, induction of lipid synthesis and outdoor mass cultivation in a low-cost photobioreactor. *Biotechnology and Bioengineering* 102(1):100–112.

Roncallo, O. P., García Freites, S., Paternina Castillo, E., Bula Silvera, A., Cortina, A., and Acuña, F. (2013). Comparison of two different vertical column photobioreactors for the cultivation of *Nannochloropsis* sp. *Journal of Energy Resources Technology* 135(1):011201.

Sharma, K. K., Schuhmann, H., and Schenk, P. M. (2012). High lipid induction in microalgae for biodiesel production. *Energies* 5(5):1532–1553.

Shen, Y., Gao, J., and Li, L. (2017). Municipal wastewater treatment via co-immobilized microalgal-bacterial symbiosis: Microorganism growth and nutrients removal. *Bioresource Technology* 243:905–913.

Siaut, M., Cuiné, S., Cagnon, C., Fessler, B., Nguyen, M., Carrier, P., and Peltier, G. (2011). Oil accumulation in the model green alga *Chlamydomonas reinhardtii*: Characterization, variability between common laboratory strains and relationship with starch reserves. *BMC Biotechnology* 11(1):7.

Srivastava, G., Nishchal, and Goud, V. V. (2017). Salinity induced lipid production in microalgae and cluster analysis (ICCB 16-BR_047). *Bioresource Technology* 242:244–252.

Sydney, E. B., Sturm, W., de Carvalho, J. C., Thomaz-Soccol, V., Larroche, C., Pandey, A., and Soccol, C. R. (2010). Potential carbon dioxide fixation by industrially important microalgae. *Bioresource Technology* 101(15):5892–5896.

Takagi, M., Karseno, and Yoshida, T. (2006). Effect of salt concentration on intracellular accumulation of lipids and triacylglyceride in marine microalgae *Dunaliella* cells. *Journal of Bioscience and Bioengineering* 101(3):223–226.

Tang, D., Han, W., Li, P., Miao, X., and Zhong, J. (2011). CO_2 biofixation and fatty acid composition of *Scenedesmus obliquus* and *Chlorella pyrenoidosa* in response to different CO_2 levels. *Bioresource Technology* 102(3):3071–3076.

Ugwu, C. U., Aoyagi, H., and Uchiyama, H. (2008). Photobioreactors for mass cultivation of algae. *Bioresource Technology* 99:4021–4028.

Voegele, E. (2014). Biodiesel Magazine - The Latest News and Data about Biodiesel Production. *Aemetis Inc.* http://www.biodieselmagazine.com/articles/7767/co2-giant-partners-with-sapphire-to-develop-open-pond-system

Wang, X. W., Liang, J. R., Luo, C. S., Chen, C. P., and Gao, Y. H. (2014). Biomass, total lipid production, and fatty acid composition of the marine diatom *Chaetoceros muelleri* in response to different CO_2 levels. *Bioresource Technology* 161:124–130.

Watanabe, M., Suzuki, Y., Sasaki, K., Nakashimada, Y., and Nishio, N. (1999). Flocculating property of extracellular polymeric substance derived from a marine photosynthetic bacterium, *Rhodovulum* sp. *Journal of Bioscience and Bioengineering* 87(5):625–629.

Wei, L., Huang, X., and Huang, Z. (2014). Temperature effects on lipid properties of microalgae *Tetraselmis subcordiformis* and *Nannochloropsis oculata* as biofuel resources. *Chinese Journal of Oceanology and Limnology* 33(1):99–106.

Xia, L., Rong, J., Yang, H., He, Q., Zhang, D., and Hu, C. (2014). NaCl as an effective inducer for lipid accumulation in freshwater microalgae *Desmodesmus abundans*. *Bioresource Technology* 161:402–409.

Xiong, W., Li, X., Xiang, J., and Wu, Q. (2008). High-density fermentation of microalga *Chlorella protothecoides* in bioreactor for microbio-diesel production. *Applied Microbiology and Biotechnology* 78(1):29–36.

Yeh, K. L., and Chang, J. S. (2012). Effects of cultivation conditions and media composition on cell growth and lipid productivity of indigenous microalga *Chlorella vulgaris* ESP-31. *Bioresource Technology* 105:120–127.

Yen, H. W., Ho, S. H., Chen, C. Y., and Chang, J. S. (2015). CO_2, NO_x and SO_x removal from flue gas via microalgae cultivation: A critical review. *Biotechnology Journal* 10(6):829–839.

Yen, H. W., Hu, I. C., Chen, C. Y., Ho, S. H., Lee, D. J., and Chang, J. S. (2013). Microalgae-based biorefinery - From biofuels to natural products. *Bioresource Technology* 135:166–174.

3 Strategic approaches to enhance lipid accumulation in microalgae for bioprospecting

Khalifa S.H. Eldiehy, Velentina Das, Doljit Borah, Manabendra Mandal, and Dhanapati Deka

Introduction

With the continual increase of the global population and limited existed resources, the search for unconventional resources to overcome these crises has become a critical issue. Bioprospecting has some exciting features, such as the sustainability and renewability of its sources, which make it one of the leading solutions for several long-term problems. Recently, microalgae have gained massive attention as a promising source of alternative fuels, pharmaceutical products, nutritional supplements, livestock feeds and highly valuable molecules.

Microalgae are a varied group of eukaryotic and prokaryotic photosynthetic cellular organisms that show different unicellular to colonial morphologies (Bellinger and Sigee, 2015; Khan et al., 2018). Besides, they grow rapidly because of their simple structure as well as short doubling time and produce at least 30 times more oil per acre than seeds of plants. In addition, algae may grow under extreme conditions; they can flourish on non-arable coastal lands that are unsuitable to traditional farming, using wastewater or seawater (Ullah et al., 2014; Reese et al., 2019). Microalgae can be found in aquatic habitats such as lakes, seawater, freshwater, ponds and even wastewater with a photosynthesis capability of producing around 50% of all oxygen on the planet (Zuccaro et al., 2020). It has been estimated that around 40,000 microalgal species have been described and that most of them are capable of generating and accumulating lipid content of up to 20–50% of their dry weight with the possibility of raising this percentage in some species when applying abiotic stress regime (Khan et al., 2018; Arif et al., 2020).

Lipid accumulation in microalgae has always drawn worldwide importance in the production of algal biofuels and other related products. It is important to study the cell cycle of algae and, at the same time, the metabolic pathways of algae before studying the lipid accumulation. Microalgae can grow under different metabolic pathways including the following:

i autotrophic, i.e., algae capable of self-nourishment by using inorganic materials as nutrient and energy source;

ii mixotrophic, i.e., algae that can grow either phototrophic or heterotrophic by using sunlight and organic or inorganic carbon; and
iii heterotrophic, i.e., algae that use organic compounds as their primary energy source as well as carbon (Chen et al., 2011; Pankratz et al., 2017).

Starch and lipids are two key components of algal cells and responsible for buffering variable supplies of energy and carbon that are vital for cell growth and reproduction, particularly DNA replication and nuclear and cellular division (Vitova et al., 2015). There is a competition for the formation of lipid and starch during metabolism, which begins with the similar 3-carbon such as glyceraldehyde 3-phosphate (GAP) and 3-phosphoglycerate (3PG) (Berg et al., 2002; de Jaeger et al., 2014). When the route toward starch formation is blocked, the production of lipid is increased (Kong et al., 2018).

Lipids present in microalgae can be divided into two categories. One is a polar lipid, which forms the membranes and organelles of the microalgae, and the second is a nonpolar lipid, which is the reservoir of energy. Polar lipids are also known as structural lipids such as phospholipids and glycolipids, and they are attached to the membranes of the organelles, such as chloroplast thylakoid membranes (Hu et al., 2008). Nonpolar lipids are neutral lipids and are also known as storage lipids such as hydrocarbons, pigments, free fatty acids, monoglycerides, diglycerides and triglycerides. The form of stored lipids differs according to strain, surrounding growth conditions and growth phases. The ratio of polar and nonpolar lipids may alter the metabolic rate along with different algal growth phases.

Triacylglycerol (TAG) plays an essential role in stocking energy within the microalgal cells. Three primary methods for the synthesis of triglycerides – cytosol, chloroplast and endoplasmic reticulum (ER) pathways (Aratboni et al., 2019) – are described in Figure 3.1. Generally, the structural lipid fraction contains mainly polyunsaturated fatty acids (PUFAs), whereas the monounsaturated fatty acids (MUFAs) and the saturated fatty acids (SAFAs) involve the storage lipid fraction (Olofsson et al., 2012; Torres et al., 2013).

Microalgal lipids are widely utilized in many industrial fields such as cosmetics, fuel and food and pharmaceutical industries. Some algal lipids such as phosphatidylcholine, phosphatidylglycerol and various galactosyl triglycerides can be used as surfactants with different properties rather than synthetic ones since they are degradable (Hayes and Smith, 2019). Furthermore, omega-3, long-chain n-3 PUFAs, PUFA-rich triacylglycerides, eicosanoids and derivatives of 20-carbon PUFAs are widely consumed as pharmaceutical products (Weintraub, 2013; Jha et al., 2017; Gautam and Mannan, 2020). Phospholipids are also known to be surface-active and are used as emulsifiers in pharmaceuticals, food and cosmetics (Foley et al., 2011). Hydrocarbons are another type of microalgal lipids that could be transformed into biofuels by catalytically breaking lipids of microalgae into thin alkenes (olefins) or thermochemical liquefaction (Goyal et al., 2018; Shakya et al., 2018). Microalgal lipids have been considered as one of the most promising feedstocks for

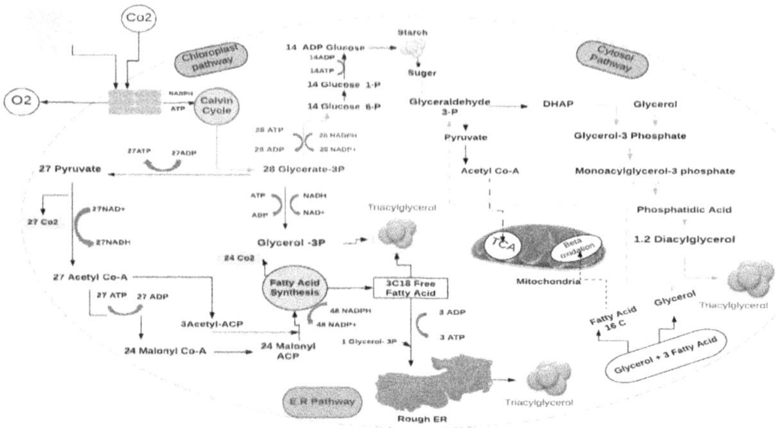

Figure 3.1 A simplified view of the main biochemical pathways for triglyceride synthesis in microalgal cells. Adapted from Aratboni et al. (2019).

biodiesel production. Microalgae biodiesel is typically produced by extracting lipids and then converting them to fatty acid methyl esters (FAME) through the transesterification process (Shirazi et al., 2017).

Generally, the type and amount of lipid accumulated in algae play a critical role in determining the quality of biodiesel produced. Some microalgal species such as *Ankistrodesmus fusiformis*, *Kirchneriella lunaris*, *Chlamydocapsa bacillus* and *Ankistrodesmus falcatus* with high levels of polyunsaturated FAME are generally preferred for biodiesel production. This preference is due to it being a much cleaner fuel than that of non-renewable energy sources, being ecologically friendlier and tending to be utilized in diesel vehicles without modifications (Nascimento et al., 2013; Bušić et al., 2018). The increase in microalgal lipid content and the reducing production costs are the main criteria in biodiesel's commercial applicability. For a considerable time, the classic approach for enhancing lipid content has been based on limited nutrients and applying stress conditions to the culture growth medium. Lately, some new strategies to resolve the problems of conventional approaches and achieve maximum possible results have been explored concerning sustainability, lipid yields and economic efficiency.

Classical approaches to enhance microalgal lipid accumulation

The culture conditions and strategies applied to microalgae cultivations are varying according to the valuable metabolites targeted for production. Under normal conditions, the microalgae tend to use their biological processes to flourish and increase the biomass. In contrast, under unfavorable conditions, the microalgae resist such circumstances by altering their metabolism pathways to produce some valuable metabolites such as lipids and carotenoids that are needed to keep their survival. Induced lipid accumulation by manipulating nutrient regime and physical surroundings has been widely studied as a traditional strategy to increase microalgal lipid content. Several theories have been forwarded to explain the mechanism behind the enhanced lipid accumulation under stress conditions. These theories can be summarized as follows:

i Stress conditions cause a reduction in growth rates that induce microalgae to store up extra lipid reserves as a consequence of the absence of an increasing number of cells (Singh et al., 2011). It also possesses the ability to store energy in the form of TAGs that serve as an essential carbon source (Merchant et al., 2012), thus helping them to resist such conditions (Fan et al., 2014).

ii Another assumption claimed that the enhanced lipid accumulation is a result of rearranging the cellular components where stress conditions cause rearrangement of membrane lipids, in particular chloroplast lipids (Srinophakun et al., 2017).

iii Other hypotheses claimed that during unfavorable conditions, cells induce the accumulation of lipids to maintain a homeostatic cellular redox situation (Dunyaporn et al., 2008).

Inducing lipid accumulation through manipulating nutrient regime

Manipulation of nutritional strategies is described as a modification of essential ingredients requirement in culture growth media (nitrogen, phosphorus, carbon source, magnesium, potassium, iron, etc.). The alteration in the concentration of macronutrients present in medium culture contributes to a change in biomass, lipid, carbohydrate and other valuable metabolites (Markou et al., 2012; Zhu, 2015; Chu, 2017).

Nitrogen

Nitrogen is one of the essential elements of culture medium constituents, representing 1–10% of the total microalgae's dry weight (Minhas et al., 2016). Nitrogen contributes to the structural, functional and cellular processes of living cells, where it is present in protein-forming amino acids and nucleic acids that make up genetic content for live organisms, along with many other organic and inorganic compounds. Manipulation of nitrogen concentration directly affects the biomass, the lipid and other metabolites' content of the algal cell. Moreover, the lipid accumulation efficiency is also affected by the type and concentration of nitrogen used.

The lipid content of *Tetraselmis* sp. was significantly affected by different forms of nitrogen, such as ammonium, nitrate, glycine, urea and yeast extract when used as a nutrient supplement for growth (Kim et al., 2016). In the same context, Wu et al. (2013) found potassium nitrate as the most appropriate nitrogen source for biomass and lipid productivity of *Monoraphidium* sp. SB2 compared to ammonium chloride and ammonium nitrate, respectively.

As reported by Altın et al. (2018), the highest lipid content (16.4%) and biomass productivity (4.21 mg/L/d) for *Chlorella variabilis* were recorded when urea was used as a sole nitrogen source followed by sodium nitrate with 15.2% and 4.15 mg/L/d, respectively, and ammonium chloride with 13.9% and 1.9 mg/L/d, respectively. Many recent studies indicate the positive impact of the nitrogen limitation simultaneously with inducing lipid accumulation of microalgae. Three prospective reasons behind lipid accumulation under nitrate starving conditions are the stimulation of phospholipids hydrolysis, activation of acyl hydrolase and a decrease in the cellular content of the thylakoid membrane (De Bhowmick et al., 2015).

Wang et al. (2019) conducted a study on the impact of nitrogen deficiency on four marine microalgae, namely *Nannochloropsis oceanica, Phaeodactylum tricornutum, Rhodomonas baltica* and *Isochrysis aff. galbana clone T-Iso*, and demonstrated that under nitrogen-limited conditions, the TAG was the primary form for the lipid accumulation in all studied strains, where, under moderate nitrogen limitation, the lipid accumulated in *N. oceanica* was induced by 51% and around 87% of fatty acid in the form of TAG. Fawzy (2017) reported that the higher lipid content and productivity of *Asteromonas gracilis* under nitrogen starvation were 29.3% and 10.21 mg/L/d, respectively.

In another study, Can et al. (2017) reported a 4.7% enhancement of the lipid accumulation in the dry weight of *Spirulina platensis* grown under nitrogen-free medium. Growing of *Parachlorella kessleri* TY02 for 7 days under both nitrogen-rich and nitrogen-limited media showed a remarkable increase in the lipid accumulation from 11.2% to 33.2% and from 12.3% to 45.5%, respectively (Gao et al., 2019). Negi et al. (2016) found that the oil accumulation in *Chlorella sorokiniana* was promoted under nitrogen-depleted growth conditions by ~20-fold. Zhu et al. (2016) also noted that *Chlorella* sp. A2 promoted the neutral lipid buildup without a significant impact on cell growth when grown with a minimum supply of nitrogen. Upon the consumption of nitrogen from the culture medium, cell division was suppressed and reached the stationary phase. Consequently, the accumulated carbons were converted into lipid substances (Adenan et al., 2016).

Phosphorus

Phosphorus has a critical role in most processes of cell metabolism, especially those involved in energy transfer, signal transduction, respiration, photosynthesis and macromolecular biosynthesis (Lei and Lu, 2011). Phosphorus constitutes significantly less than 1% of total algal biomass (0.03–0.06%) and yet is a critical ingredient for sustainable microalgae prosperity and development (Hannon et al., 2010). The absence of phosphorus in the growth medium has severe consequences on the growth of microalgae, repression of photosynthesis (Belotti et al., 2014) and impairment in phospholipid membrane (Liang et al., 2013). Microalgae consume phosphorus in the form of inorganic orthophosphate; consequently, the organic phosphates are converted into orthophosphates on the cell surface by phosphatase enzymes to facilitate the use of it by cells. This process happens especially when there is a short supply of inorganic phosphate, and this process requires a quantity of energy (active transport) to be completed (Tripathi et al., 2019). On the other hand, in high supply, the excess of phosphorus is stored in the form of polyphosphate granules, which can be consumed later when phosphorus is depleted. In converse to the immediate responses of microalgae to temperature and light, their growth does not instantly respond to environmental changes in phosphorus concentration (Sajjadi et al., 2018).

Severe reduction in phospholipid membrane has been reported in *Chlorella* sp. under phosphorus starvation conditions, and non-phosphorus sulfolipids and glycolipids substitute this shortage. This process offers an effective mechanism to preserve phosphorus (Liang et al., 2013). Induction of lipid accumulation in microalgal cells through the limitation of phosphorus concentrations has been widely explored. For instance, under low phosphorus concentrations (22 and 55 mg/L), the tropical green microalga *Scenedesmus dimorphus* KMITL showed the maximum lipid content and production amounting to 19.4% and 20.65 mg/L/d, respectively (Ruangsomboon et al., 2013). The maximum lipid content for *Isochrysis galbana* U4 was recorded at phosphorus-free media (0% P) and phosphorus-limited media (1–12.5% P). The lipid accumulations went up to approximately 50% w/w lipid by day 14 (Roopnarain et al., 2014).

Another study showed the enhancement in lipid content of *Chlorella* sp. and *Chaetoceros calcitrans* by 32.0% and 35%, respectively, when cultured under a 75% decrease in phosphorus concentrations (Adenan et al., 2016). Liang et al. (2013) conducted a study on the impact of different phosphorus concentrations on lipid accumulations of freshwater microalga *Chlorella* sp. and reported a remarkable increase in lipid content and productivity when it was cultivated under low-phosphorus conditions. The lipid content increased approximately twofold (from 23% to 53%) in *Scenedesmus* sp. when the concentration of phosphorus was reduced from 2 to 0.1 mg/L (Xin et al., 2011). Neha and Khan (2016) noted that the highest lipid accumulation in *Chlorella vulgaris* reached four-fold compared to the control culture.

Michelon et al. (2016) indicated that the depletion of phosphorus in the culture of *Chlorella* spp. stimulated lipid content from 3% to 4.1%, while the combination of nitrogen and phosphorus deficiency showed the maximum lipid accumulation by 16.3%. Another study showed that the cyanobacterium *Arthrospira platensis* had a considerable increase in the amount of lipid and carotenoids of 128% and 64%, respectively, when cultivated under phosphorus-free medium (El-Shouny et al., 2015).

Sulfur

Sulfur is a fundamental ingredient for intermediates of primary metabolism and catalysts. It has been identified in nucleotides, nucleic acids, amino acids, proteins and vitamins. In addition, it plays an essential role in mechanisms to respond to changes in abiotic and biotic environmental circumstances. These responses comprise quantitative and qualitative alterations in the proteome and in heavy metal chelators, dimethylsulfoniopropionate and compounds such as glutathione that help in the defense against oxidative stress (Giordano and Raven, 2014). Many previous studies accomplished enhanced lipid accumulation in microalgae through cultivation under the sulfur-starved condition.

Cakmak et al. (2012) compared TAG accumulation produced from wild-type *Chlamydomonas reinhardtii* CC-124 mt (−) and CC-125 mt (+) strains cultivated under sulfur starvation conditions; they reported that the starvation condition resulted in enhancement of the lipid content by 124% and 302% in *C. reinhardtii* CC-124 and CC-125, respectively. Also, Salas-Montantes et al. (2018) stated that under sulfur-free medium, the total fatty acid content for the transgenic strain *Chlamydomonas reinhardtii* was enhanced 2.6-fold (15.6 dry cell weight) compared to the untransformed strain. The cultivation of *Chlorellaceae* spp. under sulfur-deficient conditions showed an increase in the lipid accumulation amounting to 1.5- to 2.4-fold in comparison with sulfur-sufficient conditions (Mizuno et al., 2013). Moreover, *Chlamydomonas reinhardtii* accumulated higher amounts of lipid under sulfur-starved medium (Morowvat and Ghasemi, 2016).

Metals ions

The growth and lipid content of microalgae can also be affected by the concentration of metal ions. Recent studies have demonstrated that specific strains of microalgae can increase the accumulation of lipids when grown under appropriate metal ion concentrations (Che et al., 2015). Gorain et al. (2013) conducted a study to understand a relationship between the enhanced lipid accumulation for *Chlorella vulgaris* and *Scenedesmus obliquus* with metals such as calcium and magnesium. They reported that the increased magnesium concentration (2X) in the culture medium resulted in enhanced cellular lipid content of *C. vulgaris* and *S. obliquus* up to 27% and 26% (dry cell weight), respectively. In comparison, the lipid was increased up to 40.3% and 37.0% (dry cell weight) in *C. vulgaris* and *S. obliquus*, respectively, under calcium-starved conditions. The maximum lipid productivity in *Scenedesmus* sp. amounting to 249.8, 247.8 and 275.7 mg/L/d was achieved at metal ion concentrations of 1.2×10^{-3} g/L (Fe^{3+}), 7.3×10^{-3} g/L (Mg^{2+}) and 9.8×10^{-1} g/L (Ca^{2+}), respectively (Ren et al., 2014). It has been demonstrated that the increased concentration of iron positively affects the lipid accumulation of *Scenedesmus* sp. and *Chlorella* sp., and *Chlamydomonas* sp. (Sivaramakrishnan and Incharoensakdi, 2017). *Monoraphidium* sp. FXY-10 had a higher lipid content and productivity when the concentration of iron ranged from 0 to 50 μM (Che et al., 2015). Induced lipid accumulations through heavy metal consumption in several microalgae have been widely explored.

Yang et al. (2015) noticed the positive effect of cadmium and copper on the medium culture of oleaginous microalgae *Chlorella minutissima* UTEX2341, which was followed by a significant boost in lipid content amounting to 21.07% and 93.90%, respectively. The maximum lipid content (52.78%) and productivity (96.8 mg/L/d) were recorded for grown *Monoraphidium* sp. QLY-1 under 80 μM and 40 μM of cadmium, respectively (Zhao et al., 2019). The accumulation of lipids in freshwater green alga *Scenedesmus* sp. was significantly enhanced to 23.6% when copper was added at a concentration of 5 mg/L (Pham, 2019).

Napan et al. (2015) conducted a study to assess the effect of different concentrations (1, 2, 5 and 10X) for ten heavy metals based on metal composition in coal-derived flue gas (Zn, Cd, Cu, Cr, Ni, Hg, Se, Co, As and Pb) on the lipid content of microalgae *Scenedesmus obliquus.* They noticed that the excess heavy metal concentration (5 and 10X) had negatively affected and reduced lipid content by 52%, while under the baseline concentration (1X), the lipid was increased by 61% when compared to a control containing no heavy metals.

Inducing lipid accumulation through manipulation of culture conditions

Manipulation of culture conditions is defined as a variation in surrounding environmental parameters and operating conditions that overlap with the growth of microalgae like salinity, pH, light and growth temperature.

Salinity

Salinity is a complex stress factor that affects several biochemical and physiological mechanisms related to microalgal growth and development. The ability of microalgae to survive under osmotic stress conditions and their ability to alter their pattern of growth as well as to accumulate vast amounts of lipids under such circumstances have received the attention of researchers in recent years. When microalgal cells have existed to a saline stress condition, the regulation of absorption and release of ions via cell membranes, turgor pressure restoration, stress protein and osmoprotectant solutes get triggered (Alkayal et al., 2010; Talebi et al., 2013).

According to many recent studies, the cultivation of microalgal cells under salinity stress conditions is followed by the enhancement of lipid accumulations. Ji et al. (2018) reported the positive correlation between increased NaCl concentration and enhanced lipid accumulation in freshwater microalga *Scenedesmus obliquus* XJ002, and the maximum lipid content (32.3%) was recorded when they were cultured under 0.20 M NaCl. The maximum lipid content (36% dry tissue mass) for *Nannochloropsis salina* was gained when it grew under high salinity concentration (34 PSU) (Bartley et al., 2013). The increased NaCl concentration from zero to 30 g/L in the growth medium for *Chlorella sorokiniana* HS1 resulted in an increase in the lipid content from 0.22 to 0.36 g/L (Kim et al., 2016).

Kaewkannetra et al. (2012) observed that the maximum lipid content (36%) in *Scenedesmus obliquus* was recorded in the culture that was supplemented with 0.3 M NaCl at day 15 after salt stress. The cellular lipid content for *Chlorella* sp. was enhanced by culturing the studied strain under various concentrations of salt, and the highest lipid content (21.4%) was noticed at 0.5 M NaCl, which is 1.6-fold higher than that of cells cultured in a control medium (Rai et al., 2015).

pH value

pH is one of the most important environmental factors that have a strong effect on all microalgae metabolisms. pH can determine the solubility and thus the availability of CO_2 for microalgal growth. Each microalgal species has an optimal pH range for growth as well as lipid production. The optimum pH values were recorded with light acidity (pH 6) on *Chlorella sorokiniana* DOE1412 for cell growth and lipid accumulation (Qiu et al., 2017).

Almutairi and Toulibah (2017) noticed that until the fourth week for the culture of *Tetraselmis suecica*, an enhancement follows the increase in pH values in neutral lipid values and the maximum neutral lipid level was recorded when it was grown at pH 9 for 2 weeks. Moheimani (2013) studied the effect of different pH values on the growth and lipid productivity for *Tetraselmis suecica* and *Chlorella* sp. They reported that pH 7 was the optimum value for biomass and lipid productivities by 407 and 99 mg/L/d and pH 7.5 by 320 and 92 mg/L/d for *Chlorella* sp. and *T. suecica*, respectively. On the same context, he reported that, in general, lipid productivities and biomass were pH 7.5 > 7 > 8 > pH 6.5 and pH 7 > 8-7.5 > 6.5 > 6 > pH 5.5 for *T. suecica* and *Chlorella* sp., respectively. The maximum values of eicosapentaenoic acid (23.1% of the total fatty acids) and total PUFAs (38.75%) were recorded at a pH level of 7 in *Pinguiococcus pyrenoidosus* (Sang et al., 2012).

Light

In the field of algal biofuel, among the four modes of algal cultivation, the photoautotrophic mode of cultivation, which uses inorganic carbon as a food source and light as the energy source, is the most chosen approach (Chojnacka and Rocha, 2004; Chen et al., 2011). This is because the other three modes, namely heterotrophic, mixotrophic and photoheterotrophic modes, require an inorganic source of carbon, which adds to the cost of the final product making algal biofuel economically unsustainable. In the phototrophic mode, characteristics of light play an important role in algal biomass growth and its lipid accumulation (Wang et al., 2007; Das et al., 2011). Three characteristics of light play a significant role in the photosynthesis reaction: light intensity, light wavelength and light cycle (Carvalho et al., 2011).

The exposed light intensity to algae plays a significant role in the photosynthesis reaction, which is responsible for the growth of algae and their lipid accumulation (Ryther, 1956). In general, the light intensity is measured in candela (cd), but in the field of biology, the light intensity is measured in terms of micromoles/m^2 per second ($\mu mol/m^2s$). For photosynthesis reaction, photosynthetically active radiation (PAR) plays a major role (Das et al., 2011; Wang et al., 2012). To measure this radiation, a special type of radiation sensor known as a quantum sensor is used. The quantum sensor detects the light differently from the lux meter. Hence, a light source having a high candela reading might have low PAR reading and vice versa. Thus, for efficient detection and use of the light source for algal cultivation, it is essential to have a properly calibrated quantum sensor light meter (Acién et al., 2017).

In general, the relation between light intensity and growth of microalgae is shown in Figure 3.2. The light intensity of algal growth can be divided into three regions. In the first region, known as the light-limited region, with increasing light intensity, the growth of microalgae increases, starting from a light intensity below which there is no algal growth due to insufficient light. The second region, known as the light saturation region, begins from the point where there is no increase in algal growth even after increasing the light intensity. The excess light is wasted in the form of heat. After the light saturation region comes the photo-inhibition region, where the algal growth declines with an increase in light intensity (Carvalho et al., 2011).

Different microalgae have different light requirements. Hence, the values of light intensity separating these regions for different species are different.

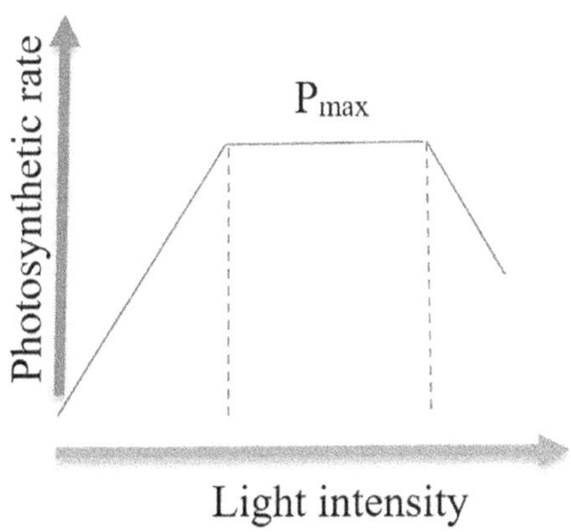

Figure 3.2 Algal growth corresponding to the incident light intensity. Adapted from Carvalho et al. (2011).

The effect of light intensity on the lipid accumulation of algal species depends on other factors as well, such as nitrogen and phosphorus supply in the cultivation media. Lipid accumulation increases at a high light intensity and under nitrogen-deficient conditions. However, excessive light intensity also declined the growth and lipid accumulation due to photo-inhibition (Yeesang and Cheirsilp, 2011; Rai et al., 2015). George et al. (2014) reported that moderately high light intensity gives better lipid accumulation and increasing the intensity further results in higher cell division and less lipid accumulation.

The wavelength of the incident light plays a significant role in the lipid accumulation of algae. The visible wavelength having a wavelength of 400–750 nm is referred to be PAR. Different pigments in microalgae use light as the energy source for its photosynthesis reaction, and various pigments react to a different wavelength of light. Different pigments in microalgae use light as the energy source for photosynthesis reaction; these pigments react to varying wavelengths of light. For instance, microalgae absorb light energy by chlorophylls in the range of 450–475 nm and 630–675 nm and by non-chlorophyll accessory such as carotenoids, which absorb light with wavelengths of 400–550, and phycobilins, which absorb light in the range of 500–650 nm (Carvalho et al., 2011).

Atta et al. (2013) reported that *Chlorella vulgaris* showed a high growth rate and lipid content using blue LED compared to white fluorescent light. Red light gives the highest algal growth rate compared to other colored lights (Wang et al., 2007; Zhao et al., 2013). As reported by Kim et al. (2014), red light gave a high rate of cell division resulting in high algal growth, but blue light increased the cell size, meaning the high accumulation of lipids. Thus, switching the light wavelength during the growth phase can result in high lipid productivity.

The light intensity and light cycle, a combination of light and dark (L:D), play a significant role in the lipid accumulation in algal cultivation. Atta et al. (2013) reported that the light intensity of 100 $\mu mol/m^2/s$ resulted in the highest lipid production (20.8%) in *Chlorella vulgaris* when the light cycle was 24:0 (L:D). However, at 200 $\mu mol/m^2/s$ the highest lipid accumulation (23%) was recorded at 12:12 (L:D) cycle, and increasing the photoperiod resulted in a decrease in lipid content. Atta et al. (2013) indicated the negative effect of excessive light on the lipid accumulation of *Chlorella vulgaris* due to their direct impact on the activity of chloroplast, which plays an essential role in lipid formation.

Temperature

Different algal species prefer a different range of temperature for their optimal growth, and different species behave differently to change in temperature. Algal species *Anabaena variabilis,* having desaturated fatty acids of 18 and 16 carbon atoms, altered the composition of fatty acid with increasing temperature, while *Anacystis nidulans* containing monounsaturated and saturated fatty acids were not affected by the change in temperature (Naoki et al., 1979; Sato and Murata, 1980). The amount of unsaturated fatty acid is inversely proportional to the growing temperature. Thus at a lower temperature, algae accumulate more lipids, compared to a higher temperature when the growth rate is fast, as during the fast growth phase algae use the energy to develop more structural components rather than producing lipids (Van Wagenen et al., 2012).

Converti et al. (2009) reported about the doubling of lipid in *Nannochloropsis oculata* algal species by increasing the growth temperature from 20°C to 25°C. However, the increase or decrease in temperature less than the mentioned degree resulted in a steep decline of lipid. On the other hand, *Chlorella vulgaris* lipid content decreased from 14.7% to 5.9% with an increase in temperature from 25°C to 30°C (Converti et al., 2009). Different stresses are given to enhance the growth and lipid accumulation for algae culture. However, incorporating more than one stressor, i.e., limiting the chemical supply accompanied by temperature stress, does not give any additive effect on lipid accumulation (Roleda et al., 2013). In a similar observation, Fakhry and Maghraby (2015) found that the effect of temperature stress combined with nitrogen stress on lipid and triglyceride accumulation was not as significant as applying only nitrogen limitation stress.

New approaches for enhancing microalgal lipid accumulation

The major obstacles that face conventional methods include rise in cost per production, limited scalability and the reduction in biomass productivity associated with reduced lipid accumulations. To overcome those dilemmas, approaches that are more efficient are recently used. These approaches include two-stage cultivation, combined abiotic stress, co-cultivation of microalgae with other microorganisms, and molecular and genetic engineering techniques for expressing foreign genes that can alter the biosynthetic pathway for lipid accumulation in microalgae. Application of the enzyme acetyl-CoA carboxylase (ACCase), which catalyzes an essential metabolic step in the synthesis of lipids in algae; the use of nanoparticles for lipid enhancement; supplementation of plant growth regulators; and the use of chemical additives are also reported.

Coupling of abiotic stress

Coupling of two or more abiotic stresses is one of the most essential recently used methods of inducing lipid accumulations in microalgae. The main feature of this approach is that it can be used to produce microalgae biodiesel on a commercial scale. The maximum lipid content and productivity for *Chlamydomonas reinhardtii* amounting to 73.4% and 10.9 mg/L/d, respectively, were recorded when cultured under an incorporated 100 mM $CaCl_2$/100 mM NaCl (Mori et al., 2020).

Li et al. (2019) conducted a study to evaluate the effect of integrating salinity stress with fulvic acid (FA) in *Monoraphidium* sp. QLY-1, and they reported that combining high salinity stress (20 g/L NaCl) and FA treatment resulted in the maximum lipid content and productivity by 59.5% and 83.67 mg/L/d, respectively. The highest lipid productivity reached 69 and 86 mg/L/d for *Chlorella sorokiniana* DPK-5 when it was grown under a minimal dose of phytohormones indole-3-acetic acid (IAA) and diethyl aminoethyl hexanoate (DA-6) combined with 50% nitrogen concentration, respectively (Babu et al., 2017).

Burch and Franz (2016) examined the impact of exposing the early stationary phase of *Phaeodactylum tricornutum* to combined stress from hydrogen peroxide and nitrogen limitation and the effect of such circumstances on enhancing the neutral lipid accretion. They noticed the synergistic effect for combined nitrogen depletion and H_2O_2 (2 mM) treatment in improving neutral lipid production, where the neutral lipid accumulation was enhanced from 0.29 to 0.56 mg/mg, which represented a 92% increase in total neutral lipid.

Moll et al. (2014) observed the maximum lipid accumulation in novel isolated diatom strain RGd-1 because of the incorporated influence of 25 mM bicarbonate with nitrogen and silicon limitations. The lipid productivity of 1.7 g/m^2/d and lipid content of up to 73% were achieved as a result of grown marine diatoms *Chaetoceros gracilis* under combined conditions including silicon stress, moderate sodium chloride and light intensity of 17.3 mol photons/m^2/d (Adams and Bugbee, 2014). Li et al. (2014) carried out a study to examine the potential effect of combination cultivation mode of nitrogen (N) and phosphorus (P) on lipid accumulations for oleaginous strain *Chlorella protothecoides*. They reported that the maximum lipid content was accomplished under co-deficiency of N and P (55.8%), compared to 52.5% and 32.77% in the case of independent nitrogen and phosphorus deficiency, respectively.

Two-stage culture system

To overcome the challenges associated with the application of nutrient deprivation in monoculture, several studies focused on using two stages of cultivation to increase biomass production along with a simultaneous improving lipid buildup. In this technique, all favorable conditions required for microalgal growth are provided at the first stage to ensure the maximum biomass production. In contrast, the appropriate stress strategies are applied in the second stage to stimulate lipid accumulation. This technique is used in its defined form by either using the hybrid cultivation strategy or using the sequential nutrient regime. The hybrid cultivation system has been used to overcome the problems related to classical monoculture systems (open-culture systems and photobioreactors (PBRs)) and obtain the advantages of both systems at the same time.

The hybrid system comprises two separate consecutive phases to improve the lipid accumulations without reducing biomass production. The first stage begins with the use of the PBR, where it is easy to control all the surrounding conditions; therefore, higher microalgae densities and increased volumetric productivity can be achieved. This stage is followed by open-culture systems where algal strains are exposed to nutrient stress to induce lipid accumulations (Figure 3.3). The efficiency of the hybrid two-stage cultivation system in enhancing biomass and lipid production in comparison with cultures grown in PBRs or open ponds has been noted by using marine microalga *Tetraselmis* sp. M8. In this study, the first stage involved cultivating the studied strain in a pressure airlift-driven PBR. In contrast, the second stage was achieved by using open raceway ponds, which included the nutrient-depleted conditions (Narala et al., 2016).

Recent studies have presented the positive effect of the sequential nutrient regime to enhance the productivity of lipid in microalgae. For instance, the additional stress of iron and light intensity can increase the impact of nitrogen stress on carbohydrate and triglyceride accumulation of *Neochloris oleoabundans* HK-129 by using two-stage cultivation (Sun et al., 2014). In this study, the first stage involved cultivating the studied strain in nitrogen- and ferric chloride-rich medium (2.94 mM and 30 μM, respectively) under light intensity (100 μmol/m/s), and it resulted in high biomass productivity amounting to 292.8 mg/L/d. The biomass that was achieved in the first stage was introduced to inoculate the second stage under specific stress conditions to increase triglyceride and carbohydrate accumulation. The mentioned conditions involved growth in nitrogen starvation medium, under iron concentrations (0.037 mM), and 200 μmol/m^2/s of light intensity. Thus, the highest lipid content and carbohydrate productivity (52 mg/L/d and 91 mg/L/d, respectively) were achieved.

By using *Nannochloropsis oculata* for salinity stress-based two-stage cultivation, the lipid content was remarkably improved (Su et al., 2011). In this study, in the first stage, the studied strain was cultivated in optimum conditions for biomass production, while in the second stage, the microalgal strain has grown under salinity stress (35 g/L) along with high light irradiance (500

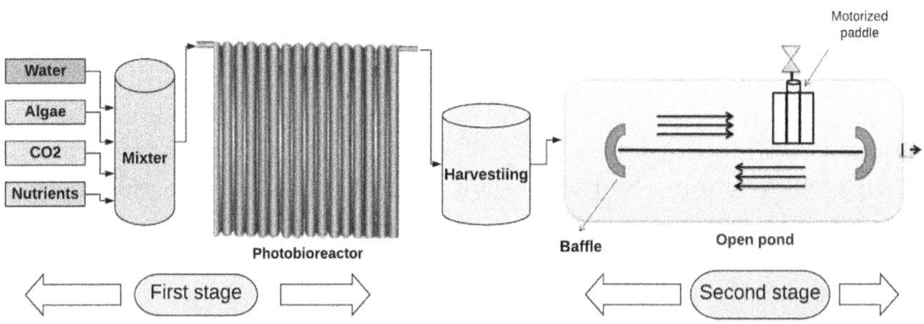

Figure 3.3 Constituents of a two-stage hybrid microalgal cultivation system.

μmol/m^2/s) and an inoculum concentration of 2.3 g/L. These conditions resulted in maximum lipid productivity of 0.324 g/L/d. In addition, this study demonstrated that the two-stage cultivation increased the final yield of lipid 2.82-fold compared with classic single cultivation systems.

Álvarez-Díaz et al. (2014) successfully used a two-stage cultivation system to enhance lipid content in the *Ankistrodesmus falcatus* from 36.5% in the case of single cultivation phase to 45.9% in the two-stage approach when cultivated in the phosphorus-free medium along with nitrogen and light presence. The two-stage cultivation using light and nutrient stress was conducted in an experiment using *Chlorella vulgaris* (Rashid et al., 2015). In this study, the studied strain was exposed to a low level from nutrients (one regular dose of BG-11) and light (2,000 lux). The biomass in this stage was used to inoculate the second stage. In the second stage, two different levels of light and nutrients were applied (2,000 and 4,000 lux), with two regular doses of BG-11. The result showed that the growth rate and lipid yield were increased in the two-stage system compared to the single-stage system.

Supplementing phytohormones

Phytohormones (also known as plant hormones) are small organic molecules and naturally occur in plants that influence physiological processes at very low concentrations (Davies, 2004). In other words, plant hormones are chemical messengers that coordinate plant cell activity (Fleet and Williams, 2011). So far, nine main phytohormones that regulate plant cell activity have been identified, including gibberellins, cytokinins, auxins, abscisic acid (ABA), jasmonates, ethylene, strigolactones, salicylic acid and brassinosteroids (Li et al., 2017). Most of the well-known phytohormones were detected in various algal taxa at concentrations comparable to the content of higher plants (Tarakhovskaya et al., 2007). The stimulant effect of phytohormones and their derivatives on the microalgal growth and enhancement of the production of valuable metabolites such as lipid, carbohydrate, carotenoid and proteins is well demonstrated in many recent studies.

Singh et al. (2020) described the impact of exogenously supplemented auxin derivatives such as indole-3-propionic acid (IPA), indole-3-butyric acid (IBA) and IAA and cytokinin derivatives such as thidiazuron (TDZ) and benzylaminopurine (BAP) on the biomass, lipid content and fatty acid profile of *Desmodesmus* sp. JS07. The result showed that among auxins, the biomass and lipid content of IBA increased by 1.96 ± 0.11 g/L and 34.9 ± 3.9%, respectively. In contrast, the biomass and lipid contents were increased by BAP of cytokinins to 1.88 ± 0.06 g/L and 31.8 ± 1.3%, respectively. Moreover, the synergistic effect of the combination of BAP (5 mg/L) and IBA (10 mg/L) on promoting the biomass and lipid content up to 2.34 ± 0.032 g/L and 42.4 ± 1.9%, respectively, was shown.

Kozlova et al. (2017) performed a study to evaluate the physiological effect of various phytohormones such as brassinolide (BL), 24-epibrassinolide

(EBL), IAA, and ABA on the microalga *Scenedesmus quadricauda*. The results appeared that all phytohormones that were studied caused stimulating impacts on *S. quadricauda* biomass production, cell growth, lipid biosynthesis and intracellular concentrations of chlorophyll a as well as a carotenoid. The accumulation of the major fatty acids was shown to have increased most pronouncedly in 100 μM ABA, where palmitoleic (C16:1), oleic (C18:1n9c) and palmitic (C16:0) acids reached maximum concentrations of 12.1, 42.1 and 34.2 of lipid, respectively. Wu et al. (2018) recorded that treatment of *Chlorella vulgaris* ZF with 1.0 mg/L ABA, 10 mg/L salicylic acid and 0.5 mg/L jasmonic acid resulted in enhanced lipid content by 45%, 42% and 49%, respectively. On the other hand, *Chlorella vulgaris* FACHB-31, which has grown under the same mentioned conditions, had an increase in lipid content by 33%, 30% and 38%, respectively. The lipid productivity of *Scenedesmus abundans* and *Chlorella ellipsoidea* was improved more than threefold and sevenfold, respectively, under auxin and gibberellin concentrations between 40 and 60 μM (González-Garcinuño et al., 2016). Likewise, all studied concentrations of ABA showed a positive stimulatory effect on the accumulation of TAGs in *Chlorella saccharophila* (Contreras-Pool et al., 2016).

The maximum lipid content and productivity amounting to 64.5% DW and 445.7 mg/L/d, respectively, in *Chromochloris zofingiensis* were noticed when it was treated with indoleacetic acid (Chen et al., 2020). On the other perspective, the use of phytohormones has been reported to induce microalgal growth and lipid aggregation under unfavorable circumstances such as high light intensity and nutrient deficiency. For instance, the addition of phytohormone to the culture of *Scenedesmus* SDEC-8 and *Chlorella sorokiniana* SDEC-18 grown under nitrogen-deficient conditions was followed by enhanced lipid productivity (SDEC-8: 26.7 and SDEC-18: 25.9 mg/L/d) and biomass concentrations. However, the two algal strains were grown poorly under the same phytohormones (Yu et al., 2018).

Renuka et al. (2018) studied the collaborative impact of cultivating microalga *Acutodesmus obliquus* under nitrogen-limited conditions with a combination of phytohormones that include indoleacetic acid, gibberellic acid and zeatin. The result showed that under nitrogen-limited conditions, the combination of indoleacetic acid (1 mg/L), gibberellic acid (5 mg/L) and zeatin (0.5 mg/L) resulted in promoted lipid and biomass productivity by 143 mg/L/d and 66.6 mg/L/d, respectively, which represent increases of 77.20% and 49.07%, respectively, compared with the control. The lipid content of *Monoraphidium* sp. QLY-1 increased with the combination of melatonin and high light intensity by 1.32 times while biomass production was maintained (Li et al., 2017).

Co-cultivation of microalgae with microorganisms

In the wild habitat, microalgae share their surroundings with a myriad of microorganisms that coexist together in complex ecological relationships that may be commensal, mutualistic or parasitic. These relationships, especially mutualistic, inspired the researchers for a long time to exciting assumptions: If microalgae could be cultured with another microorganism, the maximum benefits of such a combination would be derived. Another assumption was the ability of such co-culture to overcome the obstacles that are existing from cultivating algae individually, specifically the reduction in the growth rates and lipid productivity, which accompanies the increased lipid accumulation. Therefore, several recent studies have concentrated on enhancing lipid productivity by stimulating the increase in the accumulation of lipid and biomass side by side via microalgae co-culture with other microorganisms, including yeast, fungi, bacteria and microalgae. The exact mechanisms behind the increase in lipid yield in such circumstances, however, remain poorly intelligible.

Co-culture of microalgae with bacteria

The co-culturing of microalgae with bacteria is one of the most promising technologies that have been explored to improve the lipid yield in microalgae. This importance is due to the complementary relationship that occurs between microalgae and bacteria when they grow in a shared environment that ultimately results in their flourish together, which may be better than the growth of each of them individually. In general interactions, microalgae supply bacteria with dissolved organic matter that is needed for their biological processes. In return, the bacteria to support the growth of microalgae remineralize nitrogen, phosphorus and sulfur (Buchan et al., 2014; Yao et al., 2019). In specific interactions, microalgae supply bacteria with the fixed carbon as dissolved organic carbon. In contrast, bacteria produce siderophores to bind iron or provide vitamin B as an organic cofactor utilized by microalgae (Magdouli et al., 2016). Figure 3.4 illustrates a simplified scheme of the specific and general interactions that occur between microalgae and bacteria.

Recently, *Chlorella pyrenoidosa* and three heterotrophic bacterial strains JN1, FN3 and FN5 having efficient ammonia oxidizing abilities were examined in the co-culture system to enhance lipid productivity (Zhou et al., 2020). The cultivated *C. pyrenoidosa* with FN5 strain resulted in promoted biomass and lipid content by 14.8% and 13.6%, respectively, compared with blank culture without FN5. Berthold et al. (2019) noticed the improvement in lipid and biomass content of green algal strain *Characium* sp. when co-cultured with *Pseudomonas composti,* whether it was added in the form of bacterial cells or cell-free filtrate. A cooperative strategy using green alga *Chlamydomonas reinhardtii* and bacterial culture *Azotobacter chroococcum* was examined. This approach resulted in the highest lipid content and productivity in the studied algal strain by 65.85% and 141.86 mg/L/d, respectively, which were 2.3- and 19.4-fold more elevated than the control's levels (Xu et al., 2018).

Ryu et al. (2014) studied the effect of co-culture of microalgae *Chlorella protothecoides, Ettlia* sp. and a consortium of mixed bacteria (activated sludge) on lipid production under photoheterotrophic/photoautotrophic conditions.

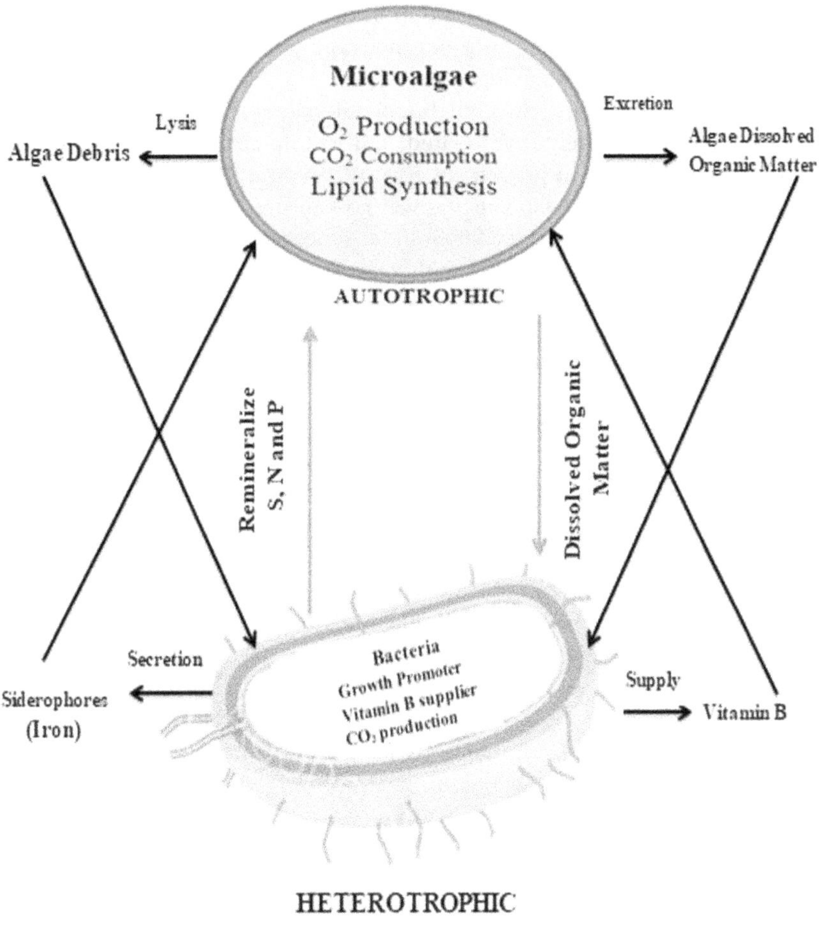

Figure 3.4 Symbiotic relationship that occurs between algae and bacteria. Adapted from Yao et al. (2019).

The results indicated that the symbiotic co-culture of the algal–bacterial system containing *Ettlia* sp. and *Chlorella protothecoides* had achieved higher lipid productivity under photoheterotrophic conditions compared to photo-autotrophic pattern by 17.3- and 28.7-fold, respectively.

Co-culture of microalgae with yeast

The oil-rich microorganisms including algae, yeast, bacteria and molds are believed as the most compatible alternatives to use as alternative primary feedstocks for biodiesel production due to their high lipid content and high growth rate (Delrue et al., 2012). In addition to algae, yeasts are able to accumulate vast amounts of lipid within their cells by using a variety of carbon sources including glycerol and molasses. Besides, yeasts have a shorter generation time than algae; this gives them the potential to grow faster and thus produce more lipid than algae.

Promoting the lipid yield through the co-culture system between microalgae and yeast has been widely investigated. For example, the total lipid content and the biomass have improved up to 60–70% and 40–50%, respectively, when green alga *Scenedesmus obliquus* was co-cultured with oleaginous yeast *Rhodotorula glutinis* (Yen et al., 2015). The synergetic effect appeared when *Chlorella vulgaris* was co-cultured with yeast strain *Rhodotorula glutinis* under photoautotrophic conditions, where the highest number of cells, biomass and lipid yield amounting to 27.73×10^5 cells/mL, 0.808 g/L and 117.73 mg/L, respectively, were recorded under the previously mentioned circumstances (Zhang et al., 2017).

In the same context, Liu et al. (2018) indicated that maximum lipid productivity and biomass concentration were achieved in the mixed culture of *Rhodotorula glutinis* and *Chlorella pyrenoidosa* at a ratio of 1:3 compared with the monoculture and other studied consortia. Jiang et al. (2018) studied the impact of co-culture of yeast *Xanthophyllomyces dendrorhous* and microalga *Chromochloris zofingiensis* on cell growth, lipid and astaxanthin production. The mutual culture promoted lipid yield and astaxanthin up to 2.37 g/L and 5.50 mg/L, respectively, at the ratio of 1:3 (yeast to algae) as compared to the individual culture of *C. zofingiensis*. During co-cultivation, microalgae provide the yeast with high oxygen levels that are needed for their growth. In return, CO_2 and some minerals are supplied by the yeast to support the growth and produce lipids (Magdouli et al., 2016).

Co-culture of microalgae with other microalgae or cyanobacteria

The co-culture systems have been used as an effective and safer alternative to reduce the cost of raw materials that are used for biodiesel production. Microalgae grow as a consortium or colony in a system of microalgae–microalgae co-culture, and the relationship that exists between them can be competitive or mutualistic. A symbiotic relationship allows each individual to benefit without a negative impact from one on the other. In contrast, in the competitive relation, one strain benefits at the expense of the other by exploiting or parasitism on the competitor (Zhu, 2015).

Gonçalves et al. (2016) studied the impact of co-culture between *Synechocystis salina* and *Pseudokirchneriella subcapitata*, *Chlorella vulgaris* and *Microcystis aeruginosa* on lipid and biomass production and nutrient removal. The results revealed that in the three studied co-culture systems, the biomass productivities were promoted compared to individual culture. On the other hand, the maximum lipid productivity was achieved when *S. salina* was grown in consortia with *M. aeruginosa* and *P. subcapitata*. The consortium containing ten strains from microalgae and cyanobacteria isolated from dairy farm wastewater achieved the maximum biomass and lipid productivity by 219.87 $\times 10^3$/ha/year and 51.37 $\times 10^3$ L of algal oil/ha/year when cultured in treated wastewater and non-treated wastewater, respectively (Hena et al., 2015). Mahapatra et al. (2014) reported that the microalgae consortia (collected from wastewater-fed urban lakes) resulted in the highest lipid content by 28.5%, which is 16% higher than *Chlorella minutissima* UTEX2341 cultivated in optimized media.

Chemical additives

Ethylenediaminetetraacetic acid

Ethylenediaminetetraacetic acid (EDTA) is a polyprotic acid comprising six functional groups (two amine groups and four carboxylic acid groups). Therefore, it has been thoroughly studied in wastewater phytoremediation (Xu et al., 2011). EDTA is widely used for enhancing metal solubility in microalgal growth media to facilitate its uptake in the microalgal cells (Ren et al., 2014).

Lipid accumulation in different microalgal strains through the supplementation of EDTA has been investigated in many recent studies. Satpati et al. (2016) noticed the enhancement of lipid accumulation of *Chlorococcum infusionum* when it is subjected to a high concentration of EDTA more than control. In contrast, the lipid content of *Chlorella ellipsoidea* was enhanced under the low level of EDTA. Also, the result revealed a 1.9-fold increase in the lipid accumulation of *C. infusionum* (20 days of incubation) and 1.7-fold improvement in the lipid content of *C. ellipsoidea* (22 days of incubation) with supplementation of 0.08 g/L and 0.05 g/L EDTA, respectively, into the medium. Ren et al. (2014) reported that compared to the control, about 28% of total lipid content and 29% of lipid productivity of green microalga *Scenedesmus* sp. could be improved by adding 1.0×10^{-3} g/L EDTA. The lipid productivity of *Acutodesmus obliquus* was enhanced from 73.2 to 80.2 mg/L/d when integrating the EDTA concentration and optimizing combined metal stress under nitrogen-deficient conditions (Singh et al., 2016).

Brefeldin A

Brefeldin A (BFA) is a chemical substance known for its inhibitory effect on vesicular transport from the ER to the Golgi, which disrupts the transportation of newly synthesized molecules (protein and lipid), and thus accumulates lipids and proteins (Kim et al., 2013). Kato et al. (2013) reported that all BFA concentrations (1.0, 2.5 or 5.0 μM) resulted in enhanced TAG accumulation of *Chlamydomonas reinhardtii*, which was cultured in the TAP medium for 2 days; moreover, TAG levels were related positively to the BFA levels up to 5 μM. Kim et al. (2013) noticed that the treatment cells of *Chlamydomonas reinhardtii* 75 μg/mL BFA enhanced the TAGs by 34% compared to the control.

Triethylamine

The addition of chemical inducers to a culture medium of microalgae is an efficient pattern of enhancing oil accumulation. Recently, attention to the use of triethylamine as a chemical inducer to trigger the biosynthesis of TAGs has been drawn. Chen et al. (2020) found that the lipid content of oleaginous *Dunaliella tertiolecta*, which was treated with 100 ppm trimethylamine, was found to be enhanced to 2.73 ± 0.03 mg/L/cell, which was 1.7-fold higher as compared to the normal condition.

Liang et al. (2019) demonstrated the positive effect of combining glycerol and triethylamine for enhancing the biomass and lipid content *of Dunaliella tertiolecta* using a two-stage cultivation system. The results indicated that at the first stage with 1 g/L glycerol addition, the biomass was improved because of rapid cell growth, leading to the highest cell biomass (1.30 g/L), while the maximum lipid weight content and lipid concentration amounting to 37.7% (dry cell weight) and 383.6 mg/L were noticed at the second stage by the cultivation of the studied strain with 100 ppm triethylamine. Adding 100 ppm of triethylamine to the medium culture of *Dunaliella tertiolecta* resulted in boosting the lipid content per cell and the total lipid production by 80% and 20%, respectively, in comparison with the group control (Xue and Jiang, 2017).

Azide

The azide addition to microalgal medium to induce the lipid accumulation within cells was extensively studied in many recent studies. The single-cell lipid content and lipid yield of *Dunaliella tertiolecta* were increased by 21.7% and 10.4%, respectively, when it was treated with <50 μM of sodium azide (Az). In contrast, the synergistic effect of combined 2.5 M salt stress and 50 μM Az on the lipid accumulation appeared when the studied strain was cultured under two-step culture conditions, where the single-cell lipid content and total lipid productivity were enhanced by 70.5% and 10%, respectively, compared to the control (Chen et al., 2019).

Yahya et al. (2018) noticed that the supplementation of 2 μM Az to the growth medium of *Acutodesmus obliquus* and *Desmodesmus maximus* resulted in an increase in the lipid content by 40% and 20%, respectively. Zalogin and Pick (2014a) reported that compared to the nitrogen deprivation condition, the addition of Az resulted in improved triglyceride yield of marine species *Chlorella desiccata* by 60–80%. Moreover, the maximum induction of TAGs by azide depends strictly on high CO_2 levels and light intensity. Zalogin and Pick (2014b) indicated that Az inhibits the assimilation of nitrogen through a reduction in the activity of nitrogen reductase enzyme and leads to a high accumulation of the lipid. They explained the mechanism behind the boost in the triglyceride accumulation in microalgae by using azide.

Nanoparticles

The particles with a maximum size of 100 nm are usually known as nanoparticles, and these particles show new properties in comparison with larger bulk material particles; novel features are derived because of variations in the characteristics of particular materials such as size, distribution and morphology (Khanna et al., 2019). Many recent studies indicated the positive effect of using nanoparticles to induce lipid buildup in microalgae.

Ren et al. (2020) demonstrated that oxidative stress caused by the combination of xenon lamp radiation and nanoparticles could help to improve the lipid production of *Scenedesmus* sp. The highest total lipid content and biomass concentration reached 40.26% and 3.18 g/L, respectively, which were triggered by cultivating the studied strain with 150 mg/L SiC nanoparticles and a photoperiod of 6:18 h.

He et al. (2017) reported that the lipid productivity of *Scenedesmus obliquus* was increased to 8.9%, 39.6% and 18.5% when it was exposed to a low dose of nanoparticles (5 mg/L carbon nanotubes, 5 mg/L nano-Fe_2O_3 and 40 mg/L nano-MgO, respectively), while the limited increase in lipid productivity was observed when it was exposed to higher nanoparticle doses. The lipid production of *Chlorella vulgaris* was improved up to 118.2 ± 5.7% by supplementation of $MgSO_4$ nanoparticles (Sarma et al., 2014). It has been reported that the lipid content of *Desmodesmus subspicatus*, *Nannochloropsis limnetica*, *Parachlorella kessleri* and *Trachydiscus minutus* was increased up to 38%, 48%, 30.9% and 35%, respectively, when they were exposed to 5.1 mg/L of zero-valent iron nanoparticles (nZVI) (Pádrová et al., 2015).

Application of Acetyl CoA Carboxylase (ACCase) enzyme

In this category, we will learn about the various processes involved in lipid accumulation by genetic manipulation of certain molecules. First, we will discuss the isolation and characterization of the enzyme, and secondly, we will discuss the cloning of the enzyme. The enzyme acetyl-coenzyme A (acetyl-CoA) is the instant precursor of fatty acid synthesis. Microalgae can produce acetyl-CoA in chloroplasts and mitochondria from pyruvate by pyruvate dehydrogenase (Dunahay, 1993). In mitochondria, when acetyl-CoA cannot enter through membranes, then acetyl-CoA hydrolase will break down to acetate and free CoA, after which acetate would enter the chloroplast and assimilate into acetyl-CoA by the action of acetyl-CoA synthetase (Dunahay, 1993; Dunahay et al., 1995). The produced acetyl-CoA is used as a substrate by ACCase to produce malonyl-CoA. In this reaction, malonyl-CoA is a substrate for fatty acid synthesis (Roessler, 1990; Dunahay et al., 1997). The enzyme ACCase has biotin which helps in the carboxylation of acetyl-CoA to form malonyl-CoA.

This section tells us about the isolation and characterization of ACCase from *C. cryptica* to identify the role of this enzyme precisely in lipid accumulation, which takes place in a silicon starvation environment. By the incorporation of bicarbonate-14C into malonyl-CoA, the activity of this enzyme can be studied. The enzyme requires a moderately alkaline pH for its optimum activity, mostly stable at pH 6.5. Citrate, sulfhydryl reductants, NaCl, KCl and divalent metal cations can stabilize the enzyme (Roessler, 1990). After having the knowledge of ACCase for lipid accumulation now, it becomes the major target for isolation and gene manipulation to enhance the lipid in a microalgal species.

Roessler and Ohlrogge (1993) studied lipid biosynthetic pathways by cloning and characterizing the ACCase gene from *C. cryptica*. In this work, ACCase protein was purified to analyze the amino acid sequence, which was used to form oligonucleotide primers that were used in a polymerase chain reaction (PCR) to increase an ACCase gene fragment from the overall DNA of *C. cryptica*. Furthermore, various genetic modifications have been done to study the activity of ACCase to increase the TAG synthesis (Roessler and Ohlrogge, 1993; Roessler et al., 1994). However, these works were only the starting phase for the research in this field. These works were not that successful because of lesser-known facts for the regulatory paths of carbon flow in the plastids of the microalgae. Now, the recent studies say that the application of starchless mutant and exogenous acetyl boost can lead to enhancement of lipid in microalgae (Ramanan et al., 2013). When it has been accepted that acetyl-CoA is the main pioneer for the de novo synthesis of fatty acid in the chloroplast of oleaginous microalgae, the major problem was the conversion to malonyl-CoA via plastidic ACCase for TAG biosynthesis (Huerlimann and Heimann, 2013). For most of the oleaginous plants, the carbon flux study shows the major metabolic control at the fatty acid biosynthesis, which includes many enzymatic stages (Guschina et al., 2014). Thus, all those studies together show that carbon precursors in the form of acetyl-coenzyme can be a strategic approach for identifying the oleaginous microalgae.

Molecular and genetic engineering techniques

Here, we will discuss genetic engineering or mutagenesis by manipulating the algal biosynthetic pathways to produce high levels of lipid in microalgal strains. Bioengineering techniques were adopted on repressing key genes that induce lipid synthesis in microalgal species. The reproduction of microalgal species also plays a vital role in these types of techniques. Mostly they reproduce asexually by the simple fission method, but certain microalgal strains do reproduce sexually. One feasible way for high lipid accumulation in algal strains would be the selection of natural genetic variants with desired traits.

Strategies are made to induce genetic variation in a pool of cells of microalgal strains using mutagenesis (Zhao et al., 2009). However, there is a limitation in this method. The strain with the desired trait is important for the production of large numbers of mutants. Hence, from both the techniques, we can conclude that lipid biosynthesis in algae is a great success in the area of microalgal genetic engineering. The enzyme ACCase plays the most regulatory role for lipid accumulation. The cloning of the gene that codes for this enzyme was done, which was beneficial because regulatory sequences from this gene were used to develop a genetic transformation and this gene allowed more areas of studies for the researchers for the effects of overexpression of this enzyme on lipid accumulation (Schneider and Roessler, 1995). These methods were further modified by another method, which is by decreasing lipid catabolism. In this method, inactivation of genes involved in the activation of both TAG and β-oxidation of fatty acids takes place, thereby enhancing the lipid content.

To avoid the absence of homologous recombination in microalgae, gene inactivation is done by random RNA silencing or mutagenesis (Molnar et al., 2009; Zhao et al., 2009). There are some recent works related to genetic engineering techniques that were applied for improving the resistance in microalgae, which also helped in lipid enhancement of microalgae. The modifications include overexpression of small heat shock protein (ch-sHSP) in *Synechococcus elongatus*, which made the species thermotolerant (Beacham et al., 2017). Overexpression of the homogentisate phytyltransferase vitamin E2 (VTE2) can make the microalgal species tolerant to oxidative stress (Li et al., 2010).

Conclusion

Bioprospecting of valuable metabolites from microalgae has gained extensive attention as a promising feedstock for biofuels, pharmaceutical products, nutritional supplements, livestock feeds and highly valuable molecules that are able to substitute the conventional existing resources. The main dilemmas that face the industrial scale for lipid production through classical approaches are the reduction in biomass productivity associated with lipid accumulations, cost-per-production rise and its limited scalability since some of these strategies entail additional costs and energy. Stress conditions created through manipulating the nutrient regime, which includes essential ingredients of culture growth media such as nitrogen, phosphorus, sulfur and metal ions, resulted in a severalfold increase in lipid content.

Alteration of culture conditions such as osmotic stress, pH, light and temperature also plays an important role in the lipid accumulation of microalgae. Lately, to overcome the limitation of cultural approaches to increase the lipid content, advanced strategies such as two-stage cultivation, combined abiotic stress, co-cultivation with other microorganisms, molecular and genetic engineering techniques, incorporation of chemical additives like EDTA, BFA, triethylamine and azide, and the use of nanoparticles and enzyme ACCase have been implemented. The new approaches used for inducing the microalgal lipid accumulation show favorable outcomes by tackling the problems associated with a classical approach, thus giving a promising scope for using microalgal lipid as the next-generation feedstock.

References

Acién, F. C., Molina, E., Reis, A., Torzillo, G., Zittelli, G. C., Sepúlveda, C., and Masojídek, J. (2017). Photobioreactors for the production of microalgae. In: Gonzalez-Fernandez, C., Muñoz, R. (eds.), *Microalgae-Based Biofuels and Bioproducts: From Feedstock Cultivation to End-Products*, Woodhead Publishing, pp. 1–44.

Adams, C., and Bugbee, B. (2014). Enhancing lipid production of the marine diatom *Chaetoceros gracilis*: Synergistic interactions of sodium chloride and silicon. *Journal of Applied Phycology* 26(3): 1351–1357.

Adenan, N. S., Yusoff, F. M., Medipally, S. R., and Shariff, M. (2016). Enhancement of lipid production in two marine microalgae under different levels of nitrogen and phosphorus deficiency. *Journal of Environmental Biology* 37: 669–676.

Alkayal, F., Albion, R. L., Tillett, R. L., Hathwaik, L. T., Lemos, M. S., and Cushman, J. C. (2010). Expressed sequence tag (EST) profiling in hyper saline shocked *Dunaliella salina* reveals high expression of protein synthetic apparatus components. *Plant Science* 179(5): 437–449.

Almutairi, A. W., and Toulibah, H. E. (2017). Effect of salinity and ph on fatty acid profile of the green algae *Tetraselmis suecica*. *Journal of Petroleum and Environmental Biotechnology* 8(3): 3–8.

Altın, N., Kutluk, T., Uyar, B., and Kapucu, N. (2018). Effect of different nitrogen sources on the growth and lipid accumulation of *Chlorella variabilis*. *Journal of Applied Biological Sciences* 12(2): 38–40.

Álvarez-Díaz, P. D., Ruiz, J., Arbib, Z., Barragán, J., Garrido-Pérez, C., and Perales, J. A. (2014). Lipid production of microalga *Ankistrodesmus falcatus* increased by nutrient and light starvation in a two-stage cultivation process. *Applied Biochemistry and Biotechnology* 174(4): 1471–1483.

Aratboni, H. A., Rafiei, N., Garcia-Granados, R., Alemzadeh, A., and Morones-Ramírez, J. R. (2019). Biomass and lipid induction strategies in microalgae for biofuel production and other applications. *Microbial Cell Factories* 18(1): 178–195.

Arif, M., Bai, Y., Usman, M., Jalalah, M., Harraz, F. A., Al-Assiri, M. S., and Zhang, C. (2020). Highest accumulated microalgal lipids (polar and non-polar) for biodiesel production with advanced wastewater treatment: Role of lipidomics. *Bioresource Technology* 298: 122299.

Atta, M., Idris, A., Bukhari, A., and Wahidin, S. (2013). Intensity of blue LED light: A potential stimulus for biomass and lipid content in fresh water microalgae *Chlorella vulgaris*. *Bioresource Technology* 148: 373–378.

Babu, A. G., Wu, X., Kabra, A. N., and Kim, D. P. (2017). Cultivation of an indigenous *Chlorella sorokiniana* with phytohormones for biomass and lipid production under N-limitation. *Algal Research* 23: 178–185.

Bartley, M. L., Boeing, W. J., Corcoran, A. A., Holguin, F. O., and Schaub, T. (2013). Effects of salinity on growth and lipid accumulation of biofuel microalga *Nannochloropsis salina* and invading organisms. *Biomass and Bioenergy* 54: 83–88.

Beacham, T. A., Sweet, J. B., and Allen, M. J. (2017). Large scale cultivation of genetically modified microalgae: A new era for environmental risk assessment. *Algal Research* 25: 90–100.

Bellinger, E. G., and Sigee, D. C. (eds.) (2015). *Freshwater Algae: Identification, Enumeration and Use as Bioindicators*. Chichester: Wiley-Blackwell.

Belotti, G., de Caprariis, B., De Filippis, P., Scarsella, M., and Verdone, N. (2014). Effect of *Chlorella vulgaris* growing conditions on bio-oil production via fast pyrolysis. *Biomass and Bioenergy* 61: 187–195.

Berg, A. H., Combs, T. P., and Scherer, P. E. (2002). ACRP30/adiponectin: An adipokine regulating glucose and lipid metabolism. *Trends in Endocrinology & Metabolism* 13(2): 84–89.

Berthold, D. E., Shetty, K. G., Jayachandran, K., Laughinghouse IV, H. D., and Gantar, M. (2019). Enhancing algal biomass and lipid production through bacterial co-culture. *Biomass and Bioenergy* 122: 280–289.

Buchan, A., LeCleir, G. R., Gulvik, C. A., and González, J. M. (2014). Master recyclers: Features and functions of bacteria associated with phytoplankton blooms. *Nature Reviews Microbiology* 12(10): 686–698.

Burch, A. R., and Franz, A. K. (2016). Combined nitrogen limitation and hydrogen peroxide treatment enhances neutral lipid accumulation in the marine diatom *Phaeodactylum tricornutum*. *Bioresource Technology* 219: 559–565.

Bušić, A., Kundas, S., Morzak, G., Belskaya, H., Mardetko, N., Ivančić Šantek, M., Komes, D., Novak, S., and Šantek, B. (2018). Recent trends in biodiesel and biogas production. *Food Technology and Biotechnology* 56(2): 152–173.

Cakmak, T., Angun, P., Demiray, Y. E., Ozkan, A. D., Elibol, Z., and Tekinay, T. (2012). Differential effects of nitrogen and sulfur deprivation on growth and biodiesel feedstock production of *Chlamydomonas reinhardtii*. *Biotechnology and Bioengineering* 109(8): 1947–1957.

Can, S. S., Koru, E., and Cirik, S. (2017). Effect of temperature and nitrogen concentration on the growth and lipid content of *Spirulina platensis* and biodiesel production. *Aquaculture International* 25(4): 1485–1493.

Carvalho, A. P., Silva, S. O., Baptista, J. M., and Malcata, F. X. (2011). Light requirements in microalgal photobioreactors: An overview of biophotonic aspects. *Applied Microbiology and Biotechnology* 89: 1275–1288.

Che, R., Huang, L., and Yu, X. (2015). Enhanced biomass production, lipid yield and sedimentation efficiency by iron ion. *Bioresource Technology* 192: 795–798.

Chen, C. Y., Yeh, K. L., Aisyah, R., Lee, D. J., and Chang, J. S. (2011). Cultivation, photobioreactor design and harvesting of microalgae for biodiesel production: A critical review. *Bioresource Technology* 102(1): 71–81.

Chen, H. H., Xue, L. L., Liang, M. H., and Jiang, J. G. (2019). Sodium azide intervention, salinity stress and two-step cultivation of *Dunaliella tertiolecta* for lipid accumulation. *Enzyme and Microbial Technology* 127: 1–5.

Chen, H. H., Xue, L. L., Liang, M. H., and Jiang, J. G. (2020). Intervention of triethylamine on *Dunaliella tertiolecta* reveals metabolic insights into triacylglycerol accumulation. *Algal Research* 47: 2211–9264.

Chen, J. H., Wei, D., and Lim, P. E. (2020). Enhanced coproduction of astaxanthin and lipids by the green microalga *Chromochloris zofingiensis*: Selected phytohormones as positive stimulators. *Bioresource Technology* 295: 122242.

Chojnacka, K., and Rocha, F. J. M. (2004). Kinetic and stoichiometric relationships of the energy and carbon metabolism in the culture of microalgae. *Biotechnology* 3(1): 21–34.

Chu, W. L. (2017). Strategies to enhance production of microalgal biomass and lipids for biofuel feedstock. *European Journal of Phycology* 52(4): 419–437.

Contreras-Pool, P. Y., Peraza-Echeverria, S., Ku-González, Á. F., and Herrera-Valencia, V. A. (2016). The phytohormone abscisic acid increases triacylglycerol content in the green microalga *Chlorella saccharophila* (Chlorophyta). *Algae* 31(3): 267–276.

Converti, A., Casazza, A. A., Ortiz, E. Y., Perego, P., and Borghi, M. D. (2009). Effect of temperature and nitrogen concentration on the growth and lipid content of *Nannochloropsis oculata* and *Chlorella vulgaris* for biodiesel production. *Chemical Engineering and Processing* 48: 1146–1151.

Das, P., Lei, W., Aziz, S. S., and Obbard, J. P. (2011). Enhanced algae growth in both phototrophic and mixotrophic culture under blue light. *Bioresource Technology* 102: 3883–3887.

Davies, P. J. (ed.) (2004). *Plant Hormones: Biosynthesis, Signal Transduction, Action!.* Dordrecht: Springer.

De Bhowmick, G., Koduru, L., and Sen, R. (2015). Metabolic pathway engineering towards enhancing microalgal lipid biosynthesis for biofuel application a review. *Renewable and Sustainable Energy Reviews* 50: 1239–1253.

de Jaeger, L., Verbeek, R. E., Draaisma, R. B., Martens, D. E., Springer, J., Eggink, G., and Wijffels, R. H. (2014). Superior triacylglycerol (TAG) accumulation in starchless mutants of *Scenedesmus obliquus*: (I) mutant generation and characterization. *Biotechnology for Biofuels* 7(1): 69–80.

Delrue, F., Setier, P. A., Sahut, C., Cournac, L., Roubaud, A., Peltier, G., and Froment, A. K. (2012). An economic, sustainability, and energetic model of biodiesel production from microalgae. *Bioresource Technology* 111: 191–200.

Dunahay, T. G. (1993). Transformation of *Chlamydomonas reinhardtii* with silicon carbide whiskers. *Biotechniques* 15(3): 452–455.

Dunahay, T. G., Adler, S. A., and Jarvik, J. W. (1997). Transformation of microalgae using silicon carbide whiskers. In: Tuan, R. S. (ed.), *Recombinant Gene Expression Protocols.* Totowa: Humana Press, pp. 503–509.

Dunahay, T. G., Jarvis, E. E., and Roessler, P. G. (1995). Genetic transformation of the diatoms *Cyclotella cryptica* and *Navicula saprophila*. *Journal of Phycology* 31(6): 1004–1012.

Dunyaporn, T., Weiqin, L., Marcia, A. O., Nilsa, R., and Peng, H. (2008). Redox regulation of cell survival. *Antioxidants & Redox Signaling* 10(8): 1343–1374.

El-Shouny, W., Sharaf, M., Abomohra, A. E. F., and Abo-Eleneen, M. (2015). Production enhancement of some valuable compounds of *Arthrospira platensis*. *Journal of Basic and Environmental Sciences* 2(3): 74–83.

Fakhry, R. M., and Maghraby, D. M. E. (2015). Lipid accumulation in response to nitrogen limitation and variation of temperature in *Nannochloropsis salina*. *Botanical Studies* 56: 1–8.

Fan, J., Cui, Y., Wan, M., Wang, W., and Li, Y. (2014). Lipid accumulation and biosynthesis genes response of the oleaginous *Chlorella pyrenoidosa* under three nutrition stressors. *Biotechnology for Biofuels* 7(1): 7–17.

Fawzy, M. A. (2017). Fatty acid characterization and biodiesel production by the marine microalga *Asteromonas gracilis*: Statistical optimization of medium for biomass and lipid enhancement. *Marine Biotechnology* 19(3): 219–231.

Fleet, C. M., and Williams, M. E. (2011). Gibberellins. Teaching Tools in Plant Biology: Lecture Notes. The Plant Cell. www.plantcell.org/cgi/doi/10.1105/tpc.110.tt0810.

Foley, P. M., Beach, E. S., and Zimmerman, J. B. (2011). Algae as a source of renewable chemicals: Opportunities and challenges. *Green Chemistry* 13(6): 1399–1405.

Gao, Y., Feng, J., Lv, J., Liu, Q., Nan, F., Liu, X., and Xie, S. (2019). Physiological changes of *Parachlorella Kessleri* TY02 in lipid accumulation under nitrogen stress. *International Journal of Environmental Research and Public Health* 16(7): 1188–1205.

Gautam, S., and Mannan, M. A. U. (2020). The role of algae in nutraceutical and pharmaceutical production. In: Singh, J., Meshram, V., Gupta, M. (eds.), *Bioactive Natural products in Drug Discovery.* Singapore: Springer, pp. 665–685.

George, B., Pancha, I., Desai, C., Chokshi, K., Paliwal, C., Ghosh, T., and Mishra, S. (2014). Effects of different media composition, light intensity and photoperiod on morphology and physiology of freshwater microalgae *Ankistrodesmus falcatus* – A potential strain for bio-fuel production. *Bioresource Technology* 171: 367–374.

Giordano, M., and Raven, J. A. (2014). Nitrogen and sulfur assimilation in plants and algae. *Aquatic Botany* 118: 45–61.

Gonçalves, A. L., Pires, J. C., and Simões, M. (2016). Biotechnological potential of *Synechocystis salina* co-cultures with selected microalgae and cyanobacteria: Nutrients removal, biomass and lipid production. *Bioresource Technology* 200: 279–286.

González-Garcinuño, Á., Sánchez-Álvarez, J. M., Galán, M. A., and Martin del Valle, E. M. (2016). Understanding and optimizing the addition of phytohormones in the culture of microalgae for lipid production. *Biotechnology Progress* 32: 1203–1211.

Gorain, P. C., Bagchi, S. K., and Mallick, N. (2013). Effects of calcium, magnesium and sodium chloride in enhancing lipid accumulation in two green microalgae. *Environmental Technology* 34(13–14): 1887–1894.

Goyal, G., Kuhn, J. N., and Philippidis, G. P. (2018). Light alkene production by cracking *Picochlorum oculatum* microalgae using aluminosilicate catalysts. *Biomass and Bioenergy* 108: 252–257.

Guschina, I. A., Everard, J. D., Kinney, A. J., Quant, P. A., and Harwood, J. L. (2014). Studies on the regulation of lipid biosynthesis in plants: Application of control analysis to soybean. *Biochimica et Biophysica Acta-Biomembranes* 1838(6): 1488–1500.

Hannon, M., Gimpel, J., Tran, M., Rasala, B., and Mayfield, S. (2010). Biofuels from algae: Challenges and potential. *Biofuels* 5: 763–784.

Hayes, D. G., and Smith, G. A. (2019). Biobased surfactants: Overview and industrial state of the art. In: Hayes, D. G., Daniel K. S., Richard D. A. (eds.), *Biobased Surfactants Synthesis, Properties, and Applications*. London: AOCS Press, pp. 3–38.

He, M., Yan, Y., Pei, F., Wu, M., Gebreluel, T., Zou, S., and Wang, C. (2017). Improvement on lipid production by *Scenedesmus obliquus* triggered by low dose exposure to nanoparticles. *Scientific Reports* 7(1): 1–12.

Hena, S., Fatimah, S., and Tabassum, S. (2015). Cultivation of algae consortium in a dairy farm wastewater for biodiesel production. *Water Resources and Industry* 10: 1–14.

Hu, Q., Sommerfeld, M., Jarvis, E., Ghirardi, M., Posewitz, M., Seibert, M., and Darzins, A. (2008). Microalgal triacylglycerols as feedstocks for biofuel production: Perspectives and advances. *The Plant Journal* 54(4): 621–639.

Huerlimann, R., and Heimann, K. (2013). Comprehensive guide to acetyl-carboxylases in algae. *Critical Reviews in Biotechnology* 33(1): 49–65.

Jha, D., Jain, V., Sharma, B., Kant, A., and Garlapati, V. K. (2017). Microalgae-based pharmaceuticals and nutraceuticals: An emerging field with immense market potential. *ChemBioEng Reviews* 4(4): 257–272.

Ji, X., Cheng, J., Gong, D., Zhao, X., Qi, Y., Su, Y., and Ma, W. (2018). The effect of NaCl stress on photosynthetic efficiency and lipid production in freshwater microalga—*Scenedesmus obliquus* XJ002. *Science of the Total Environment* 633: 593–599.

Jiang, X., Liu, L., Chen, J., and Wei, D. (2018). Effects of *Xanthophyllomyces dendrorhous* on cell growth, lipid, and astaxanthin production of *Chromochloris zofingiensis* by mixed culture strategy. *Journal of Applied Phycology* 30(6): 3009–3015.

Kaewkannetra, P., Enmak, P., and Chiu, T. (2012). The effect of CO_2 and salinity on the cultivation of *Scenedesmus obliquus* for biodiesel production. *Biotechnology and Bioprocess Engineering* 17(3): 591–597.

Kato, N., Dong, T., Bailey, M., Lum, T., and Ingram, D. (2013). Triacylglycerol mobilization is suppressed by brefeldin A in *Chlamydomonas reinhardtii. Plant and Cell Physiology* 54(10): 1585–1599.

Khan, M. I., Shin, J. H., and Kim, J. D. (2018). The promising future of microalgae: Current status, challenges, and optimization of a sustainable and renewable industry for biofuels, feed, and other products. *Microbial Cell Factories* 17(1): 36–55.

Khanna, P., Kaur, A., and Goyal, D. (2019). Algae-based metallic nanoparticles: Synthesis, characterization and applications. *Journal of Microbiological Methods* 163: 1–24.

Kim, B. H., Ramanan, R., Kang, Z., Cho, D. H., Oh, H. M., and Kim, H. S. (2016). *Chlorella sorokiniana* HS1, a novel freshwater green algal strain, grows and hyper-accumulates lipid droplets in seawater salinity. *Biomass and Bioenergy* 85: 300–305.

Kim, D. G., Lee, C., Park, S. M., and Choi, Y. E. (2014). Manipulation of light wavelength at appropriate growth stage to enhance biomass productivity and fatty acid methyl ester yield using *Chlorella vulgaris. Bioresource Technology* 159: 240–248.

Kim, G., Mujtaba, G., Lee, K., Kim, G., Mujtaba, G., and Lee, K. (2016). Effects of nitrogen sources on cell growth and biochemical composition of marine chlorophyte *Tetraselmis* sp. for lipid production. *Algae* 31(3): 257–266.

Kim, S., Kim, H., Ko, D., Yamaoka, Y., Otsuru, M., Kawai-Yamada, M., and Lee, Y. (2013). Rapid induction of lipid droplets in *Chlamydomonas reinhardtii* and *Chlorella vulgaris* by Brefeldin A. *PLoS One* 8(12): 81978.

Kong, F., Romero, I. T., Warakanont, J., and Li-Beisson, Y. (2018). Lipid catabolism in microalgae. *New Phytologist* 218(4): 1340–1348.

Kozlova, T. A., Hardy, B. P., Krishna, P., and Levin, D. B. (2017). Effect of phytohormones on growth and accumulation of pigments and fatty acids in the microalgae *Scenedesmus quadricauda. Algal Research* 27: 325–334.

Lei, Q. Y., and Lu, S. H. (2011). Molecular ecological responses of the dinoflagellate *Karenia mikimotoi* to phosphate stress. *Harmful Algae* 12: 39–45.

Li, D., Zhao, Y., Ding, W., Zhao, P., Xu, J. W., Li, T., and Yu, X. (2017). A strategy for promoting lipid production in green microalgae *Monoraphidium* sp. QLY-1 by combined melatonin and photoinduction. *Bioresource Technology* 235: 104–112.

Li, J., Li, C., and Smith, S. M. (eds.) (2017). *Hormone Metabolism and Signaling in Plants.* London: Academic Press.

Li, X., Li, X., Han, B., Zhao, Y., Li, T., Zhao, P., and Yu, X. (2019). Improvement in lipid production in *Monoraphidium* sp. QLY-1 by combining fulvic acid treatment and salinity stress. *Bioresource Technology* 294: 122–179.

Li, Y., Han, D., Hu, G., Dauvillee, D., Sommerfeld, M., Ball, S., and Hu, Q. (2010). *Chlamydomonas starchless* mutant defective in ADP-glucose pyrophosphorylase hyper-accumulates triacylglycerol. *Metabolic Engineering* 12(4): 387–391.

Li, Y., Han, F., Xu, H., Mu, J., Chen, D., Feng, B., and Zeng, H. (2014). Potential lipid accumulation and growth characteristic of the green alga Chlorella with combination cultivation mode of nitrogen (N) and phosphorus (P). *Bioresource Technology* 174: 24–32.

Liang, K., Zhang, Q., Gu, M., and Cong, W. (2013). Effect of phosphorus on lipid accumulation in freshwater microalga *Chlorella* sp. *Journal of Applied Phycology* 25(1): 311–318.

Liang, M. H., Xue, L. L., and Jiang, J. G. (2019). Two-stage cultivation of *Dunaliella tertiolecta* with glycerol and triethylamine for lipid accumulation: A viable way to alleviate the inhibitory effect of triethylamine on biomass. *Applied and Environmental Microbiology* 85(4): 1–10.

Liu, L., Chen, J., Lim, P. E., and Wei, D. (2018). Dual-species cultivation of microalgae and yeast for enhanced biomass and microbial lipid production. *Journal of Applied Phycology* 30(6): 2997–3007.

Magdouli, S., Brar, S. K., and Blais, J. F. (2016). Co-culture for lipid production: Advances and challenges. *Biomass and Bioenergy* 92: 20–30.

Mahapatra, D. M., Chanakya, H. N., and Ramachandra, T. V. (2014). Bioremediation and lipid synthesis through mixotrophic algal consortia in municipal wastewater. *Bioresource Technology* 168: 142–150.

Markou, G., Angelidaki, I., and Georgakakis, D. (2012). Microalgal carbohydrates: An overview of the factors influencing carbohydrates production, and of main bioconversion technologies for production of biofuels. *Applied Microbiology and Biotechnology* 96(3): 631–645.

Merchant, S. S., Kropat, J., Liu, B., Shaw, J., and Warakanont, J. (2012). TAG, You're it! *Chlamydomonas* as a reference organism for understanding algal triacylglycerol accumulation. *Current Opinion in Biotechnology* 23(3): 352–363.

Michelon, W., Da Silva, M. L. B., Mezzari, M. P., Pirolli, M., Prandini, J. M., and Soares, H. M. (2016). Effects of nitrogen and phosphorus on biochemical composition of microalgae polyculture harvested from phycoremediation of piggery wastewater digestate. *Applied Biochemistry and Biotechnology* 178(7): 1407–1419.

Minhas, A. K., Hodgson, P., Barrow, C. J., and Adholeya, A. (2016). A review on the assessment of stress conditions for simultaneous production of microalgal lipids and carotenoids. *Frontiers in Microbiology* 7: 546–565.

Mizuno, Y., Sato, A., Watanabe, K., Hirata, A., Takeshita, T., Ota, S., and Kawano, S. (2013). Sequential accumulation of starch and lipid induced by sulfur deficiency in *Chlorella* and *Parachlorella* species. *Bioresource Technology* 129: 150–155.

Moheimani, N. R. (2013). Inorganic carbon and pH effect on growth and lipid productivity of *Tetraselmis suecica* and *Chlorella* sp (*Chlorophyta*) grown outdoors in bag photobioreactors. *Journal of Applied Phycology* 25(2): 387–398.

Moll, K. M., Gardner, R. D., Eustance, E. O., Gerlach, R., and Peyton, B. M. (2014). Combining multiple nutrient stresses and bicarbonate addition to promote lipid accumulation in the diatom RGd-1. *Algal Research* 5: 7–15.

Molnar, A., Bassett, A., Thuenemann, E., Schwach, F., Karkare, S., Ossowski, S., and Baulcombe, D. (2009). Highly specific gene silencing by artificial microRNAs in the unicellular alga *Chlamydomonas reinhardtii*. *The Plant Journal* 58(1): 165–174.

Mori, K., Tanaka, Y., Morikawa, M., and Toyama, T. (2020). Enhanced lipid productivity of *Chlamydomonas reinhardtii* with combination of NaCl and $CaCl_2$ stresses. *Bioprocess and Biosystems Engineering* 43: 971–980.

Morowvat, M. H., and Ghasemi, Y. (2016). Rapid determination of lipid accumulation under sulfur starvation in *Chlamydomonas reinhardtii* microalga using Fourier Transform Infrared (FTIR) spectroscopy. *International Journal of Pharmaceutical and Clinical Research* 8: 1356–1360.

Naoki, S., Norio, M., Yoshiro, M., and Nobuo, U. (1979). Effect of growth temperature on lipid and fatty acid compositions in the blue-green algae, *Anabaena variabilis* and *Anacystis nidulans*. *Biochimica et Biophysica Acta - Lipids and Lipid Metabolism* 572(1): 19–28.

Napan, K., Teng, L., Quinn, J. C., and Wood, B. D. (2015). Impact of heavy metals from flue gas integration with microalgae production. *Algal Research* 8: 83–88.

Narala, R. R., Garg, S., Sharma, K. K., Thomas-Hall, S. R., Deme, M., Li, Y., and Schenk, P. M. (2016). Comparison of microalgae cultivation in photobioreactor, open raceway pond, and a two-stage hybrid system. *Frontiers in Energy Research* 4: 29–39.

Nascimento, I. A., Marques, S. S. I., Cabanelas, I. T. D., Pereira, S. A., Druzian, J. I., de Souza, C. O., and Nascimento, M. A. (2013). Screening microalgae strains for biodiesel production: Lipid productivity and estimation of fuel quality based on fatty acids profiles as selective criteria. *Bioenergy Research* 6(1): 1–13.

Negi, S., Barry, A. N., Friedland, N., Sudasinghe, N., Subramanian, S., Pieris, S., and Sayre, R. (2016). Impact of nitrogen limitation on biomass, photosynthesis, and lipid accumulation in *Chlorella sorokiniana*. *Journal of Applied Phycology* 28(2): 803–812.

Neha, K., and Khan, S. (2016). Effect of nitrogen, phosphorus concentrations, pH and salinity ranges on growth, biomass and lipid accumulation of *Chlorella vulgaris*. *International Journal of Pharmaceutical Sciences and Research* 7(1): 397–405.

Olofsson, M., Lamela, T., Nilsson, E., Bergé, J. P., Del Pino, V., Uronen, P., and Legrand, C. (2012). Seasonal variation of lipids and fatty acids of the microalgae *Nannochloropsis oculata* grown in outdoor large-scale photobioreactors. *Energies* 5(5): 1577–1592.

Pádrová, K., Lukavský, J., Nedbalová, L., Čejková, A., Cajthaml, T., Sigler, K., and Řezanka, T. (2015). Trace concentrations of iron nanoparticles cause overproduction of biomass and lipids during cultivation of cyanobacteria and microalgae. *Journal of Applied Phycology* 27(4): 1443–1451.

Pankratz, S., Oyedun, A. O., Zhang, X., and Kumar, A. (2017). Algae production platforms for Canada's northern climate. *Renewable and Sustainable Energy* 80: 109–120.

Pham, T. L. (2019). Removal and bioaccumulation of copper by the freshwater green alga *Scenedesmus* sp. *Vietnam Journal of Science, Technology and Engineering* 61(2): 65–70.

Qiu, R., Gao, S., Lopez, P. A., and Ogden, K. L. (2017). Effects of pH on cell growth, lipid production and CO_2 addition of microalgae *Chlorella sorokiniana*. *Algal Research* 28: 192–199.

Rai, M. P., Gautom, T., and Sharma, N. (2015). Effect of salinity, pH, light intensity on growth and lipid production of microalgae for bioenergy application. *On Line Journal of Biological Sciences* 15(4): 260–267.

Ramanan, R., Kim, B. H., Cho, D. H., Ko, S. R., Oh, H. M., and Kim, H. S. (2013). Lipid droplet synthesis is limited by acetate availability in starchless mutant of *Chlamydomonas reinhardtii*. *FEBS Letters* 587(4): 370–377.

Rashid, N., Rehman, M. S., and Han, J. I. (2015). Enhanced growth rate and lipid production of freshwater microalgae by adopting two-stage cultivation system under diverse light and nutrients conditions. *Water and Environment Journal* 29(4): 533–540.

Reese, K. L., Fisher, C. L., Lane, P. D., Jaryenneh, J. D., Moorman, M. W., Jones, A. D., and Lane, T. W. (2019). Chemical profiling of volatile organic compounds in the headspace of algal cultures as early biomarkers of algal pond crashes. *Scientific Reports* 9(1): 1–10.

Ren, H. Y., Dai, Y. Q., Kong, F., Xing, D., Zhao, L., Ren, N. Q., and Liu, B. F. (2020). Enhanced microalgal growth and lipid accumulation by addition of different nanoparticles under xenon lamp illumination. *Bioresource Technology* 297: 22409.

Ren, H. Y., Liu, B. F., Kong, F., Zhao, L., Xie, G. J., and Ren, N. Q. (2014). Enhanced lipid accumulation of green microalga *Scenedesmus* sp. by metal ions and EDTA addition. *Bioresource Technology* 169: 763–767.

Renuka, N., Guldhe, A., Singh, P., and Bux, F. (2018). Combined effect of exogenous phytohormones on biomass and lipid production in *Acutodesmus obliquus* under nitrogen limitation. *Energy Conversion and Management* 168: 522–528.

Roessler, P. G. (1990). Purification and characterization of acetyl-CoA carboxylase from the diatom *Cyclotella cryptica. Plant Physiology* 92(1): 73–78.

Roessler, P. G., Brown, L. M., Dunahay, T. G., Heacox, D. A., Jarvis, E. E., Schneider, J. C., Talbot. S. G., Zeiler, K. G. (1994). Genetic engineering approaches for enhanced production of biodiesel fuel from microalgae. In: Himmel, M. E., Baker, J. O., Overend, R. P. (eds.), *Enzymatic Conversion of Biomass for Fuels Production.* Washington, DC: American Chemical Society, pp. 255–270.

Roessler, P. G., and Ohlrogge, J. B. (1993). Cloning and characterization of the gene that encodes acetyl-coenzyme A carboxylase in the alga *Cyclotella cryptica. Journal of Biological Chemistry* 268(26): 19254–19259.

Roleda, M. Y., Slocombe, S. P., Leakey, R. J. G., Day, J. G., Bell, E. M., and Stanley, M. S. (2013). Effects of temperature and nutrient regimes on biomass and lipid production by six oleaginous microalgae in batch culture employing a two-phase cultivation strategy. *Bioresource Technology* 129: 439–449.

Roopnarain, A., Gray, V. M., and Sym, S. D. (2014). Phosphorus limitation and starvation effects on cell growth and lipid accumulation in *Isochrysis galbana* U4 for biodiesel production. *Bioresource Technology* 156: 408–411.

Ruangsomboon, S., Ganmanee, M., and Choochote, S. (2013). Effects of different nitrogen, phosphorus, and iron concentrations and salinity on lipid production in newly isolated strain of the tropical green microalga, *Scenedesmus dimorphus* KMITL. *Journal of Applied Phycology* 25(3): 867–874.

Ryther, J. H. (1956). Photosynthesis in the ocean as a function of light intensity. *Limnology and Oceanography* 1: 61–70.

Ryu, B. G., Kim, J., Farooq, W., Han, J. I., Yang, J. W., and Kim, W. (2014). Algal–bacterial process for the simultaneous detoxification of thiocyanate-containing wastewater and maximized lipid production under photoautotrophic/photoheterotrophic conditions. *Bioresource Technology* 162: 70–79.

Sajjadi, B., Chen, W. Y., Raman, A. A. A., and Ibrahim, S. (2018). Microalgae lipid and biomass for biofuel production: A comprehensive review on lipid enhancement strategies and their effects on fatty acid composition. *Renewable and Sustainable Energy Reviews* 97: 200–232.

Salas-Montantes, C. J., González-Ortega, O., Ochoa-Alfaro, A. E., Camarena-Rangel, R., Paz-Maldonado, L. M. T., Rosales-Mendoza, S., and Soria-Guerra, R. E. (2018). Lipid accumulation during nitrogen and sulfur starvation in *Chlamydomonas reinhardtii* overexpressing a transcription factor. *Journal of Applied Phycology* 30(3): 1721–1733.

Sang, M., Wang, M., Liu, J., Zhang, C., and Li, A. (2012). Effects of temperature, salinity, light intensity, and pH on the eicosapentaenoic acid production of *Pinguiococcus pyrenoidosus. Journal of Ocean University of China* 11(2): 181–186.

Sarma, S. J., Das, R. K., Brar, S. K., Le Bihan, Y., Buelna, G., Verma, M., and Soccol, C. R. (2014). Application of magnesium sulfate and its nanoparticles for enhanced lipid production by mixotrophic cultivation of algae using biodiesel waste. *Energy* 78: 16–22.

Sato, N., and Murata, N. (1980). Temperature shift-induced responses in lipids in the blue-green alga, *Anabaena variabilis*: The central role of diacylmonogalactosylglycerol in thermo-adaptation. *Biochimica et Biophysica Acta - Lipids and Lipid Metabolism* 619: 353–366.

Satpati, G. G., Gorain, P. C., and Pal, R. (2016). Efficacy of EDTA and phosphorous on biomass yield and total lipid accumulation in two green microalgae with special emphasis on neutral lipid detection by flow cytometry. *Advances in Biology* 2016: 1–12.

Schneider, J. C., and Roessler, P. G. (1995). A novel acyltransferase activity in an oleaginous alga. In: Kader J. C., Mazliak P. (eds.), *Plant Lipid Metabolism*. Dordrecht: Springer, pp. 105–107.

Shakya, R., Adhikari, S., Mahadevan, R., and Dempster, T. A. (2018). Catalytic upgrading of bio-oil produced from hydrothermal liquefaction of *Nannochloropsis* sp. *Bioresource Technology* 252: 28–36.

Shirazi, H. M., Karimi-Sabet, J., and Ghotbi, C. (2017). Biodiesel production from Spirulina microalgae feedstock using direct transesterification near supercritical methanol condition. *Bioresource Technology* 239: 378–386.

Singh, A., Nigam, P. S., and Murphy, J. D. (2011). Mechanism and challenges in commercialization of algal biofuels. *Bioresource Technology* 102(1): 26–34.

Singh, J., Jain, D., Agarwal, P., and Singh, R. P. (2020). Auxin and cytokinin synergism augmenting biomass and lipid production in microalgae *Desmodesmus* sp. JS07. *Process Biochemistry* 95: 223–234.

Singh, P., Guldhe, A., Kumari, S., Rawat, I., and Bux, F. (2016). Combined metals and EDTA control: An integrated and scalable lipid enhancement strategy to alleviate biomass constraints in microalgae under nitrogen limited conditions. *Energy Conversion and Management* 114: 100–109.

Sivaramakrishnan, R., and Incharoensakdi, A. (2017). Enhancement of total lipid yield by nitrogen, carbon, and iron supplementation in isolated microalgae. *Journal of Phycology* 53(4): 855–868.

Srinophakun, P., Thanapimmetha, A., Rattanaphanyapan, K., Sahaya, T., and Saisriyoot, M. (2017). Feedstock production for third generation biofuels through cultivation of *Arthrobacter* AK19 under stress conditions. *Journal of Cleaner Production* 142: 1259–1266.

Su, C. H., Chien, L. J., Gomes, J., Lin, Y. S., Yu, Y. K., Liou, J. S., and Syu, R. J. (2011). Factors affecting lipid accumulation by *Nannochloropsis oculata* in a two-stage cultivation process. *Journal of Applied Phycology* 23(5): 903–908.

Sun, X., Cao, Y., Xu, H., Liu, Y., Sun, J., Qiao, D., and Cao, Y. (2014). Effect of nitrogen-starvation, light intensity and iron on triacylglyceride/carbohydrate production and fatty acid profile of *Neochloris oleoabundans* HK-129 by a two-stage process. *Bioresource Technology* 155: 204–212.

Talebi, A. F., Tabatabaei, M., Mohtashami, S. K., Tohidfar, M., and Moradi, F. (2013). Comparative salt stress study on intracellular ion concentration in marine and salt-adapted freshwater strains of microalgae. *Notulae Scientia Biologicae* 5(3): 309–315.

Tarakhovskaya, E. R., Maslov, Y. I., and Shishova, M. F. (2007). Phytohormones in algae. *Russian Journal of Plant Physiology* 54(2): 163–170.

Torres, C. M., Ríos, S. D., Torras, C., Salvadó, J., Mateo-Sanz, J. M., and Jiménez, L. (2013). Microalgae-based biodiesel: A multicriteria analysis of the production process using realistic scenarios. *Bioresource Technology* 147: 7–16.

Tripathi, V., Kumar, P., Tripathi, P., Kishore, A., and Kamle, M. (eds.) (2019). *Microbial Genomics in Sustainable Agroecosystems* (Vol. 2). Singapore: Springer Nature.

Ullah, K., Ahmad, M., Sharma, V. K., Lu, P., Harvey, A., Zafar, M., and Anyanwu, C. N. (2014). Algal biomass as a global source of transport fuels: Overview and development perspectives. *Progress in Natural Science: Materials International* 24(4): 329–339.

Van Wagenen, J., Miller, T. W., Hobbs, S., Hook, P., Crowe, B., and Huesemann, M. (2012). Effects of light and temperature on fatty acid production in *Nannochloropsis salina*. *Energies* 5(3): 731–740.

Vitova, M., Bisova, K., Kawano, S., and Zachleder, V. (2015). Accumulation of energy reserves in algae: From cell cycles to biotechnological applications. *Biotechnology Advances* 33(6): 1204–1218.

Wang, B., Lan, C. Q., and Horsman, M. (2012). Closed photobioreactors for production of microalgal biomasses. *Biotechnology Advances* 30: 904–912.

Wang, C. Y., Fu, C. C., and Liu, Y. C. (2007). Effects of using light-emitting diodes on the cultivation of Spirulina platensis. *Biochemical Engineering Journal* 37: 21–25.

Wang, X., Fosse, H. K., Li, K., Chauton, M., Vadstein, O., and Reitan, K. I. (2019). Influence of nitrogen limitation on lipid accumulation and EPA and DHA content in four marine microalgae for possible use in aquafeed. *Frontiers in Marine Science* 6: 95–105.

Wang, Z. T., Ullrich, N., Joo, S., Waffenschmidt, S., and Goodenough, U. (2009). Algal lipid bodies: Stress induction, purification, and biochemical characterization in wild-type and starchless *Chlamydomonas reinhardtii*. *Eukaryotic Cell* 8(12): 1856–1868.

Weintraub, H. (2013). Update on marine omega-3 fatty acids: Management of dyslipidemia and current omega-3 treatment options. *Atherosclerosis* 230(2): 381–389.

Wu, G., Gao, Z., Du, H., Lin, B., Yan, Y., Li, G., and Cui, M. (2018). The effects of abscisic acid, salicylic acid and jasmonic acid on lipid accumulation in two freshwater *Chlorella* strains. *The Journal of General and Applied Microbiology* 64(1): 42–49.

Wu, L. F., Chen, P. C., and Lee, C. M. (2013). The effects of nitrogen sources and temperature on cell growth and lipid accumulation of microalgae. *International Biodeterioration and Biodegradation* 85: 506–510.

Xin, L., Hong-Ying, H., and Yu-Ping, Z. (2011). Growth and lipid accumulation properties of a freshwater microalga *Scenedesmus* sp. under different cultivation temperature. *Bioresource Technology* 102(3): 3098–3102.

Xu, L., Cheng, X., and Wang, Q. (2018). Enhanced lipid production in *Chlamydomonas reinhardtii* by co-culturing with *Azotobacter chroococcum*. *Frontiers in Plant Science* 9: 741.

Xu, M., Zhang, Y., Zhang, Z., Shen, Y., Zhao, M., and Pan, G. (2011). Study on the adsorption of Ca^{2+}, Cd^{2+} and Pb^{2+} by magnetic Fe_3O_4 yeast treated with EDTA dianhydride. *Chemical Engineering Journal* 168(2): 737–745.

Xue, L. L., and Jiang, J. G. (2017). Cultivation of *Dunaliella tertiolecta* intervened by triethylamine enhances the lipid content. *Algal Research* 25: 136–141.

Yahya, N. A., Suhaimi, N., Kaha, M., Hara, H., Zakaria, Z., Sugiura, N., and Iwamoto, K. (2018). Lipid production enhancement in tropically isolated microalgae by azide and its effect on fatty acid composition. *Journal of Applied Phycology* 30(6): 3063–3073.

Yang, J., Cao, J., Xing, G., and Yuan, H. (2015). Lipid production combined with biosorption and bioaccumulation of cadmium, copper, manganese and zinc by oleaginous microalgae *Chlorella minutissima* UTEX2341. *Bioresource Technology* 175: 537–544.

Yao, S., Lyu, S., An, Y., Lu, J., Gjermansen, C., and Schramm, A. (2019). Microalgae–bacteria symbiosis in microalgal growth and biofuel production: A review. *Journal of Applied Microbiology* 126(2): 359–368.

Yeesang, C., and Cheirsilp, B. (2011). Effect of nitrogen, salt, and iron content in the growth medium and light intensity on lipid production by microalgae isolated from freshwater sources in Thailand. *Bioresource Technology* 102: 3034–3040.

Yen, H. W., Chen, P. W., and Chen, L. J. (2015). The synergistic effects for the co-cultivation of oleaginous yeast-*Rhodotorula glutinis* and microalgae-*Scenedesmus obliquus* on the biomass and total lipids accumulation. *Bioresource Technology* 184: 148–152.

Yu, Z., Pei, H., Jiang, L., Hou, Q., Nie, C., and Zhang, L. (2018). Phytohormone addition coupled with nitrogen depletion almost tripled the lipid productivities in two algae. *Bioresource Technology* 247: 904–914.

Zalogin, T. R., and Pick, U. (2014a). Azide improves triglyceride yield in microalgae. *Algal Research* 3: 8–16.

Zalogin, T. R., and Pick, U. (2014b). Inhibition of nitrate reductase by azide in microalgae results in triglycerides accumulation. *Algal Research* 3: 17–23.

Zhang, K., Zheng, J., Xue, D., Ren, D., and Lu, J. (2017). Effect of photoautotrophic and heteroautotrophic conditions on growth and lipid production in *Chlorella vulgaris* cultured in industrial wastewater with the yeast *Rhodotorula glutinis*. *Journal of Applied Phycology* 29(6): 2783–2788.

Zhao, T., Wang, W., Bai, X., and Qi, Y. (2009). Gene silencing by artificial microRNAs in Chlamydomonas. *The Plant Journal* 58(1): 157–164.

Zhao, Y., Song, X., Yu, L., Han, B., Li, T., and Yu, X. (2019). Influence of cadmium stress on the lipid production and cadmium bioresorption by *Monoraphidium* sp. QLY-1. *Energy Conversion and Management* 188: 76–85.

Zhao, Y., Wang, J., Zhang, H., Yan, C., and Zhang, Y. (2013). Effects of various LED light wavelengths and intensities on microalgae-based simultaneous biogas upgrading and digestate nutrient reduction process. *Bioresource Technology* 136: 461–468.

Zhou, X., Jin, W., Wang, Q., Guo, S., Tu, R., Han, S. F., and Wang, Q. (2020). Enhancement of productivity of *Chlorella pyrenoidosa* lipids for biodiesel using co-culture with ammonia-oxidizing bacteria in municipal wastewater. *Renewable Energy* 151: 598–603.

Zhu, J., Chen, W., Chen, H., Zhang, X., He, C., Rong, J., and Wang, Q. (2016). Improved productivity of neutral lipids in *Chlorella* sp. A2 by minimal nitrogen supply. *Frontiers in Microbiology* 7: 557–568.

Zhu, L. (2015). Microalgal culture strategies for biofuel production: A review. *Biofuels, Bioproducts and Biorefining* 9(6): 801–814.

Zuccaro, G., Yousuf, A., Pollio, A., and Steyer, J. P. (2020). Microalgae cultivation systems. In: Yousuf, A. (ed.), *Microalgae Cultivation for Biofuels Production*. London: Academic Press, pp. 11–29.

4 Biodiesel production from algae through catalytic transesterification

Garima Srivastava, Devendra Kumar Maravi, and Ajeet Singh

Introduction

The continuously increasing population, advancement of technologies, environmental priorities and international socio-geopolitical relations are certain factors that shape the world energy scenario. There is a direct correlation between energy, financial growth and sustainable development of a country. The liquid fuel prices play a dominant role and have a huge impact on gross domestic product (GDP), consumption patterns and environmental and energy policies (Bollino et al., 2017; Snow, 2017). Rapid industrialization, as well as high population growth, is continuously raising the energy demand that largely depends on fossil fuel. Present energy demand and continuous depletion of fossil fuels require alternative fuels to be discovered. The sole dependency on conventional fossil fuel will not be able to fulfill the global energy requirement. In developing countries, the unprecedented growth of renewable energy has occurred as a chief energy source to meet global energy demand in a profitable, sustainable and eco-friendly way. Hence, exceptional growth and reconstruction in the renewable energy sector can be easily seen with the technological advancement and expanded investments that have directed reconsideration of energy transformation over the world.

The disproportionate demand and limited availability of fossil fuels have shifted the energy research paradigm toward other alternative energy resources to substitute conventional energy assets. With this advent concern for the supply of fossil fuel, the recent emergence of biodiesel production has become a forefront topic of consideration. Biodiesel is one of the most important liquid biofuels, derived from a variety of feedstocks comprising vegetable oils, animal fats and microorganisms. Biodiesel is often regarded as biofuel due to its origin from biomass (Borugadda and Goud, 2012).

Biodiesel

Biodiesel is a monoalkyl ester synthesized by the reaction of fatty acids (FA) with alcohol. The advantages of biodiesel include similar energy efficiency with petroleum, combustion properties and potential usage in existing compression-ignition engines. Moreover, the renewability, inherent lubricity, biodegradability, an absence of sulfur (S) content, a reduction in obnoxious gas emission and safer handling make biodiesel to stand alone as a prominent alternative energy resource(Knothe et al., 2015).

Based on feedstock and production technologies, biodiesel has been classified into various categories, such as first-, second- and third-generation biodiesel. The first-generation biodiesel evolved from edible oilseed crops that mainly include soybean, peanut, mustard and sunflower. However, uncertainty arises in the food sector, and due to the food versus fuel crisis, the existence of first-generation biodiesel became questionable. Further, to resolve this issue the second generation of biodiesel had come up using non-edible oil crops with Pongamia, Jatropha and Castor. The second-generation biodiesel has sorted the food versus fuel crisis, but they required huge arable land and high maintenance and have a large gestation period that leads to harvesting only once or twice a year. Due to the aforesaid limitations of first- and second-generation biodiesel feedstocks, the third generation of biodiesel has evolved as a promising alternative, i.e., microalgae biodiesel (Chuah et al., 2017; Piemonte et al., 2016). Microalgae belong to a class of photosynthetic organisms that have the potential to convert radiation energy into chemical energy and utilize it in various forms of biomolecules. Subsequently, these metabolic end products are used to produce biofuel. The lipid molecules, especially neutral lipids, are utilized as the major source for biodiesel production (Chisti, 2007; Chng et al., 2016; Rawat et al., 2013). Recently, the emergence of fourth-generation biodiesel from genetically engineered microalgae is also under investigation.

Microalgae

Microalgae are tiny photosynthetic microorganisms like plants that use sunlight and atmospheric CO_2 to synthesize biomass. Microalgal biomass consists of proteins (30–35%), carbohydrates (25–30%), lipids (20–25%), nucleic acids (1–2%), pigments (2–5%) and ash (2–3%) (Laurens et al., 2014). In general, microalgae can accumulate 20–25% of lipid content on their dry cell biomass that can extend up to 50–70% under certain circumstances or stress conditions (BenMoussa-Dahmen et al., 2016). There are plenty of microalgal species reported in the literature for their high biomass and lipid content such as *Chlorella, Scenedesmus, Dunaliella, Haematococcus, Phaeodactylum, Isochrysis* and *Nannochloropsis.* A critical review of the potential application of microalgal species for biofuel synthesis has been debated by several researchers (Chisti, 2007; Gouveia and Oliveira, 2009; Mata et al., 2010). However, a huge disparity among microalgal strains can be easily observed based on their respective biomass and lipid content (Table 4.1). The differences in microalgae growth profile, biomass concentration and lipid content are a product of their genetic and environmental counterparts. Hence, there is a great scope to modulate their biochemical profile by altering the environmental and genetic factors.

Table 4.1 Potential microalgal strains with high biomass productivity and lipid content

Freshwater species	*Lipid content (% dry cell weight)*
Botryococcus braunii	25–75
Chlorella	18–57
Chlorella emersonii	25–63
Chlorella protothecoides	14–57
Chlorella sorokiniana	19–22
Chlorella vulgaris	5–58
Chlorella sp.	28–32
Dunaliella salina	14–20
Dunaliella tertiolecta	16–71
Haematococcus pluvialis	25
Isochrysis sp.	25–33
Nannochloris sp.	20–56
Nannochloropsis	12–53
Nannochloropsis oculata	22–30
Nitzschia sp.	45–47
Phaeodactylum tricornutum	20–30
Scenedesmus quadricauda	2–18
Schizochytrium sp.	50–77
Tetraselmis suecica	15–23
Tetraselmis sp.	13–15

References: Chisti (2007), Gouveia and Oliveira (2009) and Mata et al. (2010).

Advantages of microalgae include higher oil production capacity as compared to traditional oilseed crops. Microalgae have been reported as better feedstock choices in terms of annual productivity and oil amount as compared to oilseed crops. A high oil yield of 136,900 L/ha can be obtained from high-lipid-containing microalgae. High-lipid-containing microalgae can take up to 70% of oil content in their dry cell weight (dcw), whereas low-lipid-containing microalgae have 30% of oil in their dry cell biomass and can end up to 58,700 liters/hectare (L/ha) oil yield. Microalgae have short cultivation and harvesting cycle; therefore, multiple harvesting can be achieved in a year independent of seasonal growth. In addition, microalgae have several other advantages over oilseed feedstocks; for example, they are incompetent with land and food market, have high oil yield, minimize exhaust gas emission and are helpful in CO_2 mitigation and wastewater remediation (Chisti, 2007; Mata et al., 2010). It is evident from the above discussion that microalgae have been preferred as a prime source of biodiesel production during the present scenario. From a sustainable and economical point of view, microalgae are paramount to the available feedstock used for biodiesel production. The fast and independent seasonal growth, high productivity, the minimum requirement of the cultivation land, utilization of wastewater and no competition with the food market have made microalgae invincible raw material for biodiesel production.

Biodiesel production from microalgae

Biosynthesis of oil in microalgae

Typically, the neutral lipid in microalgae is converted into biodiesel. These tiny oil droplets store energy in the form of CO_2-derived lipid molecules, i.e., triacylglycerol (TAG). The flexibility of microalgae to synthesize raw material for biodiesel production is a systemic light-driven process called photosynthesis. Microalgae utilize atmospheric CO_2 as a carbon source and sunlight as an energy source to synthesize the organic products of photosynthesis and liberate oxygen (O_2) as a by-product. It converts radiation energy into chemical energy in the form of organic biomolecule and subsequently forms raw organic feedstock material. The light energy absorbed by chlorophyll molecules gets converted into reducing energy packets such as ATP and NADPH. These energy coins are used in the metabolic process to synthesize biomolecules in the form of sugars, starch and oils. Furthermore, these metabolic end products are extracted from microalgal cells and converted into various forms of biofuel, e.g., bioethanol, biodiesel and biomethane (Schenk et al., 2008).

The lipid metabolism in microalgae has shown distinct pathways that are governed by various physicochemical factors. The changes in physicochemical parameters (light, temperature, pH, nutrients and salt) in the culture environment of microalgae tend to change their biochemical composition (Hu et al., 2008; Huang et al., 2010). Various physical and chemical conditions have been optimized for maximum productivity of lipid. Channelizing carbon sources toward lipid synthesis instead of cell growth and multiplication is one of the most favored approaches for high lipid induction in microalgae. It is reported that nitrogen (N) deprivation and nutrient limitation are the main factors for TAG induction in microalgal cells (Sharma et al., 2012). In addition, salt-induced lipid production is the key strategy for enhanced TAG accumulation in microalgae (Srivastava et al., 2017).

The inclusion of optimization tools for modulating the external environment to stimulate high lipid production has been reported by several researchers (Difusa et al., 2015; Kirrolia et al., 2014). For instance, *Coelastrella* M-60 and *Micractinium* M-1 freshwater microalgal species were subjected to less nutrient and high salinity along with sugarcane industry effluent, citric acid, glucose and vitamin B12. The response surface methodology (RSM) was used as an optimization tool to estimate the effect of supplemented components on microalgae growth and overall lipid yield. An increase in lipid content was obtained from 12.9% to 19.3% dcw, which was 1.35- to 1.4-fold higher as compared to the control when supplemented with 25 mg/L citric acids along with 1.25 ml/L effluent in the standard BG-11 culture medium (Karpagam et al., 2015). In another study, the Plackett–Burman (PB) study design was implemented on lipid production of *Scenedesmus* sp. by altering the medium composition. Under optimal conditions, the lipid production reached 54.6%, where palmitic acid and oleic acid were predominant (Yang et al., 2014).

The fatty acid profiles of microalgae have a major role in the qualitative regulation of biodiesel production. Biodiesel quality is mainly regulated by the fatty acid composition; therefore, strain selection is very important as its significant consequences can be observed on biodiesel properties (Olmstead et al., 2013). The high unsaturated fatty acid content in algal oil reduces transesterification efficiency. However, the short-chain monounsaturated fatty acids (MUFAs) including palmitic acid (16:0) and oleic acid (18:1) are favored for microalgae biodiesel synthesis (Table 4.2). The long chain and high degree of unsaturation cause high fluidity in biodiesel, whereas short-chain unsaturated fatty acid allows the biodiesel to get frozen easily. The fatty acid composition of microalgae oil may vary with species, cultivation conditions and extraction methods that remarkably determine the overall fuel characteristics of microalgae biodiesel. Therefore, algal strain selection, cultivation condition and downstream processing methods have a huge impact on microalgae biodiesel production (Sakthivel, 2011). The fatty acid profile of microalgae is presented in Table 4.2 (Gouveia and Oliveira, 2009; Schlagermann et al., 2012). The biodiesel properties such as density, viscosity and cetane number are significantly dependent on fatty acid profile and can be estimated empirically (Knothe, 2005).

Microalgae biodiesel production is a sequential stepwise process (Figure 4.1). In this process, at the beginning, microalgae are cultivated under controlled laboratory conditions to accumulate lipid in their dry cell biomass. Once the

Table 4.2 Presence of preferred fatty acids in microalgae oil composition required for the biodiesel synthesis

Fatty acid	*Spirulina maxima*	*Chlorella vulgaris*	*Scenedesmus obliquus*	*Dunaliella tertiolecta*	*Nannochloropsis sp.*	*Neochloris oleoabundans*	*Nannochloropsis salina*	*Phaeodactylum tricornutum*	*Botryococcus braunii*
14:0	0.34	3.07	1.48	0.47	7.16	0.43	–	4.5	0.8
16:0	40.16	25.07	21.78	17.70	23.35	19.35	37.5	25.8	21.0
16:1	9.19	5.25	5.95	0.88	26.87	1.85	23.3	37.5	2.0
16:2	–	–	3.96	3.03	0.39	1.74	–	–	6.5
16:3	0.42	1.27	0.68	1.24	0.48	0.96	–	–	15.2
16:4	0.16	4.06	0.43	10.56	–	7.24	–	–	–
18:0	1.18	0.63	0.45	–	0.45	0.98	0.9	1.3	2.9
18:1	5.43	12.64	17.93	4.87	13.20	20.29	11.9	–	3.2
18:2	17.89	7.19	21.74	12.37	1.21	12.99	1.5	5.1	13.6
18:3	18.32	19.05	3.76	30.19	–	17.43	–	2.0	33.0
18:4	0.08	–	0.21	–	–	2.10	–	–	–
20:0	0.06	0.09	–	–	–	–	0.1	–	0.2

References: Gouveia and Oliveira (2009) and Schlagermann et al. (2012).

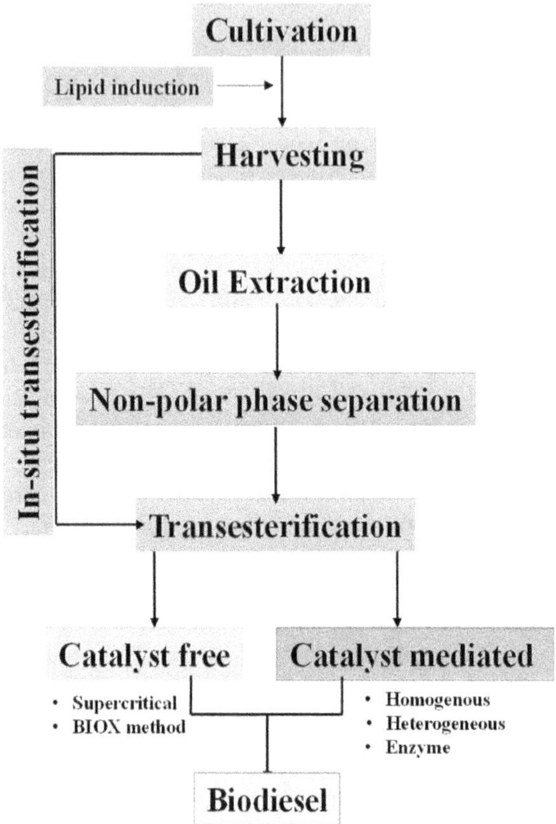

Figure 4.1 The sequential process of biodiesel production from microalgae.

cultivation period is over, the algal cells are harvested, dried and subjected to oil extraction. After the extraction, the nonpolar oil fraction is separated from water-soluble molecules and pigments. This results in a highly viscous greenish-yellow crude algae oil. Furthermore, to reduce the viscosity and other impurities and high FFA content, the crude algae oil is subjected to transesterification reaction where the conversion of oil into biodiesel takes place.

Conversion of microalgae oil to biodiesel can be carried out by catalytic or non-catalytic methods. The catalytic methods include homogenous, heterogeneous, enzymatic and nanocatalyst methods, whereas the non-catalytic methods involve supercritical and BIOX transesterification. Finally, biodiesel is separated from the glycerol phase present in the mixture. However, the purification of biodiesel is needed to be utilized in diesel engines. The multiple downstream processing steps make microalgae biodiesel uneconomically; therefore, various researchers prefer direct algal biomass transesterification. During in situ transesterifications of microalgae, an instantaneous extraction and transesterification of algal biomass take place, resulting in single-pot biodiesel synthesis.

Transesterification of microalgae oil to biodiesel

Conventionally, biodiesel can be produced via different operational methods such as blending, microemulsion, pyrolysis and transesterification. The transesterification method is the most studied and popular method among all these methods where fatty acid alkyl ester (FAAE) is produced from the reaction of fats/oil with alcohol. It can be catalyzed in the presence of alkali, acid or enzyme and can subsequently synthesize biodiesel in the most adaptive way (Borugadda and Goud, 2012).

Mechanistically, transesterification reaction is a sequential and reversible reaction, worked at equilibrium condition where one ester molecule (fat) is converted into another ester molecule to form fatty acid methyl esters (FAME). The triglyceride is converted into diglycerides, and the diglycerides are converted into monoglyceride and finally to glycerol. Stoichiometrically to synthesize one mole of alkyl esters, fatty acid and alcohol are used in a 1:3 ratio in the presence of various catalysts. However, alcohol is used in an excess amount to increase the alkyl ester synthesis and for easy phase separation among the alkyl esters and by-product glycerol molecule. In this process, the synthesis of monoglycerides is the rate-determining reaction and they are the most stable intermediate compound formed during the reaction. The transesterification process is carried out by homogeneous, heterogeneous and enzymatic catalysts (Meher et al., 2006). An efficacious transesterification reaction is indicated by the clear separation of methyl ester and glycerol moiety. The phase separation between light FAME (at the top) and heavy glycerol (at the bottom) can be easily seen in the reaction vessel. However, this crude biodiesel requires some purification before use.

Factors affecting microalgae transesterification

There are numerous external factors such as temperature, alcohol-to-oil molar ratio, reaction period, catalyst concentration and type of catalyst, purity of reactant and amount of free fatty acid (FFA) in oil that significantly affect the biodiesel yield. It was reported that high reaction temperature as well as alcohol/oil molar ratio significantly enhance the biodiesel yield, whereas high moisture content and high FFA content in oil decrease the biodiesel yield (Borugadda and Goud, 2012; Saka et al., 2006).

Presence of polar lipids

The microalgal cell wall is formed of thick parenchymatous and sclerenchymatous tissues, which is a big challenge for efficient extraction and transesterification. Microalgal cells contain phospholipids and galactolipids in their cell wall and cell membrane in the form of phosphatidylglycerols (PG), phosphatidylethanolamines (PE), phosphatidylcholines (PC), monogalactosyldiacylglycerol (MGDG) and digalactosyldiacylglycerol (DGDG). These phospholipids and glycolipids have elemental phosphorus (P) and N in their structural organization which, if they remain untreated, can cause harm to diesel engines. Conversion of these polar fatty acids into biodiesel is a challenge. It is found that there is around 30% lipid lost during esterification of polar lipid. However, neutral lipids such as TAG are highly favored over the other lipid classes as they give better biodiesel yield (>99%) as compared to phospholipids (<70%). Furthermore, the presence of phospholipids or glycolipids in oil deteriorates the biodiesel quality and reduces the low heating value (Williams and Laurens, 2010).

Moisture content

Around 80–90% of microalgal weight is due to the presence of the water molecule. Water molecule interferes with the transesterification process due to frothing or emulsification that lowers the biodiesel yield (Sathish et al., 2014). Therefore, it is recommended to use completely dried microalgal biomass or the anhydrous oil to attain maximized biodiesel yield. However, this drying process impedes an additional input cost on microalgae biodiesel production. Therefore, a novel method of direct wet algae transesterification has been developed and popularized. However, this in situ wet algae transesterification process required a high amount of reaction solvent and a strong acid catalyst to neutralize the effect of moisture content. Therefore, the development of advanced technologies such as supercritical and hydrothermal process has been designed to work with high-moisture-bearing microalgae without affecting biodiesel quality.

High FFA content

Higher amounts of FFA in oil result in high acid value. The higher acid value more than one needs a more alkaline catalyst to neutralize the FFA during the homogenous alkaline transesterification. In contrast, the presence of water molecules in such conditions causes frothing and soap formation, which is a major problem in industrial-scale biodiesel production. The increase in viscosity and the formation of foam and gels create hindrance in glycerol separation. Therefore, a two-step method is mainly used for industrial-scale biodiesel production where esterification and transesterification occur sequentially with high-FFA-containing oils (Demirbas, 2009).

Reaction temperature

Reaction temperature plays a vital role in the transesterification process. It provides sufficient collision energy to the reacting molecules for interaction and increases the mass–energy transfer for complete mixing of reactants.

However, it is important to remember that reaction should be carried out at a lower temperature as compared to the alcohol boiling point so that alcohol loss through vaporization can be minimized throughout the reaction. The optimal reaction temperature usually optimizes in the range of 50–60°C depending on the oil and solvent (Mazumdar et al., 2013). It was suggested that at higher temperatures, the intermolecular hydrogen bonding of methanol (MeOH) decreases that allows the solvation of fatty acids more significantly by increasing the interaction between fats and alcohol. At higher reaction temperatures, MeOH behaves as a nonpolar solvent, which allows easy solubilization of oil into MeOH. Moreover, increasing the temperature increases the molecular vibrations of MeOH that reduce the polarity and dielectric constant, which is a basis of supercritical transesterification (Marulanda, 2012).

Reaction time

The time needed for the reactant to completely interact with each other to form a stable product is called reaction time. A sufficient interaction time is required for completion of reaction, whereas increasing the reaction time beyond the threshold time limit reduces the product conversion efficiency due to degradation reactions and reversible nature of transesterification reaction (Srivastava et al., 2018).

Effect of alcohol/oil molar ratio

In transesterification reaction, mainly the short-chain alcohols including methanol (MeOH) and ethanol (EtOH) are usually used. MeOH and EtOH are frequently applied to the reaction among all such alcohols. MeOH usage is favored over other alcohols due to its physicochemical properties and easy catalyst solvation. The alcohol/oil molar ratio is a significant parameter that can alter the biodiesel yield. Conventionally, alcohol and oil are used in a 3:1 ratio to form fatty acid ethyl esters (FAEE). Additionally, high MeOH-to-oil molar ratio subsequently enhances overall biodiesel yield. Excess of MeOH is required basically for two purposes: first, to maintain the reaction in a forward direction, and second, for easy separation of glycerol from FAME. Moreover, a high MeOH/oil molar ratio gives a high ester yield in a shorter reaction time (Ehimen et al., 2010).

Effect of co-solvents

The co-solvents are used in transesterification due to the low solubility of oil into MeOH. Moreover, during in situ transesterification, co-solvents are used for easy permeabilization of the tough microalgal cell wall. The heterogeneous nature of transesterification reaction limits the rate of mass transfer, resulting in low solubility of oil in alcohol, especially when there is a low MeOH concentration. Therefore, the addition of co-solvent in the reaction enhances the solubility between microalgae crude lipid and MeOH that reduces the lag phase of transesterification and subsequently enhances the process efficiency. In *Chlorella vulgaris,* acid-mediated transesterification was performed to decrease the viscosity of microalgae crude oil. Under optimized conditions (MeOH/oil molar ratio of 180 and 35 wt.% of catalyst concentration), 95% of FAME was recovered in 6 h without the use of co-solvent. Further, to attain efficient and speedy recovery of biodiesel, the effect of co-solvent on transesterification was investigated using tetrahydrofuran (THF). A significant reduction in reaction time and MeOH consumption was observed. The THF yielded high FAME conversion in 3 h of reaction time with a MeOH/THF/oil molar ratio of 60:15:1 (Lam and Lee, 2013).

Algae storage condition

The fatty acid profile of microalgae drastically changes with the cultural environment. Under nitrogen or salt-stressed conditions, microalgae mainly accumulate TAG, whereas in other conditions, they synthesize polyunsaturated fatty acids (PUFAs) and other omega fatty acids. Long-term storage of wet algal biomass in ambient conditions can deteriorate the lipid quality. In different microalgal species, i.e., *Scenedesmus*, *Nannochloropsis* and heterotrophic dinoflagellates, two-step esterification followed by transesterification was investigated. The algae lipid was esterified with H_2SO_4 followed by transesterification with KOH, NaOH and potassium methoxide ($KOCH_3$). The reaction conditions were optimized for high biodiesel yield from dinoflagellates, *Scenedesmus* sp. and *Nannochloropsis* sp. The TAG content was found maximum in dinoflagellate oil and showed the highest FAME yield of 90.1%. However, the FAME yields for *Scenedesmus* sp. and *Nannochloropsis* sp. were 78.3% and 70.4%, respectively. It was observed that cellular lipid in long-term storage in wet algal biomass gets degraded due to the internal enzyme system.

The storage condition of feedstocks plays a decisive role in biodiesel quality. During the storage period, microalgae lipid may get degraded due to improper storage. In the *Scenedesmus* wet biomass paste, it was found that the sample stored at –80°C has the highest TAG (72.1%) content with negligible FFA content. However, the sample stored at room temperature (25°C) showed a high proportion of FFA (70.3%) as compared to TAG (4.4%) of the total lipid content in dry cell biomass. This degradation in TAG content significantly decreases the biodiesel yield (Chen et al., 2012).

Catalytic transesterification

Catalysts are chemical or biological entities that reduce the activation energy and enhance the rate of reaction by allowing the reaction to complete in a shorter time. Analogous to other oil feedstocks, microalgae biodiesel can be obtained by using various catalysts such as homogeneous catalyst, heterogeneous catalyst and enzyme catalyst (Table 4.3). Homogenous catalysts can be acidic or alkaline such as hydrochloric acid (HCl), sulfuric acid (H_2SO_4),

Table 4.3 Catalysts used for biodiesel production from microalgae

Algal species	*Catalyst used*	*Yield (%)*	*References*
Chlorella	Lipozyme TL IM	98	Gumbytė et al. (2018)
Chlorella protothecoides CS-41	Mg/Al–CO_3 hydrotalcite with Mg/Al	90	Zeng et al. (2014)
Chlorella vulgaris	H_2SO_4+NaOH	97	Sánchez et al. (2011)
Chlorococcum	Lipase	–	Prabakaran et al. (2019)
Freshwater algae	KOH	–	Vijayaraghavan and Hemanathan (2009)
Macroalga (kelp)	Mcgyan process	95	Krohn et al. (2011)
Scenedesmus	MW+WO_3/ZrO_2	52	Guldhe et al. (2014)
Scenedesmus	US+WO_3/ZrO_2	71	Guldhe et al. (2014)
Scenedesmus rubescens	Raw biochar, modified biochar and H_2SO_4	47 24	Tsavatopoulou et al. (2020)
Spirulina platensis	MgO/ZSM-5	92	Qu et al. (2020)
Tetraselmis	NaOH	–	Teo et al. (2014a)
Tetraselmis	Immobilized lipase	–	Teo et al. (2014a)

sodium hydroxide (NaOH), potassium hydroxide (KOH) and sodium methoxide (NaOMe). Although catalysts do not take part in the chemical reaction, there is always a loss of homogenous catalyst in every reaction.

Inactivation/reduction in catalytic activity and inefficiency in FAME conversion can be easily observed when the same catalyst is used multiple times. To reduce the catalyst loss and maintain the inherent catalytic activity of the homogenous catalyst, a novel heterogeneous catalyst has evolved. In heterogeneous catalysis, the reactant and catalyst are in a different phase and do not mix. There are several advantages of heterogeneous catalysts over homogenous catalysts such as easy recovery, high reusability and economic viability. Some of the microalgal species are represented in Table 4.4 that were used for biodiesel production by the catalytic transesterification method.

Homogenous acid catalyst

Among the homogenous acid catalysts, H_2SO_4 is the most frequently applied acid for microalgae biodiesel production. Besides, other acid catalysts used are HCl, boron trifluoride (BF_3), phosphoric acid (H_3PO_4) and some organic sulfonic acids. Mostly sulfonic acid and H_2SO_4 are preferred that give 99% conversion efficiency. In heterotrophically grown *Chlorella protothecoides*, biodiesel synthesis was reported by using H_2SO_4 as a catalyst. In this study, four different levels of catalyst loading of 25%, 50%, 60% and 100% were applied based on the oil weight of microalgae. To improve the transesterification process, various MeOH/oil molar ratios (25:1, 30:1, 45:1, 56:1, 70:1 and 84:1) and different reaction temperatures (30°C, 50°C and 90°C) were optimized during the experiment. The best biodiesel yield was reported with 100% catalyst along with a MeOH/oil molar ratio of 56:1 at 30°C (Miao and Wu, 2006).

Recently, another report was presented where H_2SO_4-catalyzed transesterification of marine microalgae *Chlorella* BDUG 91771 was found as the most efficient method for enhanced quality of biodiesel. The various homogenous and heterogeneous catalysts including NaOH, H_2SO_4 and CaO were used in transesterification, while a two-step sequential transesterification process was also performed with NaOH and H_2SO_4 or vice versa. The maximum reported yield of methyl ester was 60% with 3.5% H_2SO_4 catalyst loading in 2.5h (Mathimani et al., 2015). Certainly, the homogenous acid-catalyzed reaction yield is high, but it follows slow reaction with the requirement of high reaction time, temperature, co-solvent addition and excess of MeOH. In addition, acid catalysis is corrosive, which is a major concern for commercial biodiesel production. Therefore, to overcome such a problem, homogenous base-catalyzed reactions were favored over homogenous acid-catalyzed reactions.

Homogeneous alkali catalyst

Homogeneous alkali catalysts such as NaOH, MeONa, KOH, Na_2CO_3 and K_2CO_3 are most likely used for the transesterification process. These reagents catalyze a reaction with a high rate at a low pressure and temperature. Sodium methoxide is an extensively used catalyst in transesterification reaction having a conversion efficiency of >98% in a comparatively lesser reaction time of 30 min. The effect of alkali catalyst on quality biodiesel yield from *Neochloris oleoabundans* was investigated using NaOH and NaOMe. The effect of time, temperature and catalyst loading (dose) was optimized for high biodiesel yield. A high yield of 83.8% was reported with NaOH (3.7 g/L) with a MeOH-to-oil molar ratio of 6:1 at 60°C in 65 min (Karthikeyan et al., 2018).

In another study, 84.5% biodiesel yield was reported from marine macroalgae *Stoechospermum marginatum* in a single-stage transesterification process using NaOH as catalyst with a MeOH-to-oil molar ratio of 8:1. The physicochemical properties of algae biodiesel showed a comparable result with the existing diesel oil (Hariram et al., 2019). Nevertheless, quality biodiesel can be obtained by this method, but it is highly sensitive to FFA and moisture content; thus, this method does not support the requirement of feedstock having high FFA (<3%) and can form soaps during the reaction (Atadashi et al., 2013; Bharathiraja et al., 2014). Therefore, high-FFA-containing oil feedstocks need an extensive pretreatment before alkali-catalyzed transesterification reaction. Therefore, to resolve the problem of soap formation, a two-step esterification–transesterification reaction involving acid–base catalysis is widely adopted.

Acid and alkali catalysts

Microalgae oil consists of a high potential for qualitative biodiesel production. However, the fatty acid composition from microalgae consists of high FFA and unsaturated fatty acid that interfere with overall biodiesel yield. Therefore, to avoid such technological barriers, a dual-step transesterification process is preferred. Acid-mediated esterification and subsequent homogenous alkali-catalyzed transesterification are the most efficient and preferred way for obtaining higher biodiesel yield up to 95–99% even with high-FFA feedstocks. In a study, *Chlorella vulgaris* oil was employed and simulated for the esterification–transesterification process using Aspen Plus software coupled with pinch analysis of heat integration strategy. A two-step sequential esterification/transesterification process using H_2SO_4 (esterification) followed by NaOH was performed. The water produced during esterification was removed by flashing and further reused for catalyst washing after the transesterification to reduce the external water usage for acid catalyst (H_2SO_4) removal (Sánchez et al., 2011).

There is usually a loss of homogenous catalyst during transesterification reaction, along with chances of corrosion in the reaction vessel, difficulty in phase separation, soap formation, etc. These are some common bottlenecks in commercialization of biodiesel production (Table 4.4). To resolve the issues of homogenous catalysts and to make algae biodiesel more economical, the development of heterogeneous catalysts was made. The heterogeneous catalyst showed enormous advantages over a homogenous catalyst such as easy recovery, multiple-time reusability, low production cost and short reaction time. Identical to the homogenous catalyst, heterogeneous catalyst belongs to the class of heterogeneous acid catalyst and heterogeneous alkali catalyst. The advantages and disadvantages of the various biodiesel production processes are summarized in Table 4.4.

Heterogeneous acid catalyst

Recently, various reports on biodiesel production with a heterogeneous catalyst approach from microalgae also have been published. Conversion of *Scenedesmus obliquus* lipid to biodiesel was investigated using chromium–aluminum metal oxide (CRAL). In the study, reaction parameters were optimized by RSM for improved biodiesel yield where 98.28% FAME conversion was achieved with CRAL catalyst of 15% loading. The reusability efficiency of the CRAL catalyst was found up to four batches. A comparison between the homogenous catalyst and CRAL catalyst showed high FAME conversion efficiency of CRAL catalyst (Guldhe et al., 2017a). In another report by the same research group, tungstate zirconia (WO_3/ZrO_2) was applied as a heterogeneous acid catalyst for the conversion of *Scenedesmus obliquus* oil into FAME. A Box–Behnken design (BBD) was applied for various parameter optimizations including catalyst concentration, temperature and MeOH/oil molar ratio. WO_3/ZrO exhibited the highest conversion of 94.58% with 15% of catalyst loading. The catalyst reusability without alteration in conversion efficiency was found up to three batches. The comparative analysis with homogenous and enzyme-catalyzed reaction showed a high potential of tungsten zirconia over the homogenous and enzyme-catalyzed biodiesel conversion (Guldhe et al., 2017b).

Table 4.4 Advantages and disadvantages of catalytic transesterification

Catalyst type	*Examples*	*Advantages*	*Disadvantages*
Homogenous acids	HCl, H_2SO_4	• Perform simultaneous esterification and transesterification • Work with high-FFA-containing fatty acids • No saponification	• Highly corrosive to vessels • Difficult phase separation • Need higher reaction temperature and time • Poor reusability • Need more water for neutralization
Homogenous alkalis	NaOH, KOH, NaOMe	• Fast reaction • High catalytic efficiency • Safer and easy to handle • Cost-effective • Require moderate reaction conditions	• Need anhydrous condition • Effective on low-FFA feedstock • Saponification • Wastewater generation during washing
Heterogeneous acids	Amberlyst-31 WET, CH-A and NKC-9	• Work on high-FFA-containing oil feedstock. • No saponification	• Corrosive to vessels • Require water to remove the catalyst
Heterogeneous alkalis	CaO, MgO, Al_2O_3 Mg/Al hydrotalcite	• Non-corrosive • Eco-friendly • Reusable • Easy separation • Longer catalyst lifetime	• High cost • May require low FFA and moisture content
Enzymes	Lipase GH2, Novozyme CAL-B	• High product purity • Single-step transesterification • Require milder reaction conditions • Eco-friendly • Non-corrosive • No effect of moisture content • No saponification	• High cost • High temperatures and harsh solvent can inactivate enzymes • The excess solvent can disturb the active enzyme structure
Nanocatalysts	Fe, Ag, Au, Co, ZnO, TiO_2, SiO_2, graphene	• Cost-effective • Simple design • Recyclable • Low water consumption • High specificity • High surface • High-efficiency yield • Can be used in existing facilities	• Nanotoxicity to certain natural habitats

References: Akubude et al. (2019), Atadashi et al. (2013), Borugadda and Goud (2012), Galadima and Muraz (2014), Leung et al. (2010) and Sivaramakrishnan and Incharoensakdi (2017).

Teo et al. (2014) has reported a 92% yield of biodiesel from *Nannochloropsis oculata* with the application of calcium methoxide ($CaOCH_3$) catalyst in MeOH. The solid $Ca(OCH_3)$ catalyst was hydrothermally synthesized and was easy to recover after transesterification. In this optimization, various limiting factors including temperature, catalyst amount and MeOH/oil molar ratio were also analyzed. The fatty acid profile of *N. oculata* biodiesel also showed a high fraction of PUFAs that made it a viable biodiesel to be utilized in the colder climatic region (Teo et al., 2014b).

A high biodiesel yield was reported from esterification of *Nannochloropsis gaditana* oil using heterogeneous zeolite catalyst (H-Beta zeolite, ZSM-5, Beta, h-ZSM-5 and H-Beta). The effect of temperature and MeOH as a solvent was investigated. The best-performed conversion efficiency of oil to biodiesel was observed with H-Beta zeolite when used at a higher reaction temperature and a high MeOH concentration. The lipid profiling of *Nannochloropsis gaditana* oil was performed the presence of MUFAs along with saturated fatty acid (SFA) in relatively high concentrations (Carrero et al., 2011).

Heterogeneous alkali catalyst

There are numerous heterogeneous solid alkali catalysts including hydrotalcite, alkaline earth metal oxides and basic zeolites that have been reported with efficient yield (Atadashi et al., 2013). Umdu et al. (2009) have reported a 97.5% biodiesel yield from *Nannochloropsis oculata* in the presence of calcium oxide (CaO)- and magnesium oxide (MgO)-doped alumina (Al). It was reported that Ca and MgO were non-functional in pure form, whereas the alumina-doped CaO (CaO/Al_2O_3) was the most active catalyst in all the mixed oxide catalysts. It has shown reusability up to three batches. It was concluded that the catalytic activity depends on the basic strength of catalyst produced during gel–sol production, not on the crystallite sizes of CaO in Al_2O_3 (Umdu et al., 2009). In another report, *Nannochloropsis oculata* and *Spirulina* sp. were studied with commercially available pure catalyst such as CaO, $CaCO_3$ and $Ca(OH)_2$ and Al_2O_3. The highest catalytic activity was found in CaO-supported Al_2O_3 resulting in 90–99% biodiesel yield (Turkkul et al., 2020).

Enzyme catalyst

There are also reports available on enzyme-catalyzed transesterification of microalgae. Lipase enzyme is a biocatalyst, which is used in the conversion of fats/oil into FAAE. Lipase catalyzes the interesterification and hydrolysis of lipid molecule by a ping-pong reaction mechanism where TAG gets converted to DAG and MAG with the consumption of water (Zhang and Adhikari, 2014). Lipase is mostly extracted from various plants, microbial species and fungi. A high conversion rate of 95% was reported in *Chlorella vulgaris* oil using recombinant lipase GH2 to produce FAME and FAEE, in an n-hexane solvent system. It was also recommended that the production of FAME is more economical and convenient than that of FAEE (Huang et al., 2015). Immobilized lipase was applied to biodiesel production from *Chlamydomonas*.

It was found that a very tiny amount of lipase (0.10 g) is sufficient to catalyze the transesterification reaction that resulted in a yield of 72%

under optimized conditions (Choong et al., 2016). A photobioreactor-grown microalga *Chlorella vulgaris* ESP-31 had 63.2% lipid content. The lipid was transesterified by enzymatic transesterification where immobilized lipase was utilized. It was isolated from *Burkholderia* sp. C20. Transesterification was carried out by two methods: first, directly with the extracted microalgae lipid (M-I); and second, by the in situ transesterification (M-II). It was found that the in situ transesterification (M-II) method yielded high (97.3%) biodiesel, whereas extracted lipid conversion (M-I) to biodiesel had yielded only 72.1%. The result showed the insignificant impact of moisture content on enzyme-catalyzed reaction, whereas it was found that the wet algal biomass supported the enzyme catalysis to obtain higher biodiesel (Tran et al., 2012).

Nanocatalyst

The notable intention of economical biodiesel production lies in decreasing the catalyst amount and moderate reaction conditions. Nanocatalysts combine the advantages of both homogenous and heterogeneous catalysts and eliminate the demerits from them. However, the size, geometry, aggregation behavior, surface functionality and composition, chemical and physical environments, etc., affect the catalytic activity of nanocatalysts (Akubude et al., 2019). In *Nannochloropsis oculata*, biodiesel was synthesized via nanocatalyst $Ca(OCH_3)_2$. Reaction parameters were optimized, and the highest FAME yield of 99% was obtained with 3% of $Ca(OCH_3)_2$. This hydrothermally synthesized nanocatalyst also showed high stability and reusability (Teo et al., 2016).

Advanced technologies in catalytic transesterification

The conventional catalytic transesterification process has several limitations, such as the requirement of more catalyst, complex purification as well as a separation method, comparatively high cost, requirement of FFA content reduction and a hazardous chemical release into the surroundings. Therefore, various technological advancements such as supercritical method, BIOX process, microwave and ultrasonication and ionic liquid (IL) have been adopted for enhanced recovery, process intensification and low waste production in biodiesel production through a catalytic approach from microalgae. Among all these innovative methods, the predominant methods are in situ transesterification and non-catalytic supercritical method due to single-pot biodiesel synthesis.

In situ transesterification

The biodiesel synthesis from microalgae involves multiple downstream processing steps, which increase the production cost. Therefore, to reduce such expense, direct algal biomass transesterification processes have evolved. This is also identified as reactive extraction or in situ transesterification that is devoid of individual steps of biomass drying and oil extraction followed by transesterification. The biodiesel production from two algal species *Chlorella* and *Nannochloropsis oculata* has been investigated using a different catalyst (homogenous and heterogeneous) and with varying moisture content. It was found that the highest biodiesel yield was obtained from H_2SO_4-catalyzed transesterification when dried biomass of both species was used. *Chlorella* attained a conversion of 92%, whereas *Nannochloropsis oculata* yielded only 73%. The tough cell wall of marine microalgae *Nannochloropsis oculata* is a major hindrance for inefficacious extraction of oil from microalgal cell, thereby lowering the biodiesel yield (Velasquez-Orta et al., 2013).

In enzyme-mediated transesterification of wet *Nannochloropsis gaditana* biomass, FAME synthesis was done from saponifiable lipid. Novozyme 435 (N435) catalyzed the reaction with 99.5% conversion efficiency. However, the conversion efficiency was decreased by 40% while using the same lipase batch for the three consecutive transesterifications (López et al., 2016). In biodiesel production from *Chlorella vulgaris* biomass, both acid and alkali catalysts were used and compared for their conversion efficiency. A conversion of 77.6% was found from NaOH-catalyzed transesterification, whereas 96.8% conversion was obtained with H_2SO_4 under optimized culture conditions. However, the acid-catalyzed reaction has a higher yield with a ratio of 0.35:1 (oil:alcohol), but alkaline transesterification can complete the reaction in a shorter time (Velasquez-Orta et al., 2012). Mostly in in situ transesterification, the dried microalgal biomass is treated with MeOH having a tiny amount of H_2SO_4 as a catalyst.

MeOH serves the purpose of both extraction and esterification solvents. It was observed that using 10% (v/v) H_2SO_4 in MeOH recovered approximately 81% of biodiesel yield from 25 mg of wet algal biomass by increasing MeOH and H_2SO_4. It was found that high moisture content in microalgal biomass critically affects the biodiesel yield by lowering the conversion of oil to FAME. The moisture content of more than 20% significantly reduces the biodiesel yield of wet algal biomass (Sathish et al., 2014). Other than MeOH, some additional solvents can also assist the direct transesterification, e.g., chloroform and hexane. These typical nonpolar solvents easily solubilize the cell wall of microalgae for the smooth extraction of fats and oils from microalgal cells. The transesterification process is reversible. Therefore, more amount of MeOH is required to make the reaction in a forwarding direction; otherwise, the FAME is converted back into fatty acid and alcohol. Therefore, advanced in situ transesterification technologies such as supercritical extraction and hydrothermal liquefaction have been developed to resolve the low yield from wet microalgal biomass.

Supercritical transesterification

Supercritical extraction and transesterification is an alternative to conventional biodiesel production. It is an efficient (>99%), eco-friendly and fast process that can work with or without a catalyst. Therefore, it is also known as catalyst-free transesterification, which requires conditions of high temperature and pressure. Gases and liquids above their critical temperature and pressure condition behave as supercritical fluids (Saka et al., 2006). In a study, *Chlorella* oil was used for supercritical MeOH transesterification. Numerous limiting factors including reaction temperature, reaction time and MeOH/oil molar ratio were optimized through hybrid optimization of RSM-ANN-GA tools. It was observed that the highest conversion efficiency of 98.1% was obtained in just 30 min with a MeOH/oil molar ratio of 23:4. Moreover, the biodiesel properties of microalgae lipid were comparable to the diesel fuel (Srivastava et al., 2018).

A selective TAG extraction and transesterification process (using BF_3 catalyst) was investigated in supercritical CO_2 extraction. In this process, supercritical CO_2 selectively extracts TAG and eliminates phospholipids and pigments that interfere with transesterification efficiency. The high selectivity and tunability of supercritical CO_2 for contaminants such as N and P are advantageous and important for biodiesel integrity. Lower N content in fuel makes microalgae biodiesel environment-friendly by reducing the NOx emissions (Soh and Zimmerman, 2011). The speedy reaction, high conversion efficiency of oil to FAME and selective extraction and transesterification have made this approach the most convincing method for algae biodiesel production among all the other available methods.

Ultrasound-/microwave-assisted biodiesel production

The tough cell wall of microalgae hinders complete oil extraction and thus lowers the biodiesel yield, which is a major bottleneck in in situ microalgae transesterification. The microwave waves are high-energy waves that transfer energy through ionic conduction and dipole and interfacial polarization assisting the cell breakage and conversion of oil to FAAE by localized superheating, whereas ultrasound waves induce the cavitation bubble, which induces cell disruption and intense mixing of oil and alcohol. The cavitation process provides microturbulence in the reaction mixture and creates shock waves. The microturbulence induces intense mixing, whereas the shock waves assist in cell breakage. These continuous compression and rarefaction cycles increase reaction temperature and disrupt the bubble formation for easy mixing of reactants. Moreover, ultrasound-assisted transesterification can remove the hurdles attributed due to heterogeneous reaction conditions by enhancing the mass transfer.

Ultrasound- and microwave-assisted simultaneous extractive-transesterification was investigated on *Chlorella* sp. in the presence of NaOH in EtOH. FAEE yield and subsequent conversion with the microwave and ultrasound method were reported to be 18.8% and 18.5% (yield) and 96.2% and 95% (conversion), respectively. However, the conventional benchtop Bligh and Dyer method that consists of a two-step process including extraction and transesterification was reported with 13.9% and 78.1% yield and conversion of FAEE, respectively (Martinez-Guerra et al., 2014).

In *Botryococcus* sp., ultrasound-assisted enzyme-catalyzed transesterification was performed that showed a wide advantage over the conventional process. It was reported that ultrasound treatment enhances the yield and reduces the reaction time. The maximum yield of 88% methyl ester (w/w of oil) was reported with a reaction time of 4 h through the ultrasound method under optimum experimental conditions as compared to the previously reported maximum yield of 78% methyl esters (w/w of oil) with the reaction time of 36 h without ultrasound (Sivaramakrishnan and Incharoensakdi, 2017).

BIOX method

This is a very recent, advanced and non-catalyzed transesterification reaction developed by a Canadian research group. In this process, transmethylation of fatty acids takes place and around 99% reaction is completed in 5–10 min at ambient reaction temperature and pressure. This continuous process can run for any oil-bearing feedstock. The inert reclaimable co-solvent, e.g., THF, is used in this process to solubilize MeOH effectively in the nonpolar phase. The addition of co-solvent increases the solubility of MeOH into fatty acid for speedy reaction (Demirbas, 2008). Particularly, the BIOX process is not much popularized in microalgae biodiesel production. However, the addition of co-solvents was already investigated and established (Lam and Lee, 2013).

Ionic liquids

One of the major problems with conventional microalgae biodiesel production is the use of toxic solvents for extraction and transesterification, which significantly affect the environment. Moreover, improper solubilization of the microalgal cell wall reduces the overall biodiesel yield. Thus, the microalgae oil extraction procedure needs a highly nonpolar solvent, which is usually vaporized easily, such as chloroform, acetone and hexane, and is neurotoxic. Hence, there is a necessity for a cleaner, greener, sustainable and economic microalgae biodiesel production process, which made the invention of ILs. ILs are salts having large asymmetric cations coupled with smaller organic or inorganic anions, often known as green solvents. The unique characteristics of these compounds include highly variable viscosity and comparative high thermal stability along with low vapor pressure.

The hydrophobicity of ILs can be altered by changing the anions and cations. The impact of IL-mediated microwave heating on in situ transesterifications of wet *Nannochloropsis* sp. biomass was studied with the application of the IL 1-ethyl-3-methylimidazolium methyl sulfate [EMIM][$MeSO_4$]. Process parameters were optimized by RSM with a biodiesel yield of 42.2% with an algae/MeOH ratio of 1:4 (wt/vol%) and a MeOH-to-IL ratio of 1:0.5 (wt/wt%) with a microwave exposure time of 14 min (Wahidin et al., 2018). In another report, oil from four microalgal species – *Botryococcus braunii* (two strains, BB763 and BB764), *Chlorella vulgaris* and *Chlorella pyrenoidosa* – was isolated and transesterified by enzyme-mediated transesterification in two different solvent systems. During this study, one solvent system was IL (1-butyl-3-methylimidazolium hexafluorophosphate, [BMIm][PF_6]), while the other one was organic solvent (tert-butanol). Two different immobilized enzyme systems – *Penicillium expansum* lipase (PEL) and *Candida antarctica* lipase B (Novozym 435) – were also studied during this study. Under optimum conditions, high transesterification efficiency was reported in both the enzyme systems in the presence of IL (90.7% and 86.2%) as compared to tert-butanol (48.6% and 44.4%) (Lai et al., 2012).

Conclusions

Microalgae being a third-generation biofuel have received enough attention as a promising biodiesel feedstock due to their enormous importance over other oilseed crops. Besides, the downstream processing and storage conditions affect the environmental and genetic factors of the microalgae biodiesel property. Microalgae oil extraction and transesterification are a very crucial step for quality control. There is a huge impact of solvent systems and catalytic transesterification techniques. In microalgae biodiesel production, several factors such as reaction time, temperature, catalyst concentration, oil-to-MeOH molar ratio, type of catalyst, purity of reactants and moisture significantly alter the biodiesel quality and yield. Moreover, it was observed that the high moisture content and FFA are the main factors that can deteriorate the biodiesel quality. Therefore, various strategies have been adopted for process intensification.

Adoption and implementation of advanced technologies can overcome the bottlenecks of traditional microalgae biodiesel production. Indeed, the catalytic transesterification of microalgae yielded >98% biodiesel with some advantages and limitations of production processes; therefore, it is recommended to cautiously select the microalgal strain as well as the biodiesel production method. In situ transesterification, supercritical extraction and transesterification and ultrasound-assisted transesterification are incorporated as advanced technologies for process intensification and to work with the lower grade FA-containing feedstocks. Microalgae biodiesel is the hope of the world energy requirement. However, improvements in the production process can be made that make microalgae biodiesel a sustainable fuel source for the future.

References

Akubude, V., Nwaigwe, K. and Dintwa, E. (2019). Production of biodiesel from microalgae via nanocatalyzed transesterification process: A review. *Materials Science for Energy Technologies* 2:216–225.

Atadashi, I., Aroua, M., Aziz, A.A. and Sulaiman, N. (2013). The effects of catalysts in biodiesel production: A review. *Journal of Industrial and Engineering Chemistry* 19:14–26.

BenMoussa-Dahmen, I., Chtourou, H., Rezgui, F., Sayadi, S. and Dhouib, A. (2016). Salinity stress increases lipid, secondary metabolites and enzyme activity in *Amphora subtropica* and *Dunaliella* sp. for biodiesel production. *Bioresource Technology* 218:816–825.

Bharathiraja, B., Chakravarthy, M., Kumar, R.R., Yuvaraj, D., Jayamuthunagai, J., Kumar, R.P. and Palani, S. (2014). Biodiesel production using chemical and biological methods–A review of process, catalyst, acyl acceptor, source and process variables. *Renewable and Sustainable Energy Reviews* 38:368–382.

Bollino, C.A., Asdrubali, F., Polinori, P., Bigerna, S., Micheli, S., Guattari, C. and Rotili, A. (2017). A note on medium-and long-term global energy prospects and scenarios. *Sustainability* 9:833.

Borugadda, V.B. and Goud, V.V. (2012). Biodiesel production from renewable feedstocks: Status and opportunities. *Renewable and Sustainable Energy Reviews* 16:4763–4784.

Carrero, A., Vicente, G., Rodríguez, R., Linares, M. and Del Peso, G. (2011). Hierarchical zeolites as catalysts for biodiesel production from *Nannochloropsis* microalga oil. *Catalysis Today* 167:148–153.

Chen, L., Liu, T., Zhang, W., Chen, X. and Wang, J. (2012). Biodiesel production from algae oil high in free fatty acids by two-step catalytic conversion. *Bioresource Technology* 111:208–214.

Chisti, Y. (2007). Biodiesel from microalgae. *Biotechnology Advances* 25:294–306.

Chng, L.M., Chan, D.J.C. and Lee, K.T. (2016). Sustainable production of bioethanol using lipid-extracted biomass from *Scenedesmus dimorphus*. *Journal of Cleaner Production* 130:68–73.

Choong, W.P., Tan, C.H., Show, P.L., Lam, H.L., Annuar, M.S.B.M., Juan, J.C., Chang, J.-S. and Ling, T.C. (2016). Efficient enzyme-catalysed transesterification of microalgal biomass from *Chlamydomonas* sp. *Energy* 116:1370–1373.

Chuah, L.F., Klemeš, J.J., Yusup, S., Bokhari, A. and Akbar, M.M. (2017). A review of cleaner intensification technologies in biodiesel production. *Journal of Cleaner Production* 146:181–193.

Demirbas, A. (2008). Current technologies in biodiesel production. *Biodiesel: A Realistic Fuel Alternative for Diesel Engines*. London: Springer, pp. 161–173.

Demirbas, A. (2009). Biodiesel from waste cooking oil via base-catalytic and supercritical methanol transesterification. *Energy Conversion and Management* 50:923–927.

Difusa, A., Talukdar, J., Kalita, M.C., Mohanty, K. and Goud, V.V. (2015). Effect of light intensity and pH condition on the growth, biomass and lipid content of microalgae Scenedesmus species. *Biofuels* 6:37–44.

Ehimen, E., Sun, Z. and Carrington, C. (2010). Variables affecting the in situ transesterification of microalgae lipids. *Fuel* 89:677–684.

Galadima, A. and Muraza, O. (2014). Biodiesel production from algae by using heterogeneous catalysts: A critical review. *Energy* 78:72–83.

Gouveia, L. and Oliveira, A.C. (2009). Microalgae as a raw material for biofuels production. *Journal of Industrial Microbiology and Biotechnology* 36:269–274.

Guldhe, A., Singh, B., Rawat, I. and Bux, F. (2014). Synthesis of biodiesel from Scenedesmus sp. by microwave and ultrasound assisted in situ transesterification using tungstated zirconia as a solid acid catalyst. *Chemical Engineering Research and Design* 92:1503–1511.

Guldhe, A., Moura, C.V., Singh, P., Rawat, I., Moura, E.M., Sharma, Y. and Bux, F. (2017a). Conversion of microalgal lipids to biodiesel using chromium-aluminum mixed oxide as a heterogeneous solid acid catalyst. *Renewable Energy* 105:175–182.

Guldhe, A., Singh, P., Ansari, F.A., Singh, B. and Bux, F. (2017b). Biodiesel synthesis from microalgal lipids using tungstated zirconia as a heterogeneous acid catalyst and its comparison with homogeneous acid and enzyme catalysts. *Fuel* 187:180–188.

Gumbytė, M., Makareviciene, V., Skorupskaite, V., Sendzikiene, E. and Kondratavicius, M. (2018). Enzymatic microalgae oil transesterification with ethanol in mineral diesel fuel media. *Journal of Renewable and Sustainable Energy* 10:013105.

Hariram, V., John, J.G. and Seralathan, S. (2019). Spectrometric analysis of algal biodiesel as a fuel derived through base-catalysed transesterification. *International Journal of Ambient Energy* 40:195–202.

Hu, Q., Sommerfeld, M., Jarvis, E., Ghirardi, M., Posewitz, M., Seibert, M. and Darzins, A. (2008). Microalgal triacylglycerols as feedstocks for biofuel production: Perspectives and advances. *The Plant Journal* 54:621–639.

Huang, G., Chen, F., Wei, D., Zhang, X. and Chen, G. (2010). Biodiesel production by microalgal biotechnology. *Applied Energy* 87:38–46.

Huang, J., Xia, J., Jiang, W., Li, Y. and Li, J. (2015). Biodiesel production from microalgae oil catalyzed by a recombinant lipase. *Bioresource Technology* 180:47–53.

Karpagam, R., Raj, K.J., Ashokkumar, B. and Varalakshmi, P. (2015). Characterization and fatty acid profiling in two fresh water microalgae for biodiesel production: Lipid enhancement methods and media optimization using response surface methodology. *Bioresource Technology* 188:177–184.

Karthikeyan, S., Kalaimurugan, K. and Prathima, A. (2018). Quality analysis studies on biodiesel production of neochloris oleoabundans algae. *Energy Sources, Part A: Recovery, Utilization, and Environmental Effects* 40:439–445.

Kirrolia, A., Bishnoi, N.R. and Singh, R. (2014). Response surface methodology as a decision-making tool for optimization of culture conditions of green microalgae *Chlorella* spp. for biodiesel production. *Annals of Microbiology* 64:1133–1147.

Knothe, G. (2005). Dependence of biodiesel fuel properties on the structure of fatty acid alkyl esters. *Fuel Processing Technology* 86:1059–1070.

Knothe, G., Krahl, J. and Van Gerpen, J. (eds) (2015). *The Biodiesel Handbook*. Elsevier.

Krohn, B.J., McNeff, C.V., Yan, B. and Nowlan, D. (2011). Production of algae-based biodiesel using the continuous catalytic Mcgyan® process. *Bioresource Technology* 102:94–100.

Lai, J.-Q., Hu, Z.-L., Wang, P.W. and Yang, Z. (2012). Enzymatic production of microalgal biodiesel in ionic liquid [BMIm][PF6]. *Fuel* 95:329–333.

Lam, M.K. and Lee, K.T. (2013). Catalytic transesterification of high viscosity crude microalgae lipid to biodiesel: Effect of co-solvent. *Fuel Processing Technology* 110:242–248.

Laurens, L.M., Van Wychen, S., McAllister, J.P., Arrowsmith, S., Dempster, T.A., McGowen, J. and Pienkos, P.T. (2014). Strain, biochemistry, and cultivation-dependent measurement variability of algal biomass composition. *Analytical Biochemistry* 452:86–95.

Leung, D.Y., Wu, X. and Leung, M. (2010). A review on biodiesel production using catalyzed transesterification. *Applied Energy* 87:1083–1095.

López, E.N., Medina, A.R., Cerdán, L.E., Moreno, P.A.G., Sánchez, M.D.M. and Grima, E.M. (2016). Fatty acid methyl ester production from wet microalgal biomass by lipase-catalyzed direct transesterification. *Biomass and Bioenergy* 93:6–12.

Martinez-Guerra, E., Gude, V.G., Mondala, A., Holmes, W. and Hernandez, R. (2014). Microwave and ultrasound enhanced extractive-transesterification of algal lipids. *Applied Energy* 129:354–363.

Marulanda, V.F. (2012). Biodiesel production by supercritical methanol transesterification: Process simulation and potential environmental impact assessment. *Journal of Cleaner Production* 33:109–116.

Mata, T.M., Martins, A.A. and Caetano, N.S. (2010). Microalgae for biodiesel production and other applications: A review. *Renewable and Sustainable Energy Reviews* 14:217–232.

Mathimani, T., Uma, L. and Prabaharan, D. (2015). Homogeneous acid catalysed transesterification of marine microalga *Chlorella* sp. BDUG 91771 lipid–an efficient biodiesel yield and its characterization. *Renewable Energy* 81:523–533.

Mazumdar, P., Dasari, S.R., Borugadda, V.B., Srivasatava, G., Sahoo, L. and Goud, V.V. (2013). Biodiesel production from high free fatty acids content *Jatropha curcas* L. oil using dual step process. *Biomass Conversion and Biorefinery* 3:361–369.

Meher, L., Sagar, D.V. and Naik, S. (2006). Technical aspects of biodiesel production by transesterification- a review. *Renewable and Sustainable Energy Reviews* 10:248–268.

Miao, X. and Wu, Q. (2006). Biodiesel production from heterotrophic microalgal oil. *Bioresource Technology* 97:841–846.

Olmstead, I.L., Hill, D.R., Dias, D.A., Jayasinghe, N.S., Callahan, D.L., Kentish, S.E., Scales, P.J. and Martin, G.J. (2013). A quantitative analysis of microalgal lipids for optimization of biodiesel and omega-3 production. *Biotechnology and Bioengineering* 110:2096–2104.

Piemonte, V., Di Paola, L., Iaquaniello, G. and Prisciandaro, M. (2016). Biodiesel production from microalgae: Ionic liquid process simulation. *Journal of Cleaner Production* 111:62–68.

Prabakaran, P., Pradeepa, V., Selvakumar, G. and Ravindran, A.D. (2019). Efficacy of enzymatic transesterification of *Chlorococcum* sp. algal oils for biodiesel production. *Waste and Biomass Valorization* 10:1873–1881.

Qu, S., Chen, C., Guo, M., Lu, J., Yi, W., Ding, J. and Miao, Z. (2020). Synthesis of MgO/ZSM-5 catalyst and optimization of process parameters for clean production of biodiesel from *Spirulina platensis*. *Journal of Cleaner Production* 276:123382.

Rawat, I., Kumar, R.R., Mutanda, T. and Bux, F. (2013). Biodiesel from microalgae: A critical evaluation from laboratory to large scale production. *Applied Energy* 103:444–467.

Saka, S., Kusdiana, D. and Minami, E. (2006). Non-catalytic biodiesel fuel production with supercritical methanol technologies. *Journal of Scientific and Industrial Research* 65:420.

Sakthivel, R. (2011). Microalgae lipid research, past, present: A critical review for biodiesel production, in the future. *Journal of Experimental Sciences* 2:29–49.

Sánchez, E., Ojeda, K., El-Halwagi, M. and Kafarov, V. (2011). Biodiesel from microalgae oil production in two sequential esterification/transesterification reactors: Pinch analysis of heat integration. *Chemical Engineering Journal* 176:211–216.

Sathish, A., Smith, B.R. and Sims, R.C. (2014). Effect of moisture on in situ transesterification of microalgae for biodiesel production. *Journal of Chemical Technology and Biotechnology* 89:137–142.

Schenk, P.M., Thomas-Hall, S.R., Stephens, E., Marx, U.C., Mussgnug, J.H., Posten, C., Kruse, O. and Hankamer, B. (2008). Second generation biofuels: High-efficiency microalgae for biodiesel production. *Bioenergy Research* 1:20–43.

Schlagermann, P., Göttlicher, G., Dillschneider, R., Rosello-Sastre, R. and Posten, C. (2012). Composition of algal oil and its potential as biofuel. *Journal of Combustion* 2012:285185.

Sharma, K.K., Schuhmann, H. and Schenk, P.M. (2012). High lipid induction in microalgae for biodiesel production. *Energies* 5:1532–1553.

Sivaramakrishnan, R. and Incharoensakdi, A. (2017). Direct transesterification of Botryococcus sp. catalysed by immobilized lipase: Ultrasound treatment can reduce reaction time with high yield of methyl ester. *Fuel* 191:363–370.

Snow, N. (2017). BP energy outlook: Global energy demand to grow 30% to 2035. *Oil and Gas Journal* 115:30–31.

Soh, L. and Zimmerman, J. (2011). Biodiesel production: The potential of algal lipids extracted with supercritical carbon dioxide. *Green Chemistry* 13:1422–1429.

Srivastava, G., Nishchal, F. and Goud, V.V. (2017). Salinity induced lipid production in microalgae and cluster analysis (ICCB 16-BR_047). *Bioresource Technology* 242:244–252.

Srivastava, G., Paul, A.K. and Goud, V.V. (2018). Optimization of non-catalytic transesterification of microalgae oil to biodiesel under supercritical methanol condition. *Energy Conversion and Management* 156:269–278.

Teo, C.L., Jamaluddin, H., Zain, N.A.M. and Idris, A. (2014a). Biodiesel production via lipase catalysed transesterification of microalgae lipids from *Tetraselmis* sp. *Renewable Energy* 68:1–5.

Teo, S.H., Islam, A. and Taufiq-Yap, Y.H. (2016). Algae derived biodiesel using nanocatalytic transesterification process. *Chemical Engineering Research and Design* 111:362–370.

Teo, S.H., Islam, A., Yusaf, T. and Taufiq-Yap, Y.H. (2014b). Transesterification of Nannochloropsis oculata microalga's oil to biodiesel using calcium methoxide catalyst. *Energy* 78:63–71.

Tran, D.T., Yeh, K.L., Chen, C.L. and Chang, J.S. (2012). Enzymatic transesterification of microalgal oil from *Chlorella vulgaris* ESP-31 for biodiesel synthesis using immobilized Burkholderia lipase. *Bioresource Technology* 108:119–127.

Tsavatopoulou, V.D., Vakros, J. and Manariotis, I.D. (2020). Lipid conversion of Scenedesmus rubescens biomass into biodiesel using biochar catalysts from malt spent rootlets. *Journal of Chemical Technology and Biotechnology* 95:2421–2429.

Turkkul, B., Deliismail, O. and Seker, E. (2020). Ethyl esters biodiesel production from Spirulina sp. and Nannochloropsis oculata microalgal lipids over alumina-calcium oxide catalyst. *Renewable Energy* 145:1014–1019.

Umdu, E.S., Tuncer, M. and Seker, E. (2009). Transesterification of *Nannochloropsis oculata* microalga's lipid to biodiesel on Al_2O_3 supported CaO and MgO catalysts. *Bioresource Technology* 100:2828–2831.

Velasquez-Orta, S., Lee, J. and Harvey, A. (2012). Alkaline in situ transesterification of *Chlorella vulgaris*. *Fuel* 94:544–550.

Velasquez-Orta, S., Lee, J. and Harvey, A. (2013). Evaluation of FAME production from wet marine and freshwater microalgae by in situ transesterification. *Biochemical Engineering Journal* 76:83–89.

Vijayaraghavan, K. and Hemanathan, K. (2009). Biodiesel production from freshwater algae. *Energy and Fuels* 23:5448–5453.

Wahidin, S., Idris, A., Yusof, N.M., Kamis, N.H.H. and Shaleh, S.R.M. (2018). Optimization of the ionic liquid-microwave assisted one-step biodiesel production process from wet microalgal biomass. *Energy Conversion and Management* 171:1397–1404.

Williams, P.J.le.B. and Laurens, L.L.M. (2010). Microalgae as biodiesel and biomass feedstocks: Review and analysis of the biochemistry, energetics and economics. *Energy and Environmental Science* 3:554–590.

Yang, F., Long, L., Sun, X., Wu, H., Li, T. and Xiang, W. (2014). Optimization of medium using response surface methodology for lipid production by *Scenedesmus* sp. *Marine Drugs* 12:1245–1257.

Zeng, H.-Y., Xu, S., Liao, M.-C., Zhang, Z.-Q. and Zhao, C. (2014). Activation of reconstructed Mg/Al hydrotalcites in the transesterification of microalgae oil. *Applied Clay Science* 91:16–24.

Zhang, H. and Adhikari, P. (2014). Enzymatic interesterification. In: Dharma, R.K. (ed.) *Trans Fats Replacement Solutions*, Elsevier, pp. 187–214.

5 Production of energy-dense bio-oil from algal biomass via different thermochemical pathways

Ali Shemsedin Reshad and Venu Babu Borugadda

Introduction

Global energy consumption has increased considerably mainly because of the increase in world population as well as the rapid industrialization over the years (Borines et al., 2013; Azizi et al., 2018; Pourkarimi et al., 2019). Biomass feedstock has received more focus as a source for an alternative and clean source of energy and can be thermally, chemically or biochemically converted into biodiesel, bioethanol, bio-oil, syngas, C_1–C_4 gas mixtures and solid biofuels (Ynik et al., 2013). Consumption of renewable fuels is in demand owing to the reduction of greenhouse gas (GHG) emissions, and to meet the renewable fuel standards specified by the individual county or province (Asomaning et al., 2016). Currently, about 95% of renewable fuels (biodiesel, bioethanol and bio-oil) are produced from the first-generation biofuel feedstocks which may affect the sustainability of their process industries. The use of first-generation feedstock for biofuel production has been considered as a great challenge due to the feedstock competition with the food materials, and it leads to the food versus fuel crisis. Because of this fact, more attention was given to second- and third-generation feedstocks where the whole plant matter is used.

Biofuels produced from non-edible feedstocks such as non-edible oilseeds, municipal solid and liquid waste and lignocellulosic materials are considered as second-generation feedstocks. Even though the second-generation biofuels do not compete with the food materials directly, excessive land use restricts their use for the production of food (Pourkarimi et al., 2019). Among the various biomass resources, second- and third-generation feedstocks such as lignocellulosic and algae-derived biomasses can complement the future clean energy supply. Several authors have reported that algae are considered as a promising third-generation feedstock owing to the higher photosynthetic performance and less competitive for fertile land relative to other terrestrial plants (Chen et al., 2014a,b; Aboulkas et al., 2017). However, in the present circumstances, biofuels produced from macro- and microalgae are economically not feasible in comparison with fossil fuel production (DalmasNeto et al., 2014).

Algal biomass-derived energy could partially overcome the challenges of petroleum-derived fuels; in spite of their eco-friendly nature, for most countries, economics has been challenging to compete with conventional fuel production. The costs related to the downstream processes such as cultivation, pretreatment, purification, extraction and production on an industrial scale are the main challenges for cost-effective algal biomass-derived production of biofuels. However, culture medium is in trend to use wastewater as a partial or complete source of nutrients such as nitrogen, potassium, phosphorus and magnesium for micro- and macroalgal growth to minimize the costs for cultivation (DalmasNeto et al., 2014). However, bio-oil production from algae depends on the cultivation methods used, nature of algal strains, chemical composition, proximate and ultimate analyses of the algae and the bio-oil production technique to extract bio-oil. Among the several factors that show a significant impact on bio-oil production, this chapter is aimed to discuss the effect of chemical composition, proximate and ultimate analyses of the algae and different bio-oil production techniques to maximize bio-oil production.

Algal biomass

Algae are aquatic organisms, and photosynthetically, they can produce energy from sunlight, water and CO_2. Further, algae can be categorized into prokaryotic and eukaryotic organisms (Table 5.1). They have the ability to adjust phosphorus and nitrogen from the surrounding environments during their growth. The produced algal biomass can be converted into various chemicals, fuels and other value-added compounds through appropriate techniques. Based on their sizes, algal strains are divided into microalgae with a size range from micrometers to tens of micrometers. Microalgae occur as single cells and/or chains connected with cells without leaves, roots and stems. Further, autophytic algal strains could utilize energy through photosynthesis and transform water (H_2O), sunlight and CO_2 into organic biomass. Heterotrophic algal species consume a small quantity of carbon from the surroundings and convert it into proteins and fat. Certain species of algae take inorganic carbon in the form of CO_2 from the troposphere and natural

Table 5.1 Various characteristics of macroalgae and microalgae

Characteristics	*Macroalgae (filamentous)*	*Microalgae (phytoplankton)*
Plant type	Multicellular	Unicellular
Growing condition	Salt- and freshwater	Salt- and freshwater
Growing ability	Up to 60 m in length	Small size (1 μm)
Nutrient storage and uptake	High nutrient storage	High nutrient uptake
Growth rate	Low	Fast
Application	Production of food and hydrocolloid	Production of food, medicinal supplements and biofuels

Reference: Raheem et al. (2015).

carbon from the surroundings through the process of mixotrophy. Algal cells have a high ratio of surface to the volume which results in better nutrient and CO_2 capture efficiency related to continental plants (Kumar et al., 2016).

Biochemical composition of algae

Chemical/biochemical components such as protein, lipid and carbohydrate; proximate analysis such as fixed carbon (FC), volatile matter (VM) and ash content (AC); and elemental analysis such as sulfur, oxygen, carbon, nitrogen, hydrogen and higher heating value (HHV) could determine the characteristics of algal biomass suitability for either biofuel or value-added chemicals. Protein, lipid and carbohydrate are the most common biochemical compositions found in algae (Sudhakar et al., 2019). Biomass derived from microalgae is rich in lipids, carbohydrates and proteins; their concentration varies depending on the strain type and several physical and biological considerations, and metabolic route throughout the process of cultivation. Biochemical, elemental and proximate analyses of the various micro- and macroalgal strains are reported in Tables 5.2–5.7. Macroalgae are a promising source for carbohydrates as they synthesize very low lipids (Kumar et al., 2016). For example, the amount of protein present in *Spirulina maxima* is 60–71 wt.%, carbohydrates in *Porphyridium cruentum* are 40–60 wt.%, and lipids in *Scenedesmus dimorphus* hold 40 wt.%, whereas *Chlorella protothecoides and Botryococcus braunii* are the major sources of glyceryl terpenoids and lipids further can be converted into hydrocarbon fuels due to their high lipid content (Sudhakar et al., 2019).

The compositions of lipids, proteins, carbohydrates and other components verify the suitability of algal biomass for a specific type of bioenergy production (Figure 5.2). The biochemical contents, particularly protein, carbohydrate and lipid-rich micro-/macroalgal biomass, make them more attractive to produce various value-added products. Algae-derived biomass could be used as bioenergy hauler in the form of compact briquette for direct incineration of biomass; liquid as bioethanol, biomethanol, biodiesel and bio-oil for engine fuel; and gas as bio-hydrogen and syngas for cooking and electricity generation through chemical, biochemical and thermochemical routes.

Elemental composition (ultimate) and proximate analysis of algae

Similar to biochemical composition, elemental (oxygen, hydrogen, carbon, nitrogen and sulfur) contents of different micro- and macroalgal species vary significantly (Tables 5.4 and 5.5) based on algal species and growing conditions. Sudhakar et al. (2019) reported that the elemental composition of algal biomass is higher than woody biomass in some algal strains. The elemental concentration (S, O, C, N, H and metallic elements) differences in algae are responsible for proximate analysis (FC, AC, moisture content (MC), and VM) (Tables 5.6 and 5.7).

Table 5.2 Biochemical composition of various microalgae

Algal strain name	*Protein*	*Lipid*	*Carbohydrate*	*References*
	(wt.%)			
Botryococcus braunii	30.4	17.7	15.5	Munoz et al. (2015)
Chlorella vulgaris	51–58	9.95	12–17	Wang et al. (2015) Yuan et al. (2015) Sunganya et al. (2016) Sudhakar et al. (2019)
Lyngbya	29.9	1.39	13.2	Maddi et al. (2011)
Cladophora	24.6	5.8	24.8	
Chlorella protothecoides	53	15	11	Chen et al. (2015b)
Chlorella pyrenoidosa	71.3	22	0.1	Chen et al. (2015b)
Chlorella	29.6	15–16.5	9–13	Rizzo et al. (2013)
Chlorella	54.9	12.37	26.2	Li et al. (2017)
Chlorella sorokiniana CY1	9.9	18.81	35.7	Chen et al. (2014a,b)
Chlamydomonas reinhardtii	48	21	17	Pourkarimi et al. (2019)
Chlamydomonas	12.18	6.8	35.7	Chen et al. (2014a,b)
Desmodesmus	38–44	13–20	10–14	Chen et al. (2015b)
Dunaliella tertiolecta	29.4	11.44	13.9	Sydney et al. (2010)
Nannochloropsis oculata	42.6	24	6	Chen et al. (2015b)
Nannochloropsis sp.	52	28	12	Pourkarimi et al. (2019)
Nannochloropsis oculata	39	17	20	Du et al. (2012)
Scenedesmus almeriensis	24.6	44.2	25.2	López-Gonzalez et al. (2014a)
Nannochloropsis gaditana	30.7	12.7	21.6	Munoz et al. (2015)
Spirulina platensis	48.4	13.3	30.2	Jena and Das (2011)
Chlorella	34	7	15.5	Babich et al. (2011)
Chlamydomonas reinhardtii	47.4	18.1	15.1	Kebelmann et al. (2013)
Chlamydomonas reinhardtii CW15+	45.7	22.4	11.5	Kebelmann et al. (2013)
Dunaliella tertiolecta	27.2	22	40.5	Kim et al. (2015)
Chlorella vulgaris	54.9	15.5	9.1	Kebelmann et al. (2013)
Chlorella vulgaris	58.1	13.5	12.4	López-Gonzalez et al. (2014b)
Scenedesmus almeriensis	44.2	24.6	25.2	López-Gonzalez et al. (2014c)
Nannochloropsis gaditana	40.5	26.3	25.1	López-Gonzalez et al. (2014a)
Dunaliella tertiolecta	61.3	2.9	21.7	Shuping et al. (2010)
Chlorella vulgaris	41.5	15.7	21	Wang et al. (2013)
Chlorella vulgaris remnants	61.2	5.7	20.3	Wang et al. (2013)

The atomic ratios of carbon to oxygen (C/O), hydrogen to oxygen (H/O), carbon to nitrogen (C/N) and hydrogen to nitrogen (H/N) found in the biomass are the key factors affecting the heating value. In contrast, the higher amounts of oxygen, nitrogen, sulfur and incombustible matter (ash) have an adverse effect and reduce the HHV. The higher heating value of the micro- and macroalgae was found to be in the range of 12–27 MJ/kg and 8–16 MJ/kg, respectively. The presence of the higher oxygen content in algal biomass reduced the energy content of the algal biomass. In general, algae have a small concentration of nitrogen and tracer amount of sulfur that leads to

Table 5.3 Biochemical composition of various macroalgae

Algal strain name	*Protein*	*Lipid*	*Carbohydrate*	*References*
	(wt.%)			
Hypnea valentiae	11.8–12.6	9.6–11.6	11.8–13	Sunganya et al. (2016)
Acanthophora spicifera	12–13.2	10–12	11.6–13.2	Sunganya et al. (2016) and Pourkarimi et al. (2019)
Dictyosphaeria cavernosa	6	10.51	42.75	Sunganya et al. (2016)
Eucheuma cottonii	2.2	3.7	35.2	Tan et al. (2013)
Laminaria japonica	14.8	1.8	51.9	Kim et al. (2011)
Gelidium amansii	10.47	0.74	71.43	Pourkarimi et al. (2019)
Microdictyon agardhianum	20.93	9.4	27.0	Sunganya et al. (2016)
Sargassum fulvellum	13	1.4	39.6	Kim et al. (2011)
Sargassum spp.	10.25	0.74	41.81	Borines et al. (2013)
Boergesenia forbesii	7.43	11.42	21.38	Kaliaperumal et al. (2002) and Sunganya et al. (2016)
Cladophora fascicularis	15.53	15.7	49.5	
Hypnea valentiae	11.8–12.6	9.6–11.6	11.8–13	
Laurencia papillosa	11.8–13	8.9–10.8	12–13.3	
Ulva lactuca	11.4–12.6	9.6–11.4	11.6–13.2	
Caulerpa racemosa	11.8–12.5	9–10.5	16	
Ulva reticulata	12.83	8.5	16.88	
Enteromorpha compressa	7.26	11.45	24.75	
Chaetomorpha aerea	10.13	8.5	31.5	
Chaetomorpha antennina	10.13	11.45	27.0	
Cladophora fascicularis	15.53	15.7	49.5	
Chaetomorpha linoides	9.45	12.0	27.0	
Boergesenia forbesii	7.43	11.42	21.38	
Dictyosphaeria cavernosa	6.0	10.51	42.75	
Caulerpa peltata	6.41	11.42	45.0	
Microdictyon agardhianum	20.93	9.4	27.0	
Valoniopsis pachynema	8.78	9.09	31.5	
Caulerpa racemosa	8.78	10.63	33.75	
Caulerpa sertularioides	9.11	6.99	49.5	
Caulerpa cupressoides	7.43	10.97	51.75	
Codium adhaerens	7.26	7.4	40.5	
Caulerpa laetevirens	8.78	8.8	56.25	
Codium tomentosum	5.06	7.1	29.25	
Caulerpa fergusonii	7.76	7.15	23.63	
Halimeda macroloba	5.4	9.89	32.63	
Codium decorticatum	6.08	9.0	50.63	

less discharge of nitrogen oxides (NO_x) and sulfur dioxide (SO_2) relative to conventional fuels. Additionally, zero net emission of CO_2 can be attained by utilizing algae-derived bio-oils, because CO_2 is in a cycle where the discharged CO_2 would be used through algal biomass regeneration.

Biofuel production from algal biomass

Production of liquid transportation fuels from algal biomass is gaining huge attention owing to several advantages and availability of the feedstocks. Appropriate technique to produce bio-oil depends on the nature of

Table 5.4 Elemental composition (C, H, N, S and O) and heating values of various microalgal strains

Algal strain name	*Carbon*	*Hydrogen*	*Oxygen*	*Nitrogen*	*Sulfur*	*HHV (MJ/kg)*	*References*
	(wt.%)						
Botryococcus braunii	32.2	6.4	27.2	5.7	–	14.4	Munoz et al. (2015)
Chlorella	46.1	6.1	19.1	6.7	0.4	20.4	Rizzo et al. (2013)
Chlorella sorokiniana CY1	45.07	7.64	35.52	3.88	–	20.4	Chen et al. (2014a,b)
Chlamydomonas JSC4	40.32	7.38	44.5	2.61	–	17.41	Chen et al. (2014a,b)
Dunaliella tertiolecta	39.0	5.37	53.02	1.99	0.62	14.24	Shuping et al. (2010)
Dunaliella tertiolecta	38.23	6.19	44.49	11.12	–	11.66	Kim et al. (2015)
Nannochloropsis oculata	39.9	5.5	24.0	6.2	–	16.8	Du et al. (2012)
Chlorella vulgaris	44.8	6.8	40.4	7.0	1.0	17.72	López-Gonzalez et al. (2014a)
Chlorella vulgaris	43.9	6.2	43.3	6.7	–	18	Kebelmann et al. (2013)
Scenedesmus almeriensis	41.9	6.7	44.7	5.9	0.8		López-Gonzalez et al. (2014b)
Nannochloropsis gaditana	49.4	7.7	34.7	7	1.1		López-Gonzalez et al. (2014c)
Nannochloropsis gaditana	30.4	6.8	26.2	5.6	–	13.7	Munoz et al. (2015)
Nannochloropsis	48.2	6.9	23.2	7.5	0.4	19.2	Rizzo et al. (2013)
Spirulina platensis	46.16	7.14	35.44	10.56	0.74	20.52	Jena and Das (2011)
Chlorella	50.2	7.25	33.2	9.3	–	21.2	Babich et al. (2011)
Chlamydomonas reinhardtii	52	7.4	29.8	10.7	–	23	Kebelmann et al. (2013)
Chlamydomonas reinhardtii CW15	50.2	7.3	31.4	11.1	–	22	Kebelmann et al. (2013)
Chlorella vulgaris	42.51	6.77	27.95	6.64	–	16.8	Wang et al. (2013)
Chlorella vulgaris remnants	45.04	6.88	29.42	9.79	–	19.44	Wang et al. (2013)

Table 5.5 Elemental composition (C, H, N, S and O) and heating values of various macroalgal strains

Algal strain name	*Carbon*	*Hydrogen*	*Oxygen*	*Nitrogen*	*Sulfur*	*HHV (MJ/kg)*	*References*
	(wt.%)						
Chorda filum	39.14	4.69	37.23	1.42	1.62	15.55	Ross et al. (2008)
Enteromorpha clathrata	32.65	4.92	24.69	4.43	2.03	12.01	Wang et al. (2013)
Fucus serratus	38.8	4.9	–	1.2	0.89	13.51	Pourkarimi et al. (2019)
Fucus serratus	33.5	4.78	34.44	2.39	1.31	16.66	Ross et al. (2008)
Fucus vesiculosus	32.88	4.77	35.63	2.53	2.44	11.1	Ross et al. (2008)
Laminaria hyperborea	34.97	5.31	35.09	1.12	2.06	16.54	Ross et al. (2008)
Laminaria japonica	35.19	5.51	40.48	1.39	0.48	13	Choi et al. (2014)
Macrocystis pyrifera	27.3	4.08	34.8	2.03	1.89	16	Ross et al. (2008)
Sargassum natans	25.9	5.57	24.18	3.58	1.22	8.68	Wang et al. (2013)

Table 5.6 Proximate analysis of various microalgal strains

Algal strain name	*Moisture content*	*Ash content*	*Volatile matter*	*Fixed carbon*	*References*
	(wt.%)				
Botryococcus braunii	8.0	27.21	–	–	Munoz et al. (2015)
Chlamydomonas JSC4	3.5	5.2	75.5	15.6	Chen et al. (2014a,b)
Chlorella	6.2	21.6	–	–	Rizzo et al. (2013)
Chlorella	6.5	9.5	65.5	18.5	Babich et al. (2011)
Chlorella sorokiniana CY1	3.8	7.9	73.2	15.1	Chen et al. (2014a,b)
Chlorella vulgaris	4.4	15.9	67.2	12.4	López-Gonzalez et al. (2014a)
Chlorella vulgaris	6.18	16.8	66.56	11.62	Wang et al. (2013)
Chlorella vulgaris	4.39	8.34	72.68	14.59	Wang et al. (2013)
Dunaliella tertiolecta	4.98	13.54	54.48	27.0	Shuping et al. (2010)
Nannochloropsis gaditana	3.5	6.4	79.8	10.2	López-Gonzalez et al. (2014a)
Scenedesmus almeriensis	2.9	19.4	67.9	9.7	López-Gonzalez et al. (2014a)
Spirulina platensis	4.54	6.56	79.14	15.24	Jena and Das (2011)

Table 5.7 Proximate analysis of various macroalgal strains

Algal strain name	*Moisture content*	*Ash content*	*Volatile matter*	*Fixed carbon*	*References*
	(wt.%)				
Chorda filum	13.1	9.9	52.2	24.9	Ross et al. (2008)
Enteromorpha clathrata	10.13	21.15	57.92	10.73	Wang et al. (2013)
Fucus serratus	4.6	21.4	65.4	–	Ynik et al. (2013)
Fucus serratus	11.4	18.6	45.5	24.2	Ross et al. (2008)
Fucus vesiculosus	12.3	11.8	51.4	23.8	Ross et al. (2008)
Laminaria hyperborea	12.4	11.2	53.5	21.5	Ross et al. (2008)
Laminaria japonica	6.1	29	58	9.7	Choi et al. (2014)
Macrocystis pyrifera	8.0	18.5	42.4	33.4	Ross et al. (2008)
Sargassum natans	10.46	29.09	48.85	11.6	Wang et al. (2013)

the feedstock and desired product distribution. Several chemical, biochemical and thermochemical conversion technologies are currently in use to produce bio-oil from algae (Raheem et al., 2015). Among these conversion pathways, thermochemical technologies are promising to produce bio-oil, in which bio-oil further could be upgraded to diesel, gasoline and jet-fuel-range hydrocarbons. In thermochemical technologies, heat, decomposition rate and feedstock compositions are the three main factors that affect the product quality, yield and energy content (heating value). Thermochemical techniques such as incineration, gasification, hydrothermal liquefaction and pyrolysis were widely used to produce bio-oil from various micro- and macroalgal strains (Onay and Kockar, 2003; Ooi et al., 2004; Sensoz and Angin, 2008; Lappi and Alen, 2011; Singh and Shadangi, 2011; Chaiya and Reubroycharoen, 2013; Dickerson and Soria, 2013; Chin et al., 2014; Mohammad et al., 2014).

Direct combustion of algal biomass

Direct combustion is one of the thermochemical techniques and simplest approaches used for the utilization of algal biomass for heat and power production (Raheem et al., 2015). The stored chemical energy in the algal biomass during the photosynthesis process could be converted into heat through direct combustion in the presence of excess air or open air. The produced heat or steam could be used for household heating and electricity generation (Azizi et al., 2018). The MC of the biomass needs to be less than 50 wt.% for direct combustion of the algal biomass in a suitable manner (Raheem et al., 2015). Dehydrating, cutting and crushing were considered to be vital pretreatment processes for algal biomass before proceeding to the combustion chamber for direct combustion which results in an increase in the cost of energy production as compared to other thermochemical conversions such as gasification, pyrolysis and hydrothermal liquefaction (Sanchez-Silva et al., 2013; López-Gonzalez et al., 2014b; Azizi et al., 2018). Generally, a combustion temperature from 800°C to 1,000°C could be used for the direct combustion of algal biomass. Although algal biomass could be burned for energy generation, it is not considered as the key process in energy generation from algal biomass and their energy generation is less as compared to woody biomass and coal (Azizi et al., 2018). Co-combustion algal biomass with coal and woody biomass is another option for the justifiable process of algal biomass direct incineration which reduces releases of GHG into the atmosphere (Raheem et al., 2015).

Gasification of algal biomass

Algal biomass is considered as a high-moisture-containing feedstock with 70–90 wt.%, and moisture removal is a mandatory pretreatment process before the biomass is fed to the gasifier (Kumar et al., 2016). During gasification at a higher temperature, large molecules of algal biomass thermally degrade into lighter hydrocarbon gaseous mixtures with methane, carbon monoxide, hydrogen and other wide range of hydrocarbon gases (Kumar et al., 2009; Raheem et al., 2018). The formation of tar (black viscous liquid) and biochar is obvious in the algal biomass gasification process (Raheem et al., 2018; Hanchate et al., 2021). The procedure takes place at a high-temperature range (800–1,000°C) in the presence of steam, oxygen, air or a mixture of these as gasifying agents without combustion and incomplete oxidation (Raheem et al., 2015). The algal biomass in the gasifier is heated to the gasification temperature either directly or indirectly from an external supply of heat.

During algal biomass gasification, the type of algal species significantly influences the final product distribution, gas yield, carbon conversion, tar formation and AC. Further, the final product composition mainly depends on chemical and biochemical compositions, type and amount of catalyst used for catalytically gasification, and gasification process parameters such as reaction temperature, reaction time, particle size, gasifying agent partial pressure and type of gasifier (Sanchez-Silva et al., 2013; Hanchate et al., 2021). Gasifiers that using oxygen or air as gasifying agents produce syngas with heating value in the range of 4–7 MJ/Nm^3 and 10–12 MJ/Nm^3, respectively (Hanchate et al., 2021). A gasification process with steam as a gasifying agent can be used for thermochemical decomposition of biomass to produce gaseous products with a higher content of hydrogen gas (Duman et al., 2014). Furthermore, the main drawback of the process is the formation of tar, which is a mixture of various hydrocarbons and can be controlled by varying process parameters and using various catalysts (Anis and Zainal, 2011). Overall, multiple reactions take place during the gasification process in air, oxygen and steam gasifier, that is, primary and secondary reactions during the gasification process (López-Gonzalez et al., 2014b; Raheem et al., 2018).

Raheem et al. (2015) studied the algal biomass gasification for hydrogen-rich gas using ZnO-Ni-CaO catalyst. The result shows that the maximum composition of H_2 gas (48.95 mol%) was obtained at a gasification temperature of 851°C for 28 min of reaction time using 16.4 wt.% ZnO-Ni-CaO catalyst loading. Figueira et al. (2015) investigated the production of syngas from *Chlorella vulgaris* strain via gasification. Their gaseous product distribution was found to be 16.79% of H_2, 50.14% of CO and 20.58 wt.% of CH_4 at 850°C gasification temperature using argon and water vapor as gasification agents (Figueira et al., 2015). In a study on co-gasification of lignocellulosic materials (*Eucalyptus globulus*) with algal biomass (*Spirulina platensis*) in a fluidized bed, 19% of H_2, 40% of CO, 25% of CO_2 and 8% of CH_4 gas composition were obtained at 800°C gasification temperature using air and steam as gasification reagents (Yang et al., 2013). Guan et al. (2012) reported that a higher amount of gas products such as H_2, CH_4 and CO_2 than that of CO, C_2H_4 and C_2H_6 from gasification of *Nannochloropsis* in supercritical water at a temperature range of 450–550°C was observed. Khoo et al. (2013) investigated the gasification of *Nannochloropsis* using a fixed-bed reactor at 850°C, and the product distribution of biochar, gas and bio-oil was found to be 58.18 wt.%, 28.08 wt.% and 13.74 wt.%, respectively.

Pyrolysis of algal biomass

Pyrolysis technique is a thermochemical process where a carbonaceous material is heated over the range of 300–700°C in the lack of oxygen atmosphere that results in the formation of a solid phase in carbon-rich material (biochar) and a vapor phase comprised of condensable organic liquids (bio-oil) and non-condensable gases (CO, CO_2, H_2, CH_4 and short-chain alkanes and alkenes) (DalmasNeto et al., 2014; Raheem et al., 2015; Azizi et al., 2018; Pourkarimi et al., 2019). Many studies have reported that bio-oil produced via algal biomass pyrolysis is a promising thermal conversion technique resulting in a high yield of bio-oil, and it can be conducted at low temperatures as compared to gasification and combustion techniques (Kim et al., 2011; López-Gonzalez et al., 2014a; Chen et al., 2015b; Raheem et al., 2015; Pourkarimi et al., 2019). The operating parameters such as temperature ramping rate, residence time and the feedstock type affect the pyrolysis final product distribution. Hence, the pyrolysis biomass was categorized into flash pyrolysis, slow pyrolysis and fast pyrolysis, based on the operating temperature.

Internal rearrangement such as the appearance of free radicals, bond breakage, dehydration and fragment species takes place that makes the growth of carbonyl, carboxyl groups and primary solid residue (biochar) in the first stage of biomass pyrolysis. Further, a secondary reaction such as cracking, dehydration and polymerization takes place for further conversion of primary solid residue into gases, tar and secondary biochar (Aysu et al., 2018; Pourkarimi et al., 2019). In flash pyrolysis, a higher yield of short-chain hydrocarbon gases is produced due to the disintegration of long-chain hydrocarbon molecules that are present in the bio-oil and biochar with short residence time (Yin, 2012). In general, the product distribution and their quality depend on the pyrolysis operating conditions (pyrolysis heating rate, temperature, reaction time, reactor configuration) and physicochemical properties (moisture, ash, volatile and FC content, biochemical and elemental compositions) of the feedstock used (Table 5.8).

Chaiwong et al. (2012) studied the pyrolysis of three different algal strains – *Spirulina, Spirogyra* and *Cladophora* – and showed production of 39–46 wt.% of bio-oil, 28–31 wt.% of biochar and 23–30 wt.% of gas at the optimum conditions of 8°C/min heating rate and 550°C temperature for 60 min of reaction time. A study on pyrolysis of *Spirulina* at 450–600°C by Chaiwong et al. (2012) reported that 550°C pyrolysis temperature was optimum to obtain 46 wt.% bio-oil with a heating value of 21.68 MJ/kg. Similarly, Vardon et al. (2012) examined the impact of pyrolysis temperature (300–600°C) at a heating rate of 40°C/min on the yield of *Scenedesmus dimorphus* strain. Their result showed a maximum yield of 39.6 wt.% with a calorific value of 28.52 MJ/kg at 500°C.

Miano and Wu (2004) have reported 18 and 24 wt.% conversions of total algal biomass of *Chlorella protothecoides* and *Microcystis aeruginosa* into pyrolytic bio-oil, respectively. The heating value of bio-oil from algal biomass feedstocks' pyrolysis ranges between 29 and 45.9 MJ/kg with a viscosity of 0.04–0.2 Pa·s and density range of 0.98–1.06 kg/L, which is higher than woody bio-oil and comparable with petroleum oil (Raheem et al., 2015).

Hydrothermal liquefaction of algal biomass

Among the thermochemical conversion technologies to produce high-density bio-oil, hydrothermal liquefaction was found favorable relative to other processes owing to the freedom to use wet feedstocks, lower oxygen content (5–10 wt.%) of the produced bio-oil, compatibility with petroleum distillates and higher heating value (30–45 MJ/kg) of the produced bio-oil (Filippis et al., 2016). Consumption of wet feedstocks avoids the drying process. Therefore, the energy and expenditures needed to dry the feedstocks are relatively lower in hydrothermal processes. Hydrothermal liquefaction is also known as hydrous pyrolysis as the typical water/moisture content is almost 80%. Typical process conditions for hydrothermal liquefaction were reported to be 200–380°C temperature and 7–25 MPa pressure (sub- or supercritical state) to keep the solvent in the liquid state with a reaction time of 10–60 min (Marcilla et al., 2013). A typical reaction network (Figure 5.1) to produce bio-oil from algae was reported by Xu et al. (2018). Along with the nature of biomass (lignocellulosic biomass and algal), hydrothermal process parameters such as temperature, pressure, residence time, hydrothermal liquefaction solvent (water or alcohols) and presence of the catalyst show a significant impact on the quality of the bio-oil.

Akhtar and Amin (2011) reported that the intermediate temperature range (300–350°C) is optimum to obtain a high-quality bio-oil with a higher yield. At this temperature range and at 7–25 MPa pressure, solvents would maintain their supercritical state to keep the solvents in a liquid state; therefore, solvents could penetrate the biomass to dissolve and extract the bio-oil. In general, water acts as a solvent and catalyst in hydrous pyrolysis and a higher solvent-to-biomass ratio leads to decreased bio-oil yield. Therefore, Filippis et al. (2016) recommended a biomass-to-solvent ratio of 0.1–0.2 for higher bio-oil yield.

Shakya et al. (2017) used nine algal strains with different batches by varying the biochemical compositions (*Chlorella*: C1 and C2; *Nannochloropsis*: N1, N2, N3 and N4; *Pavlova*: P1; *Scenedesmus*: S1 and S2) to produce bio-oil via hydrothermal liquefaction at 280°C and 320°C and studied their impact on the bio-oil yields. Likewise, strains with higher nitrogen content showed higher nitrogen percentage in the bio-oils. The higher bio-oil yield was noticed at a hydrothermal liquefaction temperature of 320°C for *Nannochloropsis* species, owing to their higher lipid contents. The presence of the lipids contributes to the bio-oil, and water-soluble glycerols end up in the aqueous phase. From these studies, the bio-oil production range was found to be 33–57 wt.% and the predicted bio-oil yield model followed the trend of lipids > proteins > carbohydrates. Among the nine strains, *Nannochloropsis* had the highest bio-oil content of 57 wt.% with 33–42 wt.% of fraction in the diesel fuel range. Most of the produced bio-oils were in the range of vacuum gas oil which is feasible to upgrade to liquid transportation fuels via co-processing at the refineries.

In addition, to enhance the bio-oil content and reduce the solid residue during hydrothermal liquefaction, different homogeneous and heterogeneous catalysts (acid and base) were introduced by Xu et al. (2018). Na_2CO_3 was the widely used a homogeneous catalyst with 5 wt.% loading during hydrothermal liquefaction with enhanced bio-oil yield in most cases. Ross

Table 5.8 Pyrolysis product distribution of various strains of algal biomass

Algae type	*Algal strain name*	*Pyrolysis process conditions*			*Bio-oil*		*Gas (wt.%)*	*Biochar (wt.%)*	*References*
		Temperature (°C)	*Pressure (MPa)*	*Atmosphere*	*Yield (wt.%)*	*HHV (MJ/kg)*			
Microalgae	*Chlorella protothecoides*	450	0.101	N_2	57.9	41	32	10.1	Miano and Wu (2004)
	Chlorella protothecoides	502	0.101	NR	55.3	39.7	36.3	8.4	Demirbas (2006)
	Chlorella	450	NR	N_2	55	27	20	30	Babich et al. (2011)
	C. vulgaris remnant	520	0.101	N_2	53	24.57	16	31	Wang et al. (2013)
	Chlorella vulgaris	500	NR	N_2	60.7	~22	7.8	31.5	Gong et al. (2014)
	Dunaliella salina	500	NR	N_2	64.9	~23	7.1	28.0	Gong et al. (2014)
	C. protothecoides	500	NR	N_2	17.5	30	29.5	~53	Miao et al. (2004)
	M. aeruginosa	500	NR	N_2	23.7	29	22.3	~54	Miao et al. (2004)
	Nannochloropsis	400	NR	N_2	31.1	24.6	NR	24.2	Pan et al. (2010)
Macroalgae	*Laminaria japonica*	425	NR	N_2	35.4	29.6	NR	~39	Choi et al. (2015)
	Saccharina japonica	470	NR	N_2	37.9	35	NR	NR	Choi et al. (2015)
	Laminaria japonica	500	NR	N_2	37.5	33.57	22.5	~40	Bae et al. (2011)
	Undaria pinnatifida	500	NR	N_2	39.5	23.33	21.5	~39.5	Bae et al. (2011)
	Porphyra tenera	500	NR	N_2	47.4	29.74	35	~17.6	Bae et al. (2011)

NR: not reported.

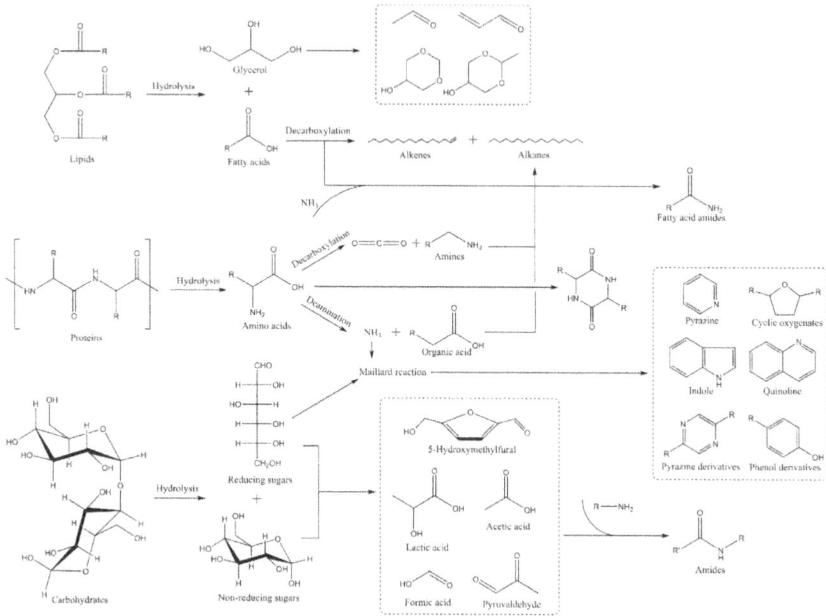

Figure 5.1 Hydrolysis of lipids, proteins and carbohydrates in hydrothermal liquefaction to produce bio-oil. Reproduced with permission from Xu et al. (2018).

et al. (2010) reported that the bio-oil yield followed the catalytic activity in the order of Na_2CO_3 > CH_3COOH > KOH > HCOOH. Likewise, organic acid catalysts such as acetic acid and formic acids were used to liquefy algal strains, and it was found that they are not effective to improve the bio-oil yield relative to the homogeneous alkali catalysts (Xu et al., 2018).

Conclusions

Production of bio-oil from algae is a smart technique to produce bio-oil via thermochemical conversion techniques that are promising such as pyrolysis and hydrothermal liquefaction. So far microalgae have received more attention relative to macroalgae for bio-oil production due to the high lipid content and favorable ultimate and proximate contents of the bio-oil. Identification of biochemical components such as lipid, carbohydrate and protein content of algal biomass is vital during the selection of algal biomass feedstock for bio-oil production. The growing conditions of algae, the type of strain, nutrients and cultivation techniques significantly affect the elemental composition and proximate and ultimate analyses of the algal biomass that have a direct impact on the bio-oil chemical composition and yield.

References

Aboulkas, A., Hammani, H., Achaby, M.E., Bilal, E., Barakat, A. and Harfi, E.K. (2017). Valorization of algal waste via pyrolysis in a fixed-bed reactor: Production and characterization of bio-oil and bio-char. *Bioresource Technology* 243: 400–408.

Akhtar, J. and Amin, N.A.S. (2011). A review on process conditions for optimum bio-oil yield in hydrothermal liquefaction of biomass. *Renewable and Sustainable Energy Reviews* 15: 1615–1624.

Anis, S. and Zainal, Z.A. (2011). Tar reduction in biomass producer gas via mechanical, catalytic and thermal methods: A review. *Renewable and Sustainable Energy Reviews* 15: 2355–2377.

Asomaning, J., Omidghane, M., Chae, M. and Bressler, D.C. (2016). Thermal processing of algal biomass for biofuel production. *Current Opinion in Green and Sustainable Chemistry* 2: 1–5.

Aysu, T., Fermoso, J. and Sanna, A. (2018). Ceria on alumina support for catalytic pyrolysis of *Pavlova* sp. microalgae to high-quality bio-oils. *Journal of Energy Chemistry* 27: 874–882.

Azizi, K., Moraveji, M.K. and Najafabadi, H.A. (2018). A review on bio-fuel production from microalgal biomass by using pyrolysis method. *Renewable and Sustainable Energy Reviews* 82: 3046–3059.

Babich, I.V., Hulst, M.V., Lefferts, L., Moulijn, J.A., Oconnor, P. and Seshan, K. (2011). Catalytic pyrolysis of microalgae to high-quality liquid bio-fuels. *Biomass and Bioenergy* 35: 3199–3207.

Bae, Y.J., Ryu, C., Jeon, J., Park, J., Suh, D.J., Suh, Y.W., Chang, D. and Park, Y.K. (2011). The characteristics of bio-oil produced from the pyrolysis of three marine macroalgae. *Bioresource Technology* 102: 3512–3520.

Borines, M.G., Leon, R.L. and Cuello, J.L. (2013). Bioethanol production from the macroalgae *Sargassum Spp. Bioresource Technology* 138: 22–29.

Chaiwong, K., Kiatsiriroat, T., Vorayos, N. and Thararax, C. (2012). Biochar production from freshwater algae by slow pyrolysis. *Maejo International Journal of Science and Technology* 6: 186–195.

Chaiya, C. and Reubroycharoen, P. (2013). Production of bio oil from para rubber seed using pyrolysis process. *Energy Procedia* 34: 905–911.

Chen, W.T., Zhang, Y., Zhang J., Yu, G., Schideman, L.C., Zhang, P. and Minarick, M. (2014a). Hydrothermal liquefaction of mixed-culture algal biomass from wastewater treatment system into bio-crude oil. *Bioresource Technology* 152: 130–139.

Chen, W.H., Huang, M.Y., Chang, J.S. and Chen, C.Y. (2014b). Thermal decomposition dynamics and severity of microalgae residues in torrefaction. *Bioresource Technology* 169: 258–264.

Chen, W.H., Lin, B.J., Huang, M.Y. and Chang, J.H. (2015a). Thermochemical conversion of microalgal biomass into biofuels: A review. *Bioresource Technology* 184: 314–327.

Chen, Y., Wu, Y., Hua, D., Li, C., Harold, M.P., Wang, J. and Yang, M. (2015b). Thermochemical conversion of low-lipid microalgae for the production of liquid fuels: Challenges and opportunities. *RSC Advances* 5: 18673–18701.

Chin, B.L.F., Yusup, S., Shoaibi, A.A., Kannan, P., Srinivasakannan, C., and Sulaiman, S.A. (2014). Kinetic studies of co-pyrolysis of rubber seed shell with high density polyethylene. *Energy Conversion and Management* 87: 746–753.

Choi, J., Choi, J.W., Suh, D.J., Ha, J.M., Hwang, J.W., Jung, H.W., Lee, K.Y. and Woo, H.C. (2014). Production of brown algae pyrolysis oils for liquid biofuels depending on the chemical pretreatment methods. *Energy Conversion and Management* 86: 371–378.

Choi, J.W., Choi, J.H., Suh, D.J. and Kim, H. (2015). Feasibility of *Laminaria Japonica* as a feedstock for fast pyrolysis in a bubbling fluidized-bed reactor. *Journal of Analytical and Applied Pyrolysis* 112: 141–149.

DalmasNeto, C.J., Sydney, E.B., Assmann, R., Neto, D.C. and Soccol, C.R. (2014). Chapter-7 'Production of biofuels from algal biomass by fast pyrolysis' in Pandey, A., Lee, D.J., Chisti, Y., SOccol, C.R. (eds.) *Biofuels from Algae*. Elsevier, Oxford, UK, 143–153.

Demirbas, A. (2006). Oily products from mosses and algae via pyrolysis. *Energy Source Part. A: Recovery Utilization and Environmental Effects* 28: 933–940.

Dickerson, T. and Soria, J. (2013). Catalytic fast pyrolysis: A review. *Energies* 6: 514–538.

Du, Z., Mohr, M., Ma, X., Cheng, Y., Lin, X., Liu, Y., Zhou, W., Cheng, P. and Ruan, R. (2012). Hydrothermal pretreatment of microalgae for production of pyrolytic bio-oil with a low nitrogen content. *Bioresource Technology* 120: 13–18.

Duman, G., Uddin, M.A. and Jale, Y. (2014). Hydrogen production from algal biomass via steam gasification. *Bioresource Technology* 166: 24–30.

Figueira, C.E., Moreira, P.F. and Giudici, R. (2015). Thermogravimetric analysis of the gasification of microalgae *Chlorella Vulgaris. Bioresource Technology* 198: 717–724.

Filippis, P.D., Caprariis, B.D., Scarsella, M., Petrullo, A. and Verdone, N. (2016). Biocrude production by hydrothermal liquefaction of olive residue. *International Journal of Sustainable Development and Planning* 11: 700–707.

Gong, X., Zhang, B., Zhang, Y., Huang, Y. and Xu, M. (2014). Investigation on pyrolysis of low lipid microalgae *Chlorella Vulgaris* and *Dunaliella Salina. Energy & Fuels* 28: 95–103.

Guan, Q., Savage, P.E. and Wei, C. (2012). Gasification of alga *Nannochloropsis* sp. in supercritical water. *The Journal of Supercritical Fluids* 61: 139–145.

Hanchate, N., Ramani, S., Mathpati, C.S. and Dalvi, V. (2021). Biomass gasification using dual fluidized bed gasification systems: A review. *Journal of Cleaner Production* 280: 123148.

Jena, U. and Das, K.C. (2011). Comparative evaluation of thermochemical liquefaction and pyrolysis for bio-oil production from microalgae. *Energy Fuels* 25: 5472–5482.

Kaliaperumal, N., Ramalingam, J.R., Kalimuthu, S. and Ezhilvalavan, R. (2002). Seasonal changes in growth, biochemical constituents and phycocolloid of some marine algae of Mandapam coast. *Seaweed Research* 24: 73–77.

Kebelmann, K., Hornung, A., Karsten, U. and Griffiths, G. (2013). Intermediate pyrolysis and product identification by TGA and Py-GC/MS of green microalgae and their extracted protein and lipid components. *Biomass and Bioenergy* 49: 38–48.

Khoo, H.H., Koh, C.Y., Shaik, M.S. and Sharratt, P.N. (2013). Bioenergy co-products derived from microalgae biomass via thermochemical conversion- life cycle energy balances and CO_2 emissions. *Bioresource Technology* 143: 298–307.

Kim, N.J., Li, H., Jung, K., Chang, H.N. and Lee, P.C. (2011). Ethanol production from marine algal hydrolysates using *Escherichia Coli* KO_{11}. *Bioresource Technology* 102: 7466–7469.

Kim, S.S., Ly, H.V., Kim, J., Lee, E.Y. and Woo, H.C. (2015). Pyrolysis of microalgae residual biomass derived from *Dunaliella tertiolecta* after lipid extraction and carbohydrate saccharification. *Chemical Engineering Journal* 263: 194–199.

Kumar, A., Jones, D.D. and Hanna, M.A. (2009). Thermochemical biomass gasification: A review of the current status of the technology. *Energies* 2: 556–581.

Kumar, K., Ghosh, S., Angelidaki, I., Holdt, S.L., Karakashev, D.B., Marales, M.A. and Das, D. (2016). Recent developments on biofuels production from microalgae and macroalgae. *Renewable and Sustainable Energy Reviews* 65: 235–249.

Lappi, H. and Alen, R. (2011). Pyrolysis of vegetable oil soaps–palm, olive, rapeseed and castor oils. *Journal of Analytical and Applied Pyrolysis* 91: 154–158.

Li, K., Zhang, L., Zhu, L. and Zhu, X. (2017). Comparative study on pyrolysis of lignocellulosic and algal biomass using pyrolysis-gas chromatography/mass spectrometry. *Bioresource Technology* 234: 48–52.

López-Gonzalez, D., Fernandez-Lopez, M., Valverde, J.L. and Sanchez-Silva, L. (2014a). Pyrolysis of three different types of microalgae: Kinetic and evolved gas analysis. *Energy* 73: 33–43.

López-Gonzalez, D., Fernandez-Lopez, M., Valverde, J.L. and Sanchez-Silva, L. (2014b). Comparison of the steam gasification performance of three species of microalgae by thermogravimetric–mass spectrometric analysis. *Fuel* 134: 1–10.

López-Gonzalez, D., Fernandez-Lopez, M., Valverde, J.L. and Sanchez-Silva, L. (2014c). Kinetic analysis and thermal characterization of the microalgae combustion process by thermal analysis coupled to mass spectrometry. *Applied Energy* 114: 227–237.

Maddi, B., Viamajala, S. and Varanasi, S. (2011). Comparative study of pyrolysis of algal biomass from natural lake blooms with lignocellulosic biomass. *Bioresource Technology* 102: 11018–11026.

Marcilla, A., Catala, L., Garcia-Quesada, J.C., Valdes, F.J. and Hernandez, M.R. (2013). A review of thermochemical conversion of microalgae. *Renewable and Sustainable Energy Reviews* 27: 11–19.

Miano, X. and Wu, Q. (2004). High yield bio-oil production from fast pyrolysis by metabolic controlling of *Chlorella Protothecoides*. *Journal of Biotechnology* 110: 85–93.

Miao, X., Wu, Q. and Yang, C. (2004). Fast pyrolysis of microalgae to produce renewable fuels. *Journal of Analytical and Applied Pyrolysis* 71: 855–863.

Mohammad, T.H., Lakhmiri, R., Azmani, A. and Hassan, I.I. (2014). Bio-oil from pyrolysis of castor shell. *International Journal of Basic and Applied Sciences IJBAS-IJENS* 14: 1–5.

Munoz, R., Navia, R., Ciudad, G., Tessini, C., Jeison, D., Mella, R., Rabert, C. and Azocar, L. (2015). Preliminary biorefinery process proposal for protein and biofuels recovery from microalgae. *Fuel* 150: 425–433.

Onay, O. and Kockar, O.M. (2003). Slow, fast and flash pyrolysis of rapeseed. *Renewable Energy* 28: 2417–2433.

Ooi, Y-S., Zakaria, R., Mohamed, A.R. and Bhatia, S. (2004). Catalytic cracking of used palm oil and palm oil fatty acid mixture for the production of liquid fuel: Kinetic modeling. *Energy Fuels* 18: 1555–1561.

Pan, P., Hu, C., Yang, W., Li, Y., Dong, L., Zhu, L., Tong, D., Qing, R. and Fan, Y. (2010). The direct pyrolysis and catalytic pyrolysis of *Nannochloropsis* sp. residue for renewable bio-oils. *Bioresource Technology* 101: 4593–4599.

Pourkarimi, S., Hallajisani, A., Alizadehdakhel, A. and Nouralishahi, A. (2019). Biofuel production through micro- and macroalgae pyrolysis – A review of pyrolysis methods and process parameters. *Journal of Analytical and Applied Pyrolysis* 142: 104599.

Raheem, A., Azlina, W.A.K.G.W., Yap, Y.H.T., Danquah, M.K. and Harun, R. (2015). Thermochemical conversion of microalgal biomass for biofuel production. *Renewable and Sustainable Energy Reviews* 49: 990–999.

Raheem, A., Ji, G., Memon, A., Sivasangar, S., Wang, W., Zhao, M. and Taufiq-Yap, Y.H. (2018). Catalytic gasification of algal biomass for hydrogen-rich gas production: Parametric optimization via central composite design. *Energy Conversion and Management* 158: 235–245.

Rizzo, A.M., Prussi, M., Bettucci, L., Libelli, I.M. and Chiaramonti, D. (2013). Characterization of microalga chlorella as a fuel and its thermogravimetric behavior. *Applied Energy* 102: 24–31.

Ross, A.B., Biller, P., Kubacki, M.L., Li, H., Lea-Langton, A. and Jones, J.M. (2010). Hydrothermal processing of microalgae using alkali and organic acids. *Fuel* 89: 2234–2243.

Ross, A.B., Jones, J.M., Kubacki, M.L. and Bridgeman, T. (2008). Classification of macroalgae as fuel and its thermochemical behaviour. *Bioresource Technology* 99: 6494–6504.

Sanchez-Silva, L., López-González, D., Garcia-Minguillan, A.M. and Valverde, J.L. (2013). Pyrolysis, combustion and gasification characteristics of *Nannochloropsis Gaditana* microalgae. *Bioresource Technology* 130: 321–331.

Sensoz, S. and Angin, D. (2008). Pyrolysis of safflower (*Charthamus tinctorius* L.) seed press cake: Part 1. The effects of pyrolysis parameters on the product yields. *Bioresource Technology* 99: 5492–5497.

Shakya, R., Adhikari, S., Mahadevan, R., Shanmugam, S.R., Nam, H., Hassan, E.B. and Dempster, T.A. (2017). Influence of biochemical composition during hydrothermal liquefaction of algae on product yields and fuel properties. *Bioresource Technology* 243: 1112–1120.

Shuping, Z., Yulong, W., Mingde, Y., Chun, L. and Junmao, T. (2010). Pyrolysis characteristics and kinetics of the marine microalgae Dunaliella tertiolecta using thermogravimetric analyzer. *Bioresource Technology* 101: 359–365.

Singh, R.K. and Shadangi, K.P. (2011). Liquid fuel from castor seeds by pyrolysis. *Fuel* 90: 2538–2544.

Sudhakar, M.P., Kumar, B.R., Mathimani, T. and Arunkumar, T. (2019). A review on bioenergy and bioactive compounds from microalgae and macroalgae-sustainable energy perspective. *Journal of Cleaner Production* 228: 1320–1333.

Sunganya, T., Varman, M., Masjuki, H.H. and Renganathan, S. (2016). Macroalgae and microalgae as a potential source for commercial applications along with biofuels production: A biorefinery approach. *Renewable and Sustainable Energy Reviews* 55: 909–941.

Sydney, E.B., Strurm, W., Carvalho, J.C., Thomaz-Soccol, V., Larroche, C., Pandey, A. and Soccol, C.R. (2010). Potential carbon dioxide fixation by industrially important microalgae. *Bioresource Technology* 101: 5892–5896.

Tan, I.S., Lam, M.K. and Lee, K.T. (2013). Hydrolysis of macroalgae using heterogeneous catalyst for bioethanol production. *Carbohydrate Polymers* 94: 561–566.

Vardon, D.R., Sharma, B.K., Blazina, G.V., Rajagopalan, K. and Strathmann, T.J. (2012). Thermochemical conversion of raw and defatted algal biomass via hydrothermal liquefaction and slow pyrolysis. *Bioresource Technology* 109: 178–187.

Wang, K., Brown, R.C., Homsy, S., Martinez, L. and Sidhu, S.S. (2013). Fast pyrolysis of microalgae remnants in a fluidized bed reactor for bio-oil and biochar production. *Bioresource Technology* 127: 494–499.

Wang, N., Tahmasebi, A., Yu, J., Xu, J., Huang, F. and Mamaeva, A. (2015). A comparative study of microwave-induced pyrolysis of lignocellulosic and algal biomass. *Bioresource Technology* 190: 89–96.

Wang, S., Wang, Q., Jiang, X., Han, X. and Ji, H. (2013). Compositional analysis of bio-oil derived from pyrolysis of seaweed. *Energy Conversion and Management* 68: 273–280.

Xu, D., Lin, G., Guo, S., Wang, S., Guo, Y. and Jing, Z. (2018). Catalytic hydrothermal liquefaction of algae and upgrading of biocrude: A critical review. *Renewable and Sustainable Energy Reviews* 97: 103–118.

Yang, K-C., Wu, K-T., Hsieh, M-H., Hsu, H-T, Chen, C-S. and Chen, H-W. (2013). Co-gasification of woody biomass and microalgae in a fluidized bed. *Journal of the Taiwan Institute of Chemical Engineers* 44: 1027–1033.

Yin, C. (2012). Microwave-assisted pyrolysis of biomass for liquid biofuels production. *Bioresource Technology* 120: 273–284.

Ynik, J., Stahl, R., Troeger, N. and Sinag, A. (2013). Pyrolysis of algal biomass. *Journal of Analytical and Applied Pyrolysis* 103: 134–141.

Yuan, T., Tahmasebi, A. and Yu, J. (2015). Comparative study on pyrolysis of lignocellulosic and algal biomass using a thermogravimetric and a fixed-bed reactor. *Bioresource Technology* 175: 333–341.

6 Technological advances in bio-crude production and lipid extraction from algae

Yulin Hu and Chunbao Charles Xu

Introduction

Over the past 50 years, rapid growth in the global population has caused a dramatic increase in energy demand. As indicated by the Organization of the Petroleum Exporting Countries (OPEC), global energy consumption is expected to rise to 17.4 billion liters daily by 2040, and this growing energy demand poses a big threat to the environment and sustainable development of global economics (Aziz et al., 2020). Biofuels can be divided into first-, second- and third-generation biofuels according to the feedstock characteristics and operational parameters. Among them, third-generation biofuels, which are commonly derived from algae, have attracted a great deal of attention since some of the technical barriers faced by the first- and second-generation biofuels can be overcome, such as the competition between food and biofuel production.

Even though algae offer several benefits like the ability to be cultivated in the wastewater and seawater, the most appealing advantage of algal biomass is related to their high content of lipids, which can be converted into biodiesel via transesterification. Under the circumstance of the rising demand for biodiesel (estimated to be 9.1 billion L/d in 2040), utilizing algae could be one of the most promising solutions to meet this growing biodiesel demand worldwide (Gebremariam and Marchetti, 2018). Several high lipid-containing algal strains such as *Chlorella vulgaris, Heterosigm akashiwo, Nannochloris, Neochloris oleoabundans, Phaeodactylum tricornutum, Schizochytrium, Scenedesmus dimorphus*, and *Scenedesmus obliquus* have been reported by Sajjadi et al. (2018).

Except for producing biodiesel from algal lipid, thermochemical methods such as fast pyrolysis and hydrothermal liquefaction (HTL) are capable of converting the whole algal biomass into bio-crude oils (or bio-oils) followed by upgrading to liquid biofuels. The yield and properties of bio-crude oil produced from fast pyrolysis and HTL are dependent on the various reaction conditions and catalyst type, which have been previously reviewed (Galadima and Muraza, 2018; Guo et al., 2015; Li et al., 2019a,b; Pourkarimi et al., 2019; Xu et al., 2018; Yang et al., 2019). In addition, the underlying reaction mechanism of biomass conversion in the process of fast pyrolysis or HTL has been fully investigated with a wide range of biomass model compounds and by studying the interaction effects between these model compounds (Anca-Couce, 2016; Hameed et al., 2019; Wang et al., 2017). Thus, this chapter aims to provide an overview of advances in technologies for bio-crude production and lipid extraction from algae, as outlined as follows:

- i Section 'Fast pyrolysis of algae' illustrates the effects of fast pyrolysis of algae with or without a catalyst on the bio-crude oil production and discusses the major differences between catalytic fast pyrolysis and non-catalytic fast pyrolysis concerning bio-crude oil yields and properties.
- ii Section 'Hydrothermal liquefaction of algae' compares pyrolysis and HTL in terms of bio-crude oil production from algae.
- iii Section 'Lipid extraction from algae' describes and analyzes the technological advances in the most recent techniques (microwave-, ultrasound- and supercritical fluid (SCF)-assisted extraction) on the lipid extraction from algae.
- iv Section 'Conclusions' summarizes the future research directions and the main conclusions drawn from this literature review work.

Fast pyrolysis of algae

Fast pyrolysis is a thermochemical conversion approach for the conversion of dried biomass at temperatures of 400–600°C, with a heating rate of >100°C/min and residence time of 0.5–2 s, producing up to 70 wt.% of pyrolysis oil. The resulting pyrolysis oil has a wide range of applications such as heating oil or a source for value-added chemicals (e.g., toluene, phenol and xylene) or for drop-in fuels after upgrading by hydrotreatment or catalytic cracking (Thangalazhy-Gopakumar et al., 2012). A summary of previous literature on the non-catalytic and catalytic fast pyrolysis of algae in different reactor configurations and reaction conditions is presented in Table 6.1.

As one of the most efficient thermochemical biomass conversion technologies, fast pyrolysis of algae with or without catalyst (non-catalytic) has attracted extensive research in recent years. For example, a parametric study of non-catalytic fast pyrolysis of algae was carried out by Ly et al. (2015) where the experiments were conducted in a bubbling fluidized-bed reactor at 350–500°C. In this work, it was reported that the yield of pyrolysis oil decreased from approx. 45 wt.% at 350°C to 27 wt.% at 500°C, accompanied by an increase in the gas yield from approx. 21 wt.% at 350°C to 34 wt.% at 500°C, which is attributed to the secondary decomposition of liquid into gaseous products. When the temperature increased from 350°C to 500°C, the higher heating value (HHV) of bio-oil obtained slightly increased from 24.8 to 27.7 MJ/kg, owing to the decrease in its oxygen content from 33.32% to 27.21%. Obviously, for biofuel applications, the bio-oil products require post-processing by deoxygenation. Compared with lignocellulose-derived bio-oil (pH: 2–4), the algal-derived bio-oils were reported to have higher pH values

Table 6.1 A summary of previous studies on the fast pyrolysis of algae

Algal species	*Reactor configuration*	*Maximum bio-oil yield (wt.%)*	*Higher heating value (MJ/kg)*	*References*
Arthrospira platensis	Py-gas chromatography–mass spectrometry (Py-GC/MS)	n.a.	30–37	Anand et al. (2016)
Chlorella protothecoides	Fluidized-bed reactor	24	29	Miao et al. (2004)
Chlorella protothecoides	Fluidized-bed reactor	57.9	41	Miao and Wu (2004)
Chlorella vulgaris	Packed-bed reactor	52.7	18.6	Thangalazhy-Gopakumar et al. (2012)
Chlorella vulgaris	Auger reactor	47.7	24–32	Sotoudehniakarani et al. (2019)
Chlorella vulgaris	Fluidized-bed reactor	53	24.6	Wang et al. (2013)
Chlorella vulgaris	Py-GC/MS	n.a.	n.a.	Adamakis et al. (2018)
Duckweed; green-blue algae; green algae; *Chlorella*	Free-fall reactor	48.9	25.9–29.9	Campanella and Harold (2012)
Enteromorpha prolifera	Free-fall reactor	n.a.	25.3	Zhao et al. (2013)
Laminaria japonica	Fluidized-bed reactor	35.4	29.6	Choi et al. (2015)
Melosira varians	Fixed-bed reactor	29	n.a.	Piloni et al. (2017)
Nannochloropsis oculata	Py-GC/MS	n.a.	33–39	Gautam and Vinu (2018)
Nannochloropsis; *Spirulina*; *Sargasso*	Double microfixed-bed reactor	n.a.	n.a.	Jia et al. (2020)
Nannochloropsis; *Tetraselmis*; *Isochrysis galbana*	Spouted-bed reactor	26.8	n.a.	Azizi et al. (2020)
Saccharina japonica	Fluidized-bed reactor	31.3	31.5	Choi et al. (2017)
Saccharina japonica	Fixed-bed reactor	40.9	33–33.4	Ly et al. (2016)
Saccharina japonica	Fluidized-bed reactor	45.0	28.3	Ly et al. (2015)
Scenedesmus sp.	Fluidized-bed reactor	55	18.4	Harman-Ware et al. (2013)
Schizochytrium limacinum	Py-GC/MS	n.a.	n.a.	Anand et al. (2017)

n.a.: not available.

(pH: 4.68–6.08), which is likely due to the differences in the textural/ash compositions of biomass feedstocks. Based on ^{1}H and ^{13}C nuclear magnetic resonance (NMR) spectroscopy, bio-oils obtained from non-catalytic fast pyrolysis of algae primarily contained short-chain hydrocarbons, alcohols, ketones and their derivatives. Among them, the most common chemical components identified in the algal-derived bio-oils were di-anhydromannitol and 2-methyl furyl ketone, at a concentration of 24.1% and 17.2%, respectively. It should be noted that di-anhydromannitol is a common additive to gasoline. Although fast pyrolysis is an industrially realized technology to convert biomass into biofuels owing to its high efficiency, it has limited application in algae conversion due to the high cost associated with feedstock dewatering/drying. The use of low-cost feedstock like field-cultivated algae could lower the cost of the entire process.

In a previous study conducted by Sotoudehniakarani et al. (2019), field-cultivated *C. vulgaris* was pyrolyzed in an auger reactor at 450–550°C for 6.4 s, compared with the same operations with commercial *C. vulgaris*. Pyrolysis of the field-cultivated algae produced a lower bio-oil yield (on dry biomass basis) at 550°C than that of the commercial *C. vulgaris*, which could be due to the high ash content (24 wt.% for the field-cultivated algae vs. 9.4 wt.% for the commercial one) and the presence of impurities in the field-cultivated algal biomass.

In recent years, the introduction of catalysts to the fast pyrolysis process has been studied to improve the quality of algal-derived bio-oil. A higher concentration of long-chain hydrocarbon and/or aromatics of bio-oil can be obtained with the addition of catalyst in the fast pyrolysis (Anand et al., 2016). Gautam and Vinu (2018) compared the fast pyrolysis of *Nannochloropsis oculata* in a single-shot micropyrolyzer at 400–600°C with or without the use of Co-Mo/γ-Al_2O_3 as the catalyst. In the non-catalytic experiments, it was observed that the major chemicals identified in the resultant bio-oils were long-chain nitriles, long-chain alkanes and alkenes, monoaromatics and polyaromatics, and the selectivity towards aliphatic/aromatic hydrocarbons reached its optimal value at a temperature of 500°C. A possible reaction mechanism was proposed, and it is illustrated in Figure 6.1. As shown in Figure 6.1, a series of reactions including decarboxylation, ammonization, aromatization and dehydration occur in the fast pyrolysis of *N. oculata*. The presence of aromatic compounds could be attributed to protein degradation. Amides and nitriles might be generated by the reaction between carboxylic acid and ammonia from protein decomposition.

Catalysts have been applied in the pyrolysis treatment with an aim of improving the quality of algal-derived bio-oils. Gautam and Vinu (2018) applied Co-Mo/γ-Al_2O_3 to catalytically pyrolyze *N. oculata* at 500°C for 60 s, at a catalyst loading of 25–67 wt.%, and weight hourly space velocities (WHSV) of 30–180/h. The HHVs of the obtained bio-oils from the catalytic fast pyrolysis were in the range of 37–39 MJ/kg, which are higher than those obtained from the non-catalytic fast pyrolysis (33.4 MJ/kg). Additionally,

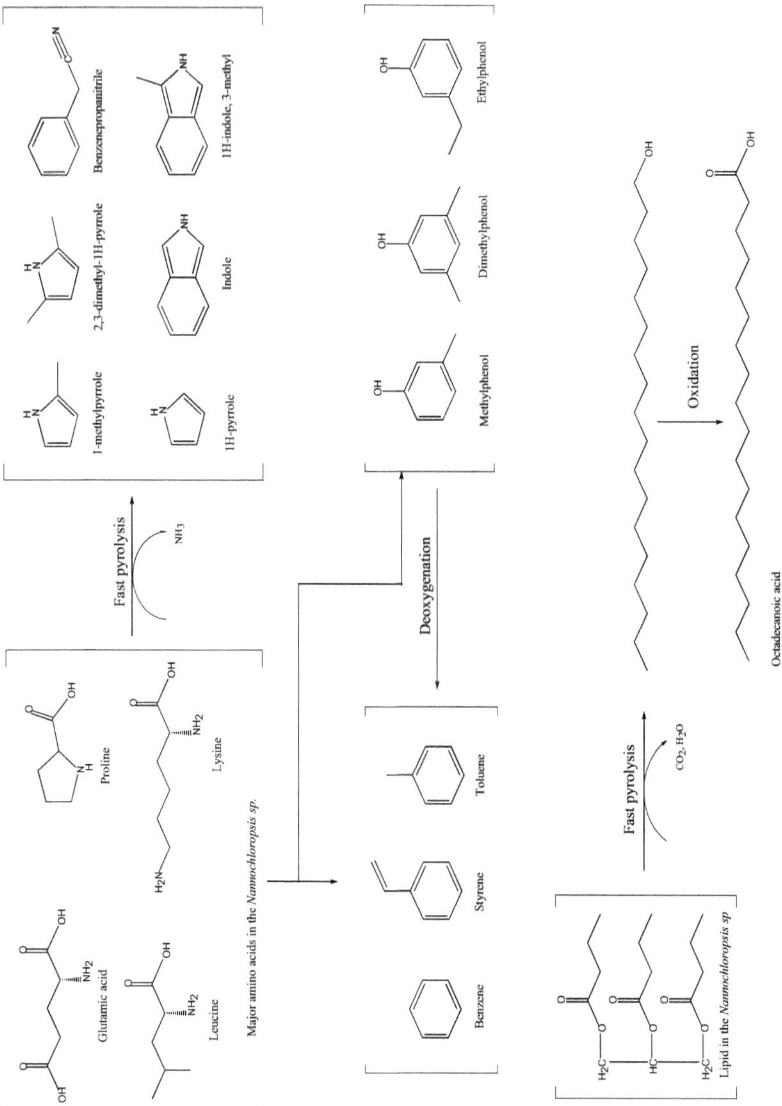

Figure 6.1 A possible reaction pathway in non-catalytic pyrolysis of microalgae. Reproduced with permission from Gautam and Vinu (2018).

compared with the bio-oil from the non-catalytic fast pyrolysis, the concentration of aromatics in the oil from the catalytic runs decreased from 11.6% to 6%, and no long-chain carboxylic acid, such as tetradecanoic acid, hexadecanoic acid and oleic acid, was identified in the oils obtained from catalytic fast pyrolysis. This absence of carboxylic acids could be due to the promoted decarboxylation and ammonization reactions in the presence of Co-Mo/γ-Al_2O_3. In catalytic fast pyrolysis, the conversion of amide compounds into nitriles could be stimulated by the Co-Mo/γ-Al_2O_3, as evidenced by a concentration of 13.8% of pentadecanenitrile in the catalytic fast pyrolysis-produced bio-oil. It should be noted that a significant amount of unsaturated hydrocarbons (e.g., $C_{20}H_{38}$) was observed in all bio-oil products from fast pyrolysis of algae with or without catalyst addition. Based on the results as discussed above, a reaction scheme for catalytic fast pyrolysis of *N. oculata* was proposed, and it is depicted in Figure 6.2.

In another study on the catalytic fast pyrolysis, Anand et al. (2016) applied various zeolite catalysts in the pyrolysis of *Arthrospira platensis* in a micropyrolyzer at 350–800°C and a heating rate greater than 2,000°C/s. The influences of temperature and catalyst type and dosage on the properties of bio-oil were thoroughly analyzed. In the case of non-catalytic fast pyrolysis, it was observed that the formation of CO_2 enhanced when increasing temperature from 350°C to 800°C, which was accompanied by the reduction in the formation of alkanes, alkenes, carboxylic acids and nitriles. For catalytic fast pyrolysis, the use of zeolite catalysts significantly affected the chemical composition of bio-oil. Specifically, acetonitrile, benzene, toluene, xylene, cyclobutene and dimethyl cyclopropane were identified as the main compounds in the bio-oils obtained with ZSM-5, zeolite-β and zeolite-Y. In addition, increasing catalyst loading from 2:1 to 50:1 wt./wt. and temperature from 350°C to 600°C resulted in the increased formation of aromatics (including monoaromatics and polyaromatics) and cycloalkanes. The results also showed that BTX and alkylbenzenes were detected as the main monoaromatic compounds in the bio-oils irrespective of the type of catalyst. The total yield of BTX followed the order of ZSM50 > ZβAm ≈ ZYAm ≈ ZYH > ZSM200 > ZβH, suggesting that the cracking tendency of zeolite catalyst is mainly determined by its surface acidity. Contrary to the non-catalytic fast pyrolysis, a series of reactions could occur or be enhanced in the catalytic fast pyrolysis of algal biomass, including the following:

- i Breakage of straight chain hydrocarbons;
- ii Dehydration of amides to form nitriles;
- iii Decarboxylation of carboxylic acids; and
- iv Cyclization, cracking, oligomerization, isomerization and aromatization.

The coke formation and its associated catalyst deactivation are the main technical barriers to the catalytic fast pyrolysis. It is essential to uncover the

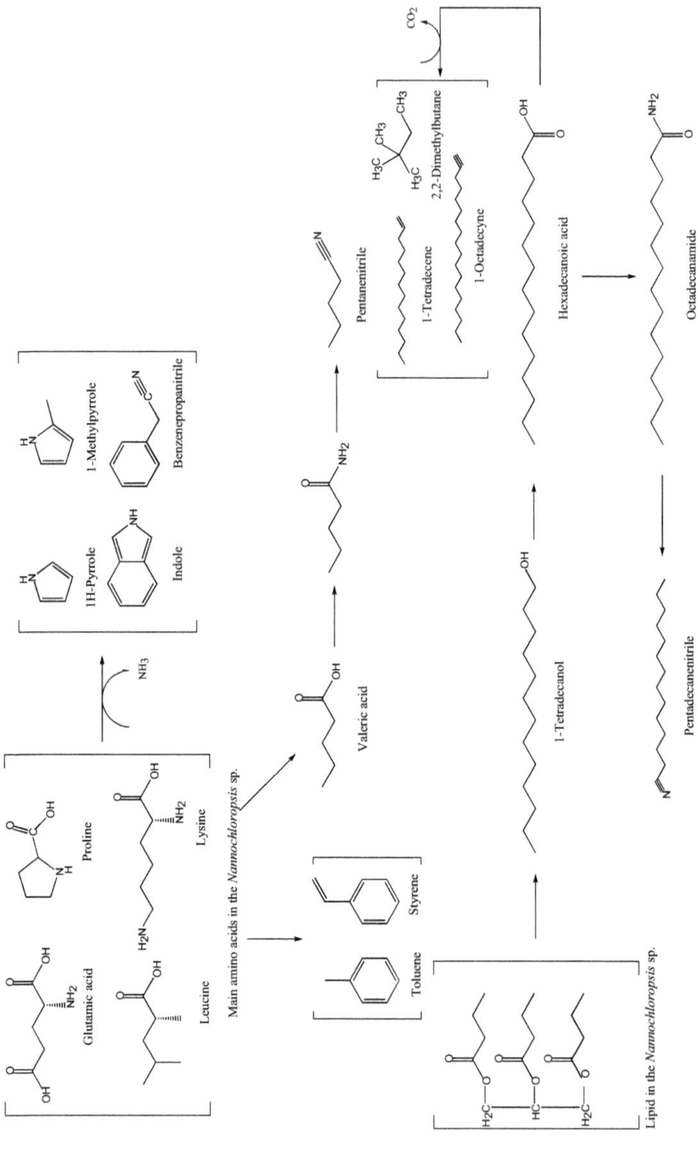

Figure 6.2 A possible reaction pathway in catalytic fast pyrolysis of microalgae. Reproduced with permission from Gautam and Vinu (2018).

underlying mechanism for the catalyst deactivation and its coke behavior in the process of catalytic fast pyrolysis of algae. To this end, Jia et al. (2020) carried out fast pyrolysis of *Nannochloropsis*, *Spirulina* and *Sargasso* in a double microfixed-bed reactor over an HZSM-5 catalyst. Generally, the deactivation rate of HZSM-5 in pyrolysis of algal biomass was found to be lower than that in pyrolysis of poplar wood, suggesting that HZSM-5 demonstrates a stronger capability to retard coke formation in the pyrolysis of algae compared to that of lignocellulosic biomass. Besides, a shorter lifetime of HZSM-5 was attained in catalytic fast pyrolysis of *Spirulina* (with a higher lipid content) than that of *Nannochloropsis*, implying that the presence of high content of lipid in the algal biomass could lead to a faster deactivation rate of the HZSM-5 catalyst. Interestingly, two different reactor configurations were applied for the pyrolysis of algal biomass mixed with a catalyst (in situ catalytic fast pyrolysis) or pyrolysis of algae without catalyst in a reactor coupled with catalyst bed (ex situ catalytic fast pyrolysis) for vapor upgrading. In the latter case, direct exposure of the catalyst to inorganic containments from ash can be prevented. Based on the results obtained, the authors speculated that one of the effective options to extend the lifetime of HZSM-5 is to perform an ex situ catalytic fast pyrolysis rather than an in situ one.

Hydrothermal liquefaction of algae

Although fast pyrolysis has been demonstrated to be an effective thermochemical technology to convert algae into bio-oils, the newly harvested algal biomass contains a significantly high-water content, which requires energy-intensive drying pretreatment before the fast pyrolysis, thus leading to a high operation cost. Hence, in recent years, HTL has been widely investigated as an alternative conversion method to fast pyrolysis for algae conversion. HTL is usually carried out in subcritical water or water–organic mixed solvent (e.g., alcohol) at moderate temperature (200–375°C) and pressure (5–28 MPa) with or without the addition of catalyst (Guo et al., 2015). The most recent studies on the HTL of algae at various reaction conditions are summarized in Table 6.2.

Due to the use of hot-compressed water as the reaction medium, the vaporization associated with high latent heat can be avoided in the HTL process, which ensures high energy efficiency. Another benefit offered by HTL is related to the variations in the physical properties of water at elevated temperature and pressure. For example, water behaves more like an organic solvent at elevated conditions. Hence, it improves the decomposition of big molecules present in the biomass and the polymerization of the degraded small molecules into bio-crude oil. This is due to the reduced hydrogen bonding of water at high temperatures and pressures compared to that at ambient temperature, leading to a lower dielectric constant of water and thus increases in the solubility of organics in the water. Besides, a sharp increase in the ionic product of water can also be observed in hot-compressed water

Table 6.2 Summary of the recent studies on the hydrothermal liquefaction of algae

Algal species	*Reaction conditions*	*Key findings*	*References*
Galdieria sulphuraria	• Temp: 310–350°C • Residence time: 30 min • Biomass loading: 5 wt.% • Initial pressure: 1.38 MPa • Solvents: ethanol, isopropanol, glycerol and crude glycerol	• The use of alcohol helped transfer protein-derived compounds from the water phase into the oily phase • The addition of ethanol, isopropanol and glycerol promoted the Maillard reaction • In comparison, crude glycerol was found to be the most effective solvent in terms of bio-crude oil yield and energy recovery	Cui et al. (2020)
Nannochloropsis sp., *Sargassum* sp.	• Temp: 260–340°C • Residence time: 30 min • Biomass loading: 15 wt.% • Initial pressure: 4 MPa	• *Nannochloropsis* sp. led to a higher bio-crude oil yield than *Sargassum* sp. • N-heterocyclics, amides and ketones were the main compounds in the algal-derived bio-crude oil • Alcohols were found to be the dominant compounds in the water phase obtained from HTL of algae • Acylation and Maillard reaction were the dominant reactions in the HTL of algae with low lipid content	He et al. (2020)
Nostoc ellipsosporum	• Temp: 240–320°C • Residence time: 60 min • Biomass loading: 4.76 wt.% • Initial pressure: 5 MPa	• The algae yield increased from 1.4 to 2.9 g/L under the optimized conditions, which was accompanied by an 88% nitrogen removal and 88% phosphate removal • A bio-crude oil yield of 24.6 wt.% was obtained at 300°C	Devi and Parthiban (2020)
Salicornia bigelovii; Cyanidioschyzon merolae	• Temp: 250–350°C • Residence time: 30 min • Biomass loading: 10 wt.% • Initial pressure: 2.9 MPa • *Cyanidioschyzon merolae*-to-*Salicornia bigelovii* mass ratio: 80/20–20/80	• A positive synergistic effect was observed between *Salicornia bigelovii* and *Cyanidioschyzon merolae* in terms of bio-crude oil yield • The highest yield of bio-crude oil of 33 wt.% was obtained at 300°C and 80/20 mass ratio of *Salicornia bigelovii/Cyanidioschyzon merolae* • Co-liquefaction led to an increase in the heating value of bio-crude oil and the energy recovery	Phani et al. (2020)

Sargassum tenerrimum	• Temp: 260–300°C • Residence time: 15 min • Catalysts: CaO/CeO_2, CaO/Al_2O_3 and CaO/ZrO_2 • Solvent: water, ethanol and methanol	• For non-catalytic HTL, the highest bio-crude oil yield was 3.3 wt.%, 23.3 wt.% and 32.0 wt.% with water, ethanol and ethanol–water mixed solvents as the reaction medium, respectively • For catalytic HTL, the maximum oil yield of 25.2 wt.% was obtained with CaO/ZrO_2 and ethanol as the reaction medium • The use of ethanol–water mixed solvents led to a higher oil yield than that obtained using ethanol alone, along with an ester selectivity of 88% • Bio-crude oil of a higher heating value was produced from catalytic HTL	Biswas et al. (2020)
Scenedesmus obliquus	• Temp: 240–320°C • Residence time: 60 min • Biomass loading: 7 wt.% • Initial pressure: 5 MPa • Catalyst: $Ca(OH)_2$ from clam shells • Catalyst loading: 0.2–1 wt.%	• The maximum bio-crude oil yield of 39.6 wt.% was obtained at 300°C for 60 min and 0.6 wt.% catalyst loading, along with 35 MJ/kg of heating value • Phenols, aromatics, carboxylic acids and aldehydes were identified as the main compounds in the bio-crude oils • The solid residue obtained from HTL of algae can be used as a catalyst for gasification to produce H_2 • The highest H_2 yield of 37 wt.% was attained at 400°C with 3 wt.% of catalyst	Arun et al. (2020a)
Spirulina platensis	• Temp: 250–320°C • Residence time: 30–60 min • Biomass loading: 10–11 wt.% • Catalyst: CeO_2 • Catalyst loading: 0.15–1.6 g	• Hydrocarbon (7.6%), amino acids (36.7%) and nitrogen-containing compounds (21.6%) were the dominant compounds in the bio-crude oil obtained from catalytic HTL • An increase in the biomass conversion from 16% to 26% was observed when using CeO_2 as the catalyst at 250°C for 30 min	Kandasamy et al. (2020)

(Continued)

Algal species	*Reaction conditions*	*Key findings*	*References*
Ulva lactuca	• Temp: 260–300°C • Residence time: 15–45 min • Biomass loading: 14.29 wt.% • Catalyst loading: 10 wt.% • Catalysts: MgO and ZSM-5 • Solvent: water, methanol and ethanol • Formic acid addition: 8.33–25 wt.%	• The highest bio-crude oil yield of 55.2 wt.% was produced at 280°C for 30 min, using MgO as the catalyst and ethanol–formic acid mixed solvents as the reaction medium • In the presence of formic acid and MgO, hydrocarbons were the main compounds in the bio-crude oils • The use of ZSM-5 and ethanol made phenolics dominant in the bio-crude oil • Both catalysts increased the boiling point distribution of bio-crude oil • The highest heating value of bio-crude oil (35.8 MJ/kg) was obtained with the presence of MgO	Li et al. (2020)
Ulva prolifera	• Temp: 260–300°C • Biomass loading: 14.29 wt.% • Solvents: water, ethanol and methanol • Residence time: 15–45 min • Catalysts: Al_2O_3, ZrO_2, CeO_2, MgO, CaO, $CaCO_3$, $MgCl_2$, $FeCl_3$ and $CuCl_2$	• The highest bio-crude oil yield of 50.6 wt.% was produced with ethanol reaction medium in the presence of MgO • The use of MgO and ethanol led to a large formation of esters, suggesting the esterification reactions between the bio-oil and ethanol solvent • MgO proved to be recyclable	Xu et al. (2020)
Ulva prolifera	• Temp: 260–300°C • Residence time: 15–45 min • Biomass loading: 14.29 wt.% • Catalysts: ZSM-5, Y-zeolite and mordenite • Catalyst loading: 10–20 wt.%	• The maximum bio-crude oil yield of 29.3 wt.% was produced at 280°C for 15 min and 15 wt.% catalyst loading of ZSM-5 • A sharp increase in the heating value of bio-crude oil was observed from 21.2 to 32.3–34.8 MJ/kg with the use of catalyst • In general, zeolite catalysts reduced the oxygen content of bio-crude oil to 17.7% • ZSM-5 exhibited a higher deoxygenation rate than zeolite-Y and mordenite	Ma et al. (2020)

at high temperatures and pressures, making water an effective acid and base catalyst for the hydrolysis of biomass molecules (Toor et al., 2011).

To choose the most effective thermochemical conversion approach for algae, many studies have compared pyrolysis and HTL in terms of product distribution and bio-crude oil properties. Previously, Jena and Das (2011) conducted HTL of *Spirulina platensis* at 250–350°C for 60 min at 5–20 MPa, with 20% of biomass loading, and the results were compared with those obtained from pyrolysis at 350–500°C for 60 min at atmospheric pressure. The comparison showed that the yield of bio-crude oil obtained from HTL (41 wt.%) was higher when compared with that obtained from pyrolysis (23–29 wt.%), accompanied by a lower yield of solid residue (6.3 wt.%) in the HTL than that in the pyrolysis treatment (28–40 wt.%). Overall, it can be stated that the HTL led to a higher biomass conversion than the pyrolysis. As for bio-crude oil properties, the pH values of all bio-oil samples obtained from both HTL and pyrolysis were >9, much higher than those of the lignocellulose-derived bio-oils, which could be attributed to the presence of high protein content in the algae.

Although the bio-oil produced by pyrolysis has a lower viscosity than that produced by HTL, the viscosity of algal bio-oil is still considerably larger than that of wood pyrolysis oils or petroleum crude oil. Thus, subsequent upgrading to lower the viscosity of the algal bio-oils should be performed before application for energy and fuels. It was also found that a higher concentration of inorganics was detected in the bio-crude oils obtained from HTL than that in the pyrolysis oil. This result can be explained as follows: During HTL, the solid and water phases remained in the reactor before bio-crude oil recovery and product separation. Hence, some inorganics might partition into the oil and/or water phase, which leads to a higher amount of inorganic compounds present in the bio-crude oil. On the contrary, the bio-oil obtained in pyrolysis was collected from the vapor condensed in a series of condensers where the solid phase was absent. A comparison of HTL and pyrolysis treatment on the physicochemical properties of bio-oil products is presented in Table 6.3.

Variation in oil viscosity with time is an indicator of the storage ability of bio-crude oil. The authors also reported that an obvious change in the viscosity was found for all bio-oil products at the first 50 days, whereas no significant change was observed in the viscosity of HTL bio-crude oil at the end of 90 days. In contrast, an evident increase in the viscosity of bio-oil samples from pyrolysis was observed, suggesting the HTL bio-crude oils exhibited a better storage ability over the pyrolysis bio-oils.

Vardon et al. (2012) compared HTL and pyrolysis in converting raw and defatted *Scenedesmus* and *Spirulina* into bio-oils at 300°C at 10–12 MPa (HTL) and at 450°C at atmospheric pressure (pyrolysis). Similar to the results reported by Jena and Das (2011), the HTL treatment produced a higher bio-crude oil yield by 7–14% than that in the pyrolysis. According to GC-MS results, cyclic nitrogenates and oxygenates, cyclic nitrogen and oxygen-containing compounds were detected as the dominant chemical species in the

Table 6.3 Physiochemical properties of bio-oil obtained from hydrothermal liquefaction and pyrolysis

Properties	*Hydrothermal liquefaction*	*Pyrolysis at 350°C*	*Pyrolysis at 500°C*
Color	Black	Reddish-brown	Reddish-brown
Odor	Smoky	Acrid smoky	Acrid smoky
pH	9.60	9.35	9.52
Density, kg/L	0.97	1.20	1.05
Viscosity at 40°C, cP	189.80	100.67	79.20
Viscosity at 60°C, cP	51.20	34.30	23.10
Elemental composition (%)^{db}			
C	73.73	67.52	74.66
H	8.90	9.82	10.57
O	10.17	11.34	6.81
N	6.30	10.71	7.13
S	0.90	0.45	0.81
H/C ratio	1.44	1.73	1.68
O/C ratio	0.10	0.13	0.06
HHV, MJ/kg	34.21	29.30	33.62
Inorganic composition (mg/kg)			
Na	14.6	14.6	14.0
Mg	69.3	11.3	2.3
Al	60.1	58.8	10.7
Si	115	54.8	15.5
P	249	63.2	39.6
Ca	116	35.4	7.6
Fe	848	135	180
Ni	72.1	6.4	24.1
Energy and mass balance			
Energy recovery, % [a]	67.9	33.9	46.7
Net energy ratio, % [b]	0.7	2.11	1.56
Carbon and hydrogen recovery, % [c]	71.7	39.3	51.3

Note: Reproduced with permission from Jena and Das (2011).

[a] $$\text{Energy recovery} = \frac{\text{HHV of bio-crude oil} \times \text{Mass of bio-crude oil}}{\text{HHV of biomass} \times \text{Mass of biomass}} \times 100\% \quad (6.1)$$

[b] $$\text{Net energy ratio} = \frac{\text{Energy consumption in HTL or pyrolysis}}{\text{Available energy of bio-crude oil produced}} \times 100\% \quad (6.2)$$

[c] $$\text{Carbon and hydrogen recovery} = \frac{\text{Weight of C and H in the bio-crude oil}}{\text{Weight of C and H in the biomass}} \times 100\% \quad (6.3)$$

HTL bio-crude oils, whereas the pyrolysis oils contained a higher concentration of cyclic oxygenates (area percentage: 16–24%) than that from HTL (area percentage: 8–12%). Detailed chemical compositions of the bio-crude oil obtained from HTL and the pyrolysis oils are shown in Table 6.4.

Table 6.4 Relative area percentage (%) of main chemical compounds identified in the bio-oils obtained in HTL and pyrolysis of algae

Feedstock	*Hydrocarbons (straight/ branched)*	*Hydrocarbons (cyclic)*	*Oxygenates (straight/ branched)*	*Oxygenates (cyclic)*	*Nitrogenates (straight/ branched)*	*Nitrogenates (cyclic)*	*N and O (straight/ branched)*	*N and O (cyclic)*
HTL								
Raw *Scenedesmus*	5 ± 2	2 ± 1	7 ± 3	12 ± 1	n.a.	22 ± 5	n.a.	9 ± 2
Defatted *Scenedesmus*	1 ± 1	1 ± 2	2 ± 3	10 ± 3	n.a.	16 ± 8	2 ± 1	9 ± 5
Spirulina	4 ± 2	2 ± 1	2 ± 2	8 ± 1	n.a.	26 ± 9	3 ± 3	8 ± 1
Pyrolysis								
Raw *Scenedesmus*	7 ± 1	3 ± 1	3 ± 1	24 ± 3	2 ± 1	13 ± 1	1 ± 1	1 ± 1
Defatted *Scenedesmus*	1 ± 2	1 ± 1	3 ± 4	19 ± 4	1 ± 1	11 ± 3	2 ± 1	2 ± 1
Spirulina	5 ± 3	4 ± 2	15 ± 4	16 ± 2	n.a.	9 ± 1	n.a.	2 ± 2

Note: Reproduced with permission from Vardon et al. (2012). n.a.: not available.

Another difference between pyrolysis and HTL is related to the variations in the boiling point distribution of the resultant bio-oil products. Precisely, pyrolysis bio-oils are mainly composed of boiling point range lower than 400°C (62–66 wt.%); in contrast, HTL bio-crude oils normally contain a majority of compounds with boiling points in the range of >400°C (46–55 wt.%) and >538°C (~ 20 wt.%). On the other hand, HTL usually has a better energy balance than pyrolysis when using algal biomass with water content more than 80% after harvesting.

Contrarily, in a more recent study, Arun et al. (2020b) reported a higher oil yield (33.5 wt.%) from pyrolysis at 460°C for 60 min compared to that of HTL (28.9 wt.%) at 320°C for 60 min. In addition, the bio-oil storage stability analysis indicated that the HTL bio-crude oil was less stable than the pyrolysis oil, suggesting the promoted secondary reactions (e.g., polymerization and condensation) in the HTL bio-crude oil during storage. In general, HTL bio-crude oil is often regarded as heavy oil due to the presence of long-chain compounds which contributes to its higher viscosity. As such, by far the results from various studies are not conclusive for which is the best thermochemical approach for algae conversion in terms of the bio-crude oil yield and quality as well as energy recovery.

Lipid extraction from algae

Conventional techniques for lipid extraction from algae such as ball milling, French press and organic solvent extraction are simple operations, but they are constrained by the requirement for a long processing time, the use of a large amount of organic solvent or high operating temperature, which could cause the degradation of target compounds and low extraction yields. Therefore, more advanced lipid extraction techniques have been recently developed, including microwave-, ultrasound- and SCF-assisted extraction (Gullón et al., 2020). Some previous studies on lipid extraction using these advanced approaches are listed in Table 6.5.

Supercritical fluid-assisted extraction

The most important parameters affecting the lipid extraction efficiency are temperature, solvent type, pressure and solvent loading. Owing to the high extraction selectivity and short processing time, SCF extraction has been considered one of the most promising techniques for heat-sensitive bioactive compounds present in algal lipids (Molino et al., 2020). SCF extraction is often performed above the critical point of the solvent; therefore, there is the nonexistence of distinct liquid and gas phases. When compared with the liquid organic solvent, SCF has the advantages of being eco-friendly, non-toxic, non-flammable and chemically inert, and non-requirement for product purification since no residual solvent is present in the target products (Khoo et al., 2020).

Table 6.5 A list of the previous studies on lipid extraction using advanced methods

Extraction methods	*Algal strains*	*References*
Supercritical CO_2-assisted		
	Botryococcus braunii	Santana et al. (2012)
	Chlorella protothecoides	Viguera et al. (2016)
	Chlorella vulgaris	Safi et al. (2014)
	Nannochloropsis oculata; Chlorella vulgaris; Spirulina platensis; Cylindrotheca closterium	Mouahid et al. (2013)
	Nannochloropsis salina; Nannochloropsis maritima	Mouahid et al. (2020)
	Pavlova sp.	Cheng et al. (2011)
	Scenedesmus sp.	Taher et al. (2014)
	Schizochytrium limacinum	Tang et al. (2011)
Microwave-assisted		
	Chlorella sorokiniana	Pan et al. (2016)
	Chlorella vulgaris	Krishnan et al. (2020)
	Nannochloropsis oculata	Biller et al. (2013)
	Nannochloropsis oculata	Ali and Watson (2015)
	Nannochloropsis sp.; *Tetraselmis* sp.	Teo and Idris (2014)
	Scenedesmus obliquus	Balasubramanian et al. (2011)
	Scenedesmus obliquus	Zhou et al. (2019)
	Scenedesmus quadricauda	Onumaegbu et al. (2019)
Ultrasound-assisted		
	Chlorella pyrenoidosa	Feng et al. (2020)
	Chlorella vulgaris	Kim et al. (2013)
	Chlorella vulgaris	Araujo et al. (2013)
	Desmodesmus sp.	González-Balderas et al. (2020)
	Nannochloropsis sp.	Yao et al. (2018)
	Scenedesmus obliquus	Ido et al. (2018)
	Scenedesmus sp.; *Chlorococcum* sp.	Keris-Sen et al. (2014)

To date, supercritical CO_2 is the most commonly used SCF applied for lipid extraction from algae, owing to its lower critical temperature of 31.1°C and pressure of 7.28 MPa compared with those of other SCFs such as ethanol (241°C; 6.3 MPa) and water (374.0°C; 22.1 MPa). Because of this nature of supercritical CO_2, it is favorable for the recovery of heat-sensitive bioactive compounds. Mouahid et al. (2020) evaluated the effects of reaction parameters on the supercritical CO_2 extraction of lipid from *Nannochloropsis maritima* and *Nannochloropsis salina*. The results showed that temperature and pressure were the most important operating conditions affecting the amount of extracted lipid. Specifically, an increase in both temperature and pressure led to an increase in the lipid extraction yield owing to a higher solubility of the SCF. The maximum yield of 14.5 wt.% was obtained at 60°C and 30 MPa. In addition, the composition of the extracted lipid from the supercritical CO_2 extraction was assessed, and palmitoleic acid (C16:1) was found to be the most abundant fatty acid in the extracted lipid. C16:1, C15:0, C18:1, C18:2, C20:0 and C20:4 were also detected. Compared to the fatty acid profile of the lipid obtained from the conventional Soxhlet extraction method, no significant difference could be observed in the composition and lipid extraction yield.

However, it should be mentioned that the nonpolar nature of supercritical CO_2 makes it only effective for extracting neutral lipid. Thus, polar co-solvents (e.g., ethanol) were always introduced into the extraction process to improve the extraction yield (Li et al., 2019). In summary, supercritical CO_2 extraction is a very promising approach for algal lipid extraction and can achieve a high lipid extraction yield, while this approach has some drawbacks, mainly the requirement for a polar solvent to improve the solubility of polar lipid and the need of high-pressure-resistant reactor (Khoo et al., 2020).

Microwave-assisted extraction

Another efficient technology to recover lipid from algae is microwave-assisted extraction. The basic working mechanism of microwave-based extraction is the use of electromagnetic irradiation to transfer the heat to the system during which the ionic conduction and dipole rotation reactions occur. Compared with conventional heated solvent extraction, microwave-assisted extraction is advantageous as it requires a lower amount of solvent, lower energy input, shorter processing time and better heat distribution and lipid extraction rate (Gullón et al., 2020).

Zhou et al. (2019) optimized the lipid extraction of *Scenedesmus obliquus* cultivated in the municipal wastewater using microwave-assisted extraction in n-hexane/isopropanol and achieved the highest extraction efficiency of 88.25% at 130°C for 15 min, n-hexane/isopropanol volume ratio of 3:2 and algae/co-solvent ratio of 50:1 (mL:g), and extraction efficiency is much higher than that of the traditional water bath-assisted extraction method.

Dai et al. (2014) studied lipid extraction from algae by microwave-assisted approach under various processing times, microwave power levels and solvent types. The maximum lipid extraction yield of 30 wt.% was attained using n-heptane/isopropanol. To further improve the lipid extraction efficiency, ionic liquid has been also introduced in the microwave-assisted extraction process (Lu et al., 2019; Pan et al., 2016). Nevertheless, the microwave-assisted extraction method could result in the formation of unstable bonds in the carbon chain structure and thus negatively affect the properties of the extracted lipid (Lee et al., 2017).

Ultrasound-assisted extraction

Ultrasound can be applied to the algal lipid extraction process, where the cell wall of the algae is collapsed and the mass transfer of the target compounds can be promoted. The primary operational variables affecting the lipid extraction efficiency by the ultrasound-assisted approach are power intensity, temperature and algae-to-solvent mass ratio (Ahmadkelayeh and Hawboldt, 2020).

Keris-Sen et al. (2014) investigated the effect of ultrasound on the cell wall disruption and lipid extraction rate with mixed microalgal cultures cultivated in either BG-11 medium or secondary effluent wastewater. The authors found that the ultrasound-assisted technique was helpful for cell wall breakage by releasing proteins and carbohydrates into the water phase, reducing the concentration of suspended solids. The maximum releasing level of the intracellular compounds from algae was obtained at the ultrasound power intensity of 0.4 kWh/L. Additionally, the effects of ultrasound on the yield of extracted lipid were investigated in two different solvent systems (e.g., n-hexane and chloroform/methanol). The application of ultrasound was found to improve the lipid extraction yield by 96% and 54% in chloroform/methanol and n-hexane, respectively. The higher lipid extraction yield attained in chloroform/methanol than that in n-hexane could be due to the nature of chloroform/methanol mixture that consists of both polar (methanol) and nonpolar (chloroform) solvents that facilitate the extraction of both storage lipid (neutral) and structural lipid (polar). On the contrary, n-hexane can only be effective for extracting nonpolar lipid. In conclusion, the main advantages of ultrasound-assisted extraction include high cell wall disruption efficiency, simple operation, less or no toxicity and short processing time. However, the need for temperature control during the treatment and high energy consumption are the main bottlenecks in the ultrasound-assisted extraction technique (Hu et al., 2019).

Conclusions

Algae have been widely accepted as a promising resource for producing liquid biofuels such as bio-crude oil (or bio-oil) and biodiesel. Fast pyrolysis and HTL are thermochemical conversion approaches to the conversion of algae to bio-oils. Although a high yield of bio-oil can be achieved by fast pyrolysis, the quality of algal-derived pyrolysis oil is poor and the algae feedstock (usually with high water content) needs dewatering/drying that is energy-intensive and costly. In contrast, in HTL of algae, the energy-intensive drying operation can be avoided, hence leading to improved overall energy efficiency. However, the properties of algal-derived HTL bio-crude oils are poor in the raw form. For example, the oils have a high viscosity, are corrosive and thermally unstable, and the presence of phase separation within the oil limits their direct application as drop-in biofuels. Consequently, the algal-derived bio-oils must be upgraded via some bio-oil upgrading techniques, e.g., solvent addition, emulsification, esterification, hydrotreatment and zeolite cracking.

Except for thermochemical conversion of algal biomass into bio-oils for drop-in biofuels after upgrading, some advanced lipid extraction approaches are also promising, e.g., ultrasound-, microwave- and SCF-assisted extraction, which can be coupled with the lipid transesterification to produce biodiesel. The ultrasound-based extraction offers benefits of high cell wall disruption efficiency, simple operation, less or no toxicity and short processing time, while the temperature must be controlled during the treatment and high energy consumption is often needed. Despite that, the microwave-assisted extraction method is effective and easy to scale up, and the formation of unstable bonds in the carbon chain structure could affect the properties of extracted lipid.

The SCF-assisted approach is an efficient and green method for algal lipid extraction, whereas a high-pressure-resistant reactor and polar solvents such as ethanol and methanol are required. Therefore, more studies are still required to develop novel lipid extraction methods to avoid the abovementioned technical challenges of the current approaches, while considering easy process scale-up and process intensification, and safety and environmental aspects. For example, solvents of low toxicity such as n-hexane and alcohols that pose a low threat to the ecosystems should be utilized. Besides, the recyclability of the solvent used must be considered to reduce the overall operational cost and achieve a greener process. Ideal algal lipid extraction methods should meet the following criteria: a shorter processing time, lower solvent consumption and the use of green or recyclable solvents.

Acknowledgments

Support for this work was provided through the Biomass Canada Cluster (BMC) that is funded through Agriculture and Agri-Food Canada's AgriScience Program and industry partners. The authors also acknowledge the financial support from the Natural Sciences and Engineering Research Council of Canada (NSERC) through the Discovery Grant awarded to Professor Charles Xu.

References

Adamakis, I.D., Lazaridis, P.A., Terzopoulou, E., Torofias, S., Valari, M., Kalaitzi, P., Rousonikolos, V., Gkoutzikostas, D., Zouboulis, A., Zalidis, G., Triantafyllidis, K.S. (2018) Cultivation, characterization, and properties of *Chlorella vulgaris* microalgae with different lipid contents and effect on fast pyrolysis oil composition. *Environmental Science and Pollution Research* 25: 23018–23032.

Ahmadkelayeh, S., Hawboldt, K. (2020) Extraction of lipids and astaxanthin from crustacean by-products: A review on supercritical CO_2 extraction. *Trends in Food Science & Technology* 103: 94–108.

Ali, M., Watson, I.A. (2015) Microwave treatment of wet algal paste for enhanced solvent extraction of lipids for biodiesel production. *Renewable Energy* 76: 470–477.

Anand, V., Gautam, R., Vinu, R. (2017) Non-catalytic and catalytic fast pyrolysis of *Schizochytrium limacinum* microalga. *Fuel* 205: 1–10.

Anand, V., Sunjeev, V., Vinu, R. (2016) Catalytic fast pyrolysis of *Arthrospira platensis (Spirulina)* algae using zeolites. *Journal of Analytical and Applied Pyrolysis* 118: 298–307.

Anca-Couce, A. (2016) Reaction mechanisms and multi-scale modelling of lignocellulosic biomass pyrolysis. *Progress in Energy and Combustion Science* 53: 41–79.

Araujo, G.S., Matos, L.J.B.L., Fernandes, J.O., Cartaxo, S.J.M., Gonçalves, L.R.B., Fernandes, F.A.N., Farias, W.R.L. (2013) Extraction of lipids from microalgae by ultrasound application: Prospection of the optimal extraction method. *Ultrasonics Sonochemistry* 20: 95–98.

Arun, J., Gopinath, K.P., SundarRajan, P.S., Malolan, R., Adithya, S., Sai Jayaraman, R., Srinivaasan Ajay, P. (2020a) Hydrothermal liquefaction of *Scenedesmus obliquus* using a novel catalyst derived from clam shells: Solid residue as catalyst for hydrogen production. *Bioresource Technology* 310: 123443.

Arun, J., Gopinath, K.P., SundarRajan, P.S., Malolan, R., AjaySrinivaasan, P. (2020b) Hydrothermal liquefaction and pyrolysis of *Amphiroa fragilissima* biomass: Comparative study on oxygen content and storage stability parameters of bio-oil. *Bioresource Technology Reports* 11: 100465.

Aziz, M.M.A., Kassim, K.A., Shokravi, Z., Jakarni, F.M., Lieu, H.Y., Zaini, N., Tan, L.S., Islam, S., Shokravi, H. (2020) Two-stage cultivation strategy for simultaneous increases in growth rate and lipid content of microalgae: A review. *Renewable and Sustainable Energy Review* 119: 109621.

Azizi, K., Keshavarz Moraveji, M., Arregi, A., Amutio, M., Lopez, G., Olazar, M. (2020) On the pyrolysis of different microalgae species in a conical spouted bed reactor: Bio-fuel yields and characterization. *Bioresource Technology* 311: 123561.

Balasubramanian, S., Allen, J.D., Kanitkar, A., Boldor, D. (2011) Oil extraction from *Scenedesmus obliquus* using a continuous microwave system - design, optimization, and quality characterization. *Bioresource Technology* 102: 3396–3403.

Biller, P., Friedman, C., Ross, A.B. (2013) Hydrothermal microwave processing of microalgae as a pre-treatment and extraction technique for bio-fuels and bio-products. *Bioresource Technology* 136: 188–195.

Biswas, B., Kumar, A., Fernandes, A.C., Saini, K., Negi, S., Muraleedharan, U.D., Bhaskar, T. (2020) Solid base catalytic hydrothermal liquefaction of macroalgae: Effects of process parameter on product yield and characterization. *Bioresource Technology* 307: 123232.

Campanella, A., Harold, M.P. (2012) Fast pyrolysis of microalgae in a falling solids reactor: Effects of process variables and zeolite catalysts. *Biomass and Bioenergy* 46: 218–232.

Cheng, C.-H., Du, T.-B., Pi, H.-C., Jang, S.-M., Lin, Y.-H., Lee, H.-T. (2011) Comparative study of lipid extraction from microalgae by organic solvent and supercritical CO_2. *Bioresource Technology* 102: 10151–10153.

Choi, J.H., Kim, S.S., Ly, H.V., Kim, J., Woo, H.C. (2017) Effects of water-washing *Saccharina japonica* on fast pyrolysis in a bubbling fluidized-bed reactor. *Biomass and Bioenergy* 98: 112–123.

Choi, J.W., Choi, J.H., Suh, D.J., Kim, H. (2015) Feasibility of *Laminaria japonica* as a feedstock for fast pyrolysis in a bubbling fluidized-bed reactor. *Journal of Analytical and Applied Pyrolysis* 112: 141–149.

Cui, Z., Cheng, F., Jarvis, J.M., Brewer, C.E., Jena, U. (2020) Roles of co-solvents in hydrothermal liquefaction of low-lipid, high-protein algae. *Bioresource Technology* 310: 123454.

Dai, Y.M., Chen, K.T., Chen, C.C. (2014) Study of the microwave lipid extraction from microalgae for biodiesel production. *Chemical Engineering Journal* 250: 267–273.

Devi, T.E., Parthiban, R. (2020) Hydrothermal liquefaction of *Nostoc ellipsosporum* biomass grown in municipal wastewater under optimized conditions for bio-oil production. *Bioresource Technology* 316: 123943.

Feng, X., Chen, Y., Lv, J., Han, S., Tu, R., Zhou, X., Jin, W., Ren, N. (2020) Enhanced lipid production by *Chlorella pyrenoidosa* through magnetic field pre-treatment of wastewater and treatment of microalgae-wastewater culture solution: Magnetic field treatment modes and conditions. *Bioresource Technology* 306: 123102.

Galadima, A., Muraza, O. (2018) Hydrothermal liquefaction of algae and bio-oil upgrading into liquid fuels: Role of heterogeneous catalysts. *Renewable Sustainable Energy Review* 81: 1037–1048.

Gautam, R., Vinu, R. (2018) Non-catalytic fast pyrolysis and catalytic fast pyrolysis of *Nannochloropsis oculata* using Co-Mo/γ-Al_2O_3 catalyst for valuable chemicals. *Algal Research* 34: 12–24.

Gebremariam, S.N., Marchetti, J.M. (2018) Economics of biodiesel production: Review. *Energy Conversion and Management* 168: 74–84.

González-Balderas, R.M., Velásquez-Orta, S.B., Valdez-Vazquez, I., Orta Ledesma, M.T. (2020) Intensified recovery of lipids, proteins, and carbohydrates from wastewater-grown microalgae *Desmodesmus* sp. by using ultrasound or ozone. *Ultrasonics Sonochemistry* 62: 104852.

Gullón, B., Gagaoua, M., Barba, F.J., Gullón, P., Zhang, W., Lorenzo, J.M. (2020) Seaweeds as promising resource of bioactive compounds: Overview of novel extraction strategies and design of tailored meat products. *Trends in Food Science & Technology* 100: 1–18.

Guo, Y., Yeh, T., Song, W., Xu, D., Wang, S. (2015) A review of bio-oil production from hydrothermal liquefaction of algae. *Renewable Sustainable Energy Review* 48: 776–790.

Hameed, S., Sharma, A., Pareek, V., Wu, H., Yu, Y. (2019) A review on biomass pyrolysis models: Kinetic, network and mechanistic models. *Biomass and Bioenergy* 123: 104–122.

Harman-Ware, A.E., Morgan, T., Wilson, M., Crocker, M., Zhang, J., Liu, K., Stork, J., Debolt, S. (2013) Microalgae as a renewable fuel source: Fast pyrolysis of *Scenedesmus* sp. *Renewable Energy* 60: 625–632.

He, S., Zhao, M., Wang, J., Cheng, Z., Yan, B., Chen, G. (2020) Hydrothermal liquefaction of low-lipid algae *Nannochloropsis* sp. and *Sargassum* sp.: Effect of feedstock composition and temperature. *Science of the Total Environment* 712: 135677.

Hu, Y., Gong, M., Feng, S., Charles, C., Bassi, A. (2019) A review of recent developments of pre-treatment technologies and hydrothermal liquefaction of microalgae for bio-crude oil production. *Renewable Sustainable Energy Review* 101: 476–492.

Ido, A.L., de Luna, M.D.G., Capareda, S.C., Maglinao, A.L., Nam, H. (2018) Application of central composite design in the optimization of lipid yield from *Scenedesmus obliquus* microalgae by ultrasound-assisted solvent extraction. *Energy* 157: 949–956.

Jena, U., Das, K.C. (2011) Comparative evaluation of thermochemical liquefaction and pyrolysis for bio-oil production from microalgae. *Energy and Fuels* 25: 5472–5482.

Jia, L., Cao, C., Cheng, Z., Wang, J., Huang, J., Yang, J., Pan, Y., Xu, M., Wang, Y. (2020) Ex Situ catalytic pyrolysis of algal biomass in a double microfixed-bed reactor: Catalyst deactivation and its coking behavior. *Energy and Fuels* 34: 1918–1928.

Kandasamy, S., Zhang, B., He, Z., Chen, H., Feng, H., Wang, Q., Wang, B., Ashokkumar, V., Siva, S., Bhuvanendran, N., Krishnamoorthi, M. (2020) Effect of low-temperature catalytic hydrothermal liquefaction of *Spirulina platensis*. *Energy* 190: 116236.

Keris-Sen, U.D., Sen, U., Soydemir, G., Gurol, M.D. (2014) An investigation of ultrasound effect on microalgal cell integrity and lipid extraction efficiency. *Bioresource Technology* 152: 407–413.

Khoo, K.S., Chew, K.W., Yew, G.Y., Leong, W.H., Chai, Y.H., Show, P.L., Chen, W.H. (2020) Recent advances in downstream processing of microalgae lipid recovery for biofuel production. *Bioresource Technology* 304: 122996.

Kim, Y.-H., Park, S., Hoo, M., Choi, Y.-K., Yang, Y.-H., Joo, H., Kim, H., Kim, H.-S., Song, K.-G., Lee, H.S. (2013) Ultrasound-assisted extraction of lipids from *Chlorella vulgaris* using [Bmim][$MeSO_4$]. *Biomass and Bioenergy* 56: 99–103.

Krishnan, S., Ghani, N.A., Aminuddin, N.F., Quraishi, K.S., Azman, N.S., Cravotto, G., Leveque, J.M. (2020) Microwave-assisted lipid extraction from *Chlorella vulgaris* in water with 0.5%–2.5% of imidazolium based ionic liquid as additive. *Renewable Energy* 149: 244–252.

Lee, S.Y., Cho, J.M., Chang, Y.K., Oh, Y.K. (2017) Cell disruption and lipid extraction for microalgal biorefineries: A review. *Bioresource Technology* 244: 1317–1328.

Li, F., Srivatsa, S.C., Bhattacharya, S. (2019a) A review on catalytic pyrolysis of microalgae to high-quality bio-oil with low oxygeneous and nitrogenous compounds. *Renewable Sustainable Energy Reviews* 108: 481–497.

Li, P., Li, H., Jing, K., David, A., Lin, J., Deng, G. (2019b) Evaluation of lipid extraction from microalgae based on different phase regions of CO_2-expanded ethanol. *Chemical Engineering and Processing - Process Intensification* 138: 1–6.

Li, Y., Zhu, C., Jiang, J., Yang, Z., Feng, W., Li, L., Guo, Y., Hu, J. (2020) Hydrothermal liquefaction of macroalgae with in-situ-hydrogen donor formic acid: Effects of process parameters on products yield and characterizations. *Industrial Crops and Products* 153: 112513.

Lu, H., Yu, X., Li, H., Tu, S.-T., Sebastian, S. (2019) Lipids extraction from wet *Chlorella pyrenoidosa* sludge using recycled [BMIM]Cl. *Bioresource Technology* 291: 121819.

Ly, H.V., Kim, S.S., Choi, J.H., Woo, H.C., Kim, J. (2016) Fast pyrolysis of *Saccharina japonica* alga in a fixed-bed reactor for bio-oil production. *Energy Conversion and Management* 122: 526–534.

Ly, H.V., Kim, S.S., Woo, H.C., Choi, J.H., Suh, D.J., Kim, J. (2015) Fast pyrolysis of macroalga *Saccharina japonica* in a bubbling fluidized-bed reactor for bio-oil production. *Energy* 93: 1436–1446.

Ma, C., Geng, J., Zhang, D., Ning, X. (2020) Hydrothermal liquefaction of macroalgae: Influence of zeolites based catalyst on products. *Journal of the Energy Institute* 93: 581–590.

Miao, X., Wu, Q. (2004) High yield bio-oil production from fast pyrolysis by metabolic controlling of *Chlorella protothecoides*. *Journal of Biotechnology* 110: 85–93.

Miao, X., Wu, Q., Yang, C. (2004) Fast pyrolysis of microalgae to produce renewable fuels. *Journal of Analytical and Applied Pyrolysis* 71: 855–863.

Molino, A., Mehariya, S., Di Sanzo, G., Larocca, V., Martino, M., Leone, G.P., Marino, T., Chianese, S., Balducchi, R., Musmarra, D. (2020) Recent developments in supercritical fluid extraction of bioactive compounds from microalgae: Role of key parameters, technological achievements and challenges. *Journal of CO_2 Utilization* 36: 196–209.

Mouahid, A., Crampon, C., Toudji, S.A.A., Badens, E. (2013) Supercritical CO_2 extraction of neutral lipids from microalgae: Experiments and modelling. *Journal of Supercritical Fluids* 77: 7–16.

Mouahid, A., Seengeon, K., Martino, M., Crampon, C., Kramer, A., Badens, E. (2020) Selective extraction of neutral lipids and pigments from *Nannochloropsis salina* and *Nannochloropsis maritima* using supercritical CO_2 extraction: Effects of process parameters and pre-treatment. *Journal of Supercritical Fluids* 165: 104934.

Onumaegbu, C., Alaswad, A., Rodriguez, C., Olabi, A. (2019) Modelling and optimization of wet microalgae *Scenedesmus quadricauda* lipid extraction using microwave pre-treatment method and response surface methodology. *Renewable Energy* 132: 1323–1331.

Pan, J., Muppaneni, T., Sun, Y., Reddy, H.K., Fu, J., Lu, X., Deng, S. (2016) Microwave-assisted extraction of lipids from microalgae using an ionic liquid solvent [BMIM][HSO_4]. *Fuel* 178: 49–55.

Phani, K., Dandamudi, R., Muhammed, K., Laideson, M., Murdock, T., Seger, M., Mcgowen, J., Lammers, P.J., Deng, S., Torr, S. (2020) Hydrothermal liquefaction of *Cyanidioschyzon merolae* and *Salicornia bigelovii Torr.*: The interaction effect on product distribution and chemistry. *Fuel* 277: 118146.

Piloni, R.V., Brunetti, V., Urcelay, R.C., Daga, I.C., Moyano, E.L. (2017) Chemical properties of biosilica and bio-oil derived from fast pyrolysis of *Melosira varians*. *Journal of Analytical and Applied Pyrolysis* 127: 402–410.

Pourkarimi, S., Hallajisani, A., Alizadehdakhel, A., Nouralishahi, A. (2019) Biofuel production through micro- and macroalgae pyrolysis – A review of pyrolysis methods and process parameters. *Journal of Analytical and Applied Pyrolysis* 142: 104599.

Safi, C., Camy, S., Frances, C., Varela, M.M., Badia, E.C., Pontalier, P.Y., Vaca-Garcia, C. (2014) Extraction of lipids and pigments of Chlorella vulgaris by supercritical carbon dioxide: Influence of bead milling on extraction performance. *Journal of Applied Phycology* 26: 1711–1718.

Sajjadi, B., Chen, W.Y., Raman, A.A.A., Ibrahim, S. (2018) Microalgae lipid and biomass for biofuel production: A comprehensive review on lipid enhancement strategies and their effects on fatty acid composition. *Renewable and Sustainable Energy Review* 97: 200–232.

Santana, A., Jesus, S., Larrayoz, M.A., Filho, R.M. (2012) Supercritical carbon dioxide extraction of algal lipids for the biodiesel production. *Procedia Engineering* 42: 1755–1761.

Sotoudehniakarani, F., Alayat, A., McDonald, A.G. (2019) Characterization and comparison of pyrolysis products from fast pyrolysis of commercial *Chlorella vulgaris* and cultivated microalgae. *Journal of Analytical and Applied Pyrolysis* 139: 258–273.

Taher, H., Al-Zuhair, S., Al-Marzouqi, A.H., Haik, Y., Farid, M., Tariq, S. (2014) Supercritical carbon dioxide extraction of microalgae lipid: Process optimization and laboratory scale-up. *Journal of Supercritical Fluids* 86: 57–66.

Tang, S., Qin, C., Wang, H., Li, S., Tian, S. (2011) Study on supercritical extraction of lipids and enrichment of DHA from oil-rich microalgae. *Journal of Supercritical Fluids* 57: 44–49.

Teo, C.L., Idris, A. (2014) Enhancing the various solvent extraction method via microwave irradiation for extraction of lipids from marine microalgae in biodiesel production. *Bioresource Technology* 171: 477–481.

Thangalazhy-Gopakumar, S., Adhikari, S., Chattanathan, S.A., Gupta, R.B. (2012) Catalytic pyrolysis of green algae for hydrocarbon production using H+ZSM-5 catalyst. *Bioresource Technology* 118: 150–157.

Toor, S.S., Rosendahl, L., Rudolf, A. (2011) Hydrothermal liquefaction of biomass: A review of subcritical water technologies. *Energy* 36: 2328–2342.

Vardon, D.R., Sharma, B.K., Blazina, G.V., Rajagopalan, K., Strathmann, T.J. (2012) Thermochemical conversion of raw and defatted algal biomass via hydrothermal liquefaction and slow pyrolysis. *Bioresource Technology* 109: 178–187.

Viguera, M., Marti, A., Masca, F., Prieto, C., Calvo, L. (2016) The process parameters and solid conditions that affect the supercritical CO_2 extraction of the lipids produced by microalgae. *Journal of Supercritical Fluids* 113: 16–22.

Wang, K., Brown, R.C., Homsy, S., Martinez, L., Sidhu, S.S. (2013) Fast pyrolysis of microalgae remnants in a fluidized bed reactor for bio-oil and biochar production. *Bioresource Technology* 127: 494–499.

Wang, S., Dai, G., Yang, H., Luo, Z. (2017) Lignocellulosic biomass pyrolysis mechanism: A state-of-the-art review. *Progress in Energy and Combustion Science* 62: 33–86.

Xu, D., Lin, G., Guo, S., Wang, S., Guo, Y., Jing, Z. (2018) Catalytic hydrothermal liquefaction of algae and upgrading of biocrude: A critical review. *Renewable Sustainable Energy Review* 97: 103–118.

Xu, J., Dong, X., Wang, Y. (2020) Hydrothermal liquefaction of macroalgae over various solids, basic or acidic oxides and metal salt catalyst: Products distribution and characterization. *Industrial Crops and Products* 151: 112458.

Yang, C., Li, R., Zhang, B., Qiu, Q., Wang, B., Yang, H., Ding, Y., Wang, C. (2019) Pyrolysis of microalgae: A critical review. *Fuel Processing Technology* 186: 53–72.

Yao, S., Mettu, S., Law, S.Q.K., Ashokkumar, M., Martin, G.J.O. (2018) The effect of high-intensity ultrasound on cell disruption and lipid extraction from high-solids viscous slurries of *Nannochloropsis* sp. biomass. *Algal Research* 35: 341–348.

Zhao, H., Yan, H.X., Liu, M., Sun, B.B., Zhang, Y., Dong, S.S., Qi, L.B., Qin, S. (2013) Production of bio-oil from fast pyrolysis of macroalgae *Enteromorpha prolifera* powder in a free-fall reactor. *Energy Sources, Part A Recovery, Utilization and Environmental Effects* 35: 859–867.

Zhou, X., Jin, W., Tu, R., Guo, Q., Han, S., Chen, C., Wang, Qing, Liu, W., Jensen, P.D., Wang, Q. (2019) Optimization of microwave assisted lipid extraction from microalga *Scenedesmus obliquus* grown on municipal wastewater. *Journal of Cleaner Production* 221: 502–508.

7 Hydrothermal gasification of biomass to produce hydrogen-rich syngas

Pankaj Kalita and Rabindra Kangsha Banik

Introduction

Industrialization and increase in per capita energy consumption have enforced the energy industry shift toward higher generation capacity and look for alternative routes for energy generation. Moreover, fossil fuel-derived energy causes severe environmental damage, and it has led the policymakers to make stringent ecological regulations (Kumar et al., 2018). Besides, the inequality in demand and supply makes it inevitable for the technological upliftment of conventional energy sources (Tekin and Karagöz, 2013). It has contributed to finding alternative energy sources and upliftment of the existing technologies for the higher generation following the environmental regulations and bringing out new technologies that are much more efficient and create less damage to the environment.

Hydrothermal processing is gaining widespread importance as it is catering toward increasing energy demands by producing energy-dense fuel and value-added chemicals as well as mitigating emergent environmental concerns. The process permits effective heat integration resulting in reduced water vaporization from hydrothermal media. Hydrothermal processing of biomass or any other solid fuel is typically a thermochemical process comprising thermal disintegration in the presence of hot-compressed water. The process converts biomass or any other solid into biochar (solid), bio-oil (liquid) or gas (e.g., hydrogen, methane, etc.) by controlling operating constraints such as pressure, temperature, catalyst and residence time (Kumar et al., 2018). This process can directly utilize biomass having high moisture content without pre-drying (Tekin et al., 2014). By-products produced during this process can be used for electricity generation and recovery of value-added chemicals. Hydrothermal liquefaction (HTL) and hydrothermal gasification (HTG) are the fundamental hydrothermal routes. In contrast, hydrothermal carbonization (HTC) is a somewhat modern process and has matured in the current years (Erlach et al., 2012).

Hydrothermal processing occurs under subcritical as well as supercritical water conditions. These conditions are defined concerning the critical point of water (i.e., T_C: 374°C; and P_C: 22.1 MPa) (Tekin et al., 2014). Water has different characteristics for each condition, as shown in Figure 7.1. Water

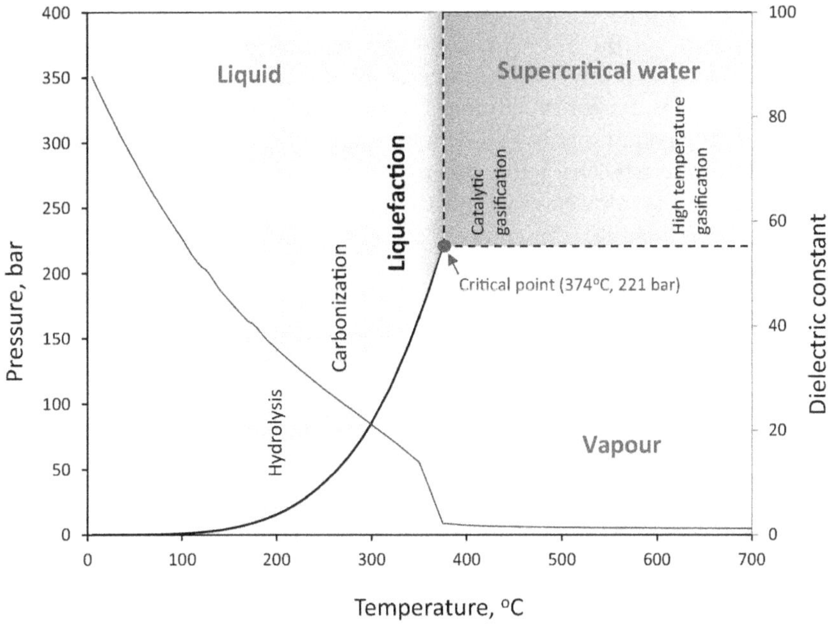

Figure 7.1 Phase diagram and dielectric constant (static) of water at 20 MPa. Reproduced with permission from Tran (2016).

plays a significant role, such as acting as a solvent, as a catalyst and as a reactant in hydrothermal processing (Tekin and Karagöz, 2013). Most of the biomass constituents (e.g., cellulose, hemicellulose and lignin) are mostly insoluble in water at room temperature. However, they are soluble at elevated temperature. In the presence of supercritical water, macromolecules of biomass are fragmented to their primary building blocks to produce gas. On the other hand, at moderate pressure and temperature, macromolecules are fragmented into bio-oil having high viscosity (Tekin et al., 2014). The hydrothermal process initiates at about 100°C with the dispersion of water-soluble fraction of the biomass feedstocks to water; subsequently, hydrolysis starts beyond 150°C with the disintegration of cellulose and hemicellulose into its monomeric chains. As the temperature rises to 200°C at a pressure of 1 MPa, slurry formation takes place and the process shifts toward either carbonization, liquefaction or gasification subject to the target product (Kumar et al., 2018).

Water condition for hydrothermal treatment

Water is widely used as an inert solvent in numerous organic reactions. The critical temperature and pressure of water are 374°C and 22.1 MPa, respectively. If the properties of water are changed below its critical point, phase transition may take place, whereas beyond its critical point, the properties are altered without phase change (Kumar et al., 2018). Under sufficient pressure and temperature (100–374°C) if the water remains in the liquid phase, it is termed as subcritical water. At temperature and pressure beyond 374°C and 22.1 MPa, the water is termed supercritical water (Tekin et al., 2014). Water at both these conditions has some advantages due to its different physicochemical properties. With the rise in temperature, viscosity and dielectric constant (ε) of water decrease, but the diffusion coefficient increases leading toward higher reaction rates. Water molecules also act as a hydrogen source and take part in hydrolysis reactions as a reactant. Table 7.1 shows the physicochemical properties of water which are a function of its pressure and temperature. Solvent characteristics of water are dependent on the dielectric constant, which is again a function of pressure and temperature of water (Guo et al., 2010). Apart from hydrothermal processing, supercritical water is used for the power generation cycle and also for segregating toxic compounds by oxidation from biological and organic waste.

History and background of hydrothermal treatment

The first thermochemical processing was applied in 1788 to convert biomass to bio-crude (Dimitriadis and Bezergianni, 2017). However, the thermochemical technologies have been abandoned due to the low fossil fuel prices, and recently, these have been reconsidered to meet the increasing energy demand. In the USA at Pittsburgh Energy Research Center, researchers have contributed toward liquefaction of lignite under hydrothermal conditions during the 1970s and 1980s (Biller and Ross, 2012).

In 1982, Schaleger and his team had performed direct liquefaction of biomass at the Lawrence Berkeley Laboratory, whereas some other scientists in the USA have performed experimentation on biomass liquefaction at the same time. The technology established at Pittsburgh Research Center used drying and grinding during biomass pretreatment, whereas Lawrence Berkeley

Table 7.1 Physicochemical properties of water

	Ambient water	*Subcritical water*	*Supercritical water*	
Temperature, T (°C)	25	250	400	400
Pressure, P (MPa)	0.1	5.0	25.0	50.0
Density, ρ (g/cm^3)	0.997	0.80	0.17	0.58
Thermal conductivity, λ (mWm^{-1} K^{-1})	606	620	160	438
Specific heat, C_p (kJ/kg/K)	4.182	4.878	13.0	6.8
Viscosity, μ (mPa s)	0.89	0.106	0.03	0.07
Dielectric constant, ε	78.5	27.1	5.9	10.5

Note: Reproduced with permission from Tekin et al. (2014).

Laboratory has treated the biomass by acid hydrolysis before processing (Elliott, 1980). Shell in the Netherlands has developed a commercial-scale biomass liquefaction unit under hydrothermal processing. In Asia, NextFuels is introducing a process for commercial HTL having a plant capacity of 1,000 barrels/d from palm oil wastes. A novel process called CatLiq was developed by a Danish company for the processing of sewage sludge comprising algae and manure (Nielsen et al., 2012). These initiatives led to the development of commercial technologies for hydrothermal processing at the industrial level in collaboration with research institutions.

Ridgeline Energy Services in Canada procured the commercial-scale HTL plant for the conversion of waste to oil using thermal depolymerization within the reactor developed by Changing World Technologies (Tran, 2016). Biller and Ross (2012) have reported in their studies that initially, HTL investigations were carried out using terrestrial biomass resources and consequently using algal feedstocks at the University of Arizona (USA), University of Toronto (Canada) and Royal Institute of Technology (Sweden). The biorefinery concept is driven by HTL for the production of fuels and valued chemicals.

In the USA, Modell and his associates from the Massachusetts Institute of Technology (MIT) witnessed the complete disintegration of glucose in supercritical water and thereby proposed the idea of HTG (Matsumura, 2015). It has influenced the group to apply the novel technology for supercritical water oxidation (SCWO) and testing high-risk organic wastes under supercritical conditions. A technology was developed at the Pacific Northwest National Laboratory (USA) utilizing metal catalysts at lower temperatures (around 400°C) (Elliott et al., 1993; Sealock et al., 1993), whereas a group at the University of Hawaii established a technology using a carbonaceous catalyst at higher temperatures (around 600°C) (Xu et al., 1996; Yu et al., 1993). It has created the platform for scientists and researchers from various institutions such as Karlsruhe Research Center (2.4 tons/d), Hiroshima University (1 ton/d), the University of Leeds and the University of Michigan to work in the field of HTG to develop pilot-scale HTG plants (Matsumura, 2015). In 1913, Friedrich Bergius suggested the concept of HTC by simulating the natural coalification technique (Kumar et al., 2018). Later, the process has been recognized as a hydrothermal deterioration of organic substances for the production of fuel and valuable chemicals (Kambo and Dutta, 2015).

Hydrothermal treatment of biomass

The term biomass indicates biological material or waste derived from living entities having potential energy sources and is regarded as abundant, clean and eco-friendly (McKendry, 2002). Moreover, due to its sustainable nature, biomass is considered as a feasible alternative energy resource and estimated to offer around 25% of the world's total energy demand in the near future (Briens et al., 2008). Although biomass is reckoned as a major alternative energy resource, there are challenges in proper process selection for effective utilization due to its wide variety. Moreover, the high moisture content and the lower carbon content in some of its variety make it difficult for economical utilization as a significant amount of energy is consumed during its drying operation. Hydrothermal processing has privilege over the conventional biomass processing technologies due to its high flexibility on feedstock selection and elimination of the drying cost of high-moisture-content biomass feed (Heidari et al., 2019).

In hydrothermal processing of biomass, the high-moisture feedstock can be converted into various products such as biochar (solid), bio-oil (liquid) and gas (H_2, CH_4, CO, etc.) by controlling the process parameters (Tekin et al., 2014). Figure 7.2 shows the various products obtained from hydrothermal processing of biomass. The process parameters such as pressure, temperature, catalyst and residence time play a significant role to achieve the desired product. This process is executed at a temperature of 180–374°C and pressure ranging from 2 to 22 MPa for solid and liquid products and at a higher temperature (400–600°C) and pressure (>22 MPa) for more gaseous products at suitable water conditions. Table 7.2 summarizes the process conditions of the thermal treatment of biomass and the products obtained by these methods. Biomass containing 30% moisture requires less energy in HTG under supercritical conditions compared to pre-drying biomass (Savage et al., 2010). Moreover, an increase in operating pressure prevents the possible phase change, thereby enhancing the yield of the process (Peterson et al., 2008).

In recent years, the focus has been given to a wide variety of biomass feedstocks, comprising agricultural wastes as well as algae for hydrothermal processing (Elliott, 2011). Algal biomass contains biogenic carbon, nitrogen,

Table 7.2 Conditions for the thermal transformations of biomass and the resulting products

Process	*Conditions*			*Products (wt.%)*		
	Temperature (°C)	*Pressure (MPa)*	*Time*	*Solid*	*Liquid*	*Gas*
Slow pyrolysis	~400	0.1	10–20 h	35	30	35
Intermediate pyrolysis	~500	0.1	10–20 s	20	50	30
Fast pyrolysis	~500	0.1	1–3 s	12	75	13
Torrefaction	200–300	2–10	1–2 h	60	20	20
Hydrothermal carbonization	180–250	2–10	10–20 h	80	17	3
Hydrothermal liquefaction	300–350	5–20	5–60 min	–	–	–
Hydrothermal gasification	400–600	>22.4	1–60 min	–	–	–

Note: Reproduced with permission from Krylova and Zaitchenko (2018).

phosphorus and micronutrients in the aqueous phase, which can further be reprocessed for algal cultivation purposes. Besides, valued chemicals such as acetone, acetic acid and ethanol can be recovered using extraction as well as catalytic processes (Maddi et al., 2017).

Apart from the benefits of hydrothermal treatment of biomass, some drawbacks are also there in the process. Several obstacles have been encountered, such as enormous water handling vessels and the capability required for expensive and complex reactors for commercialization of this technology (Zhang et al., 2010). Furthermore, it is challenging to smoothly accomplish the separation and extraction of products at the industrial level. Moreover, variability in the hydrothermal medium and complicated mass balancing make it challenging to determine the gas yield (Karagöz et al., 2004).

Hydrothermal liquefaction of biomass

It is a thermochemical conversion method performed at moderate temperature and high pressure with subcritical water. In this process, a wide range of biomass can be converted to liquid products (bio-oil) having higher energy density (Toor et al., 2011). The HTL process is similar to fossil fuel formation, although it requires lower temperature, pressure and residence time. The operating conditions for HTL are provided in Table 7.2. The HTL process has several advantages; for example, it is suitable for wet biomass and is less energy-intensive than pyrolysis (Sasaki et al., 2002). The transformation of biomass to bio-oil involves many complex reactions because of the complexity of the biomass molecules (Zhang, 2010). The main constituents of biomass are carbohydrates, proteins, fats and lignin. The decomposition steps involve depolymerization of the complex structure of the biomass followed by degradation of monomers and finally recombination of the fragmented components to yield bio-oil (Toor et al., 2011).

Large molecules of the biomass structure on hydrolysis dissociate to water-soluble monomers and oligomers, which might undergo further deterioration or re-polymerization or have reduced functional groups in the structure (Savage et al., 2010). In the case of cellulose molecules, glucose monomers are linked by β-(1–4) glycosidic bonds, which contribute toward the creation of strong intramolecular as well as intermolecular hydrogen bonds. At high temperature and pressure, subcritical water disrupts the hydrogen bond of the cellulose molecules, leading toward the formation of glucose monomers and some other oligomers. The oligomers thus produced may undergo quick hydrolysis to form glucose and subsequently form a wide variety of products such as acetaldehyde, acetic acid, glycolaldehyde, glyceraldehyde and furfural derivatives depending upon the operating conditions (Peterson et al., 2008).

Hemicellulose comprises a variety of sugar monomers, such as xylose, galactose and mannose with side chains. The complex structure of hemicellulose offers less resistance toward hydrolysis than cellulose and readily dissolves in water beyond 180°C (Bobleter, 1994). The hydrolysis of hemicellulose leads to the formation of a furanose ring, a pyranose ring or an open-chain structure by xylose molecules. The pyranose ring of xylose leads to the formation of furfural, whereas the open-chain structure leads to the formation of glyceraldehyde, formic acid, lactic acid and acetol (Peterson et al., 2008). In addition, lignin, another compound found in biomass, has higher molecular weight and is more irregular and complex in structure than hemicellulose.

The hydrothermal degradation of lignin results in the formation of low molecular weight phenolic components (Saisu et al., 2003). Most of the heavier compounds of the lignocellulosic biomass, including benzonitriles, benzyl aryl ethers, benzamides, cyclohexyl phenyl and pyridine carbonitriles, are thermally non-reactive at normal conditions. However, at around 250–350°C, water molecules turn out to be active and reaction with organic compounds proceeds. At this condition, benzamides, benzonitriles and pyridine carbonitrile swiftly hydrolyze to form ammonium carboxylates, whereas cyclohexyl phenyl dissociates forming methyl cyclopentene along with thiophenol, phenol or aniline (Zhang, 2010).

The production of bio-oil is influenced by the type of biomass feedstock chosen due to different compositions. The presence of cellulose and hemicellulose increases bio-oil yield, while lignin contributes mainly toward more residual fraction. Moreover, liquefaction of loose biomass results in higher oxygen and moisture content of bio-oil which is unfavorable as it degrades the quality and calorific value of fuel (Akhtar and Amin, 2011). Smaller feedstock particle size enhances heat penetration, which improves the conversion rate and efficiency of bio-oil production. Figure 7.2 gives a schematic of the batch hydrothermal process for biomass feedstocks.

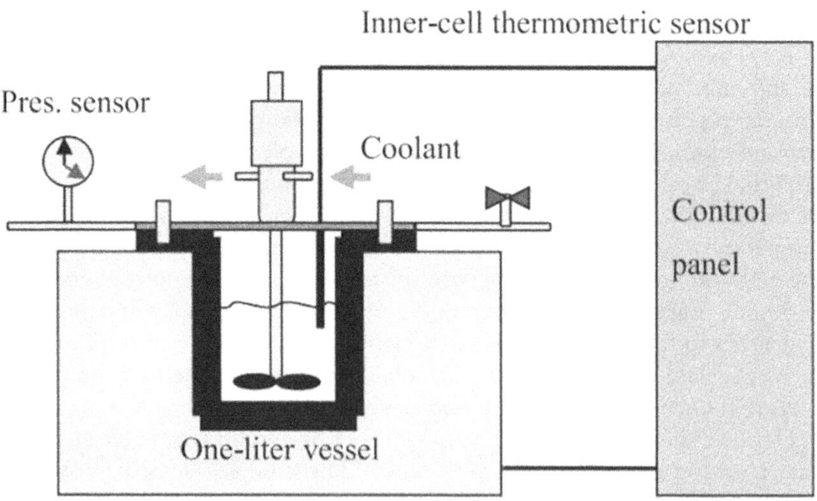

Figure 7.2 Schematic diagram of the batch hydrothermal process for biomass feedstock. Reproduced with permission from Nonaka et al. (2011).

Liquefaction of biomass is typically endothermic at a lower temperature, whereas it turns into exothermic at elevated temperatures (Akhtar and Amin, 2011). With an increase in temperature, there is a boost in bio-oil production up to a threshold point after which the gas yield increases. At a moderate temperature (<275°C), the partial breakdown of biomass components results in a lower yield of bio-oil. Studies have shown that a moderate temperature of around 300–350°C results in higher yield bio-oil. In addition to that, pressure also plays a significant role in the HTL process, maintaining water in the aqueous state and reducing the handling cost of the two-phase system. However, at supercritical conditions, the pressure becomes insignificant. Therefore, a further increase in pressure under supercritical conditions prevents C–C bond fragmentation due to higher local solvent density. Apart from pressure and temperature, residence time plays an essential role in product composition and conversion efficiency. At near-supercritical conditions, the short residence time is desirable as rapid degradation of biomass may result in dominating secondary and tertiary reactions (Akhtar and Amin, 2011).

Although bio-oils produced from HTL have lower oxygen content and higher calorific value, they cannot be utilized straightway as a transportation fuel (Huber et al., 2006). For their utilization as transportation fuel, it is crucial to improve certain physical properties. Improved bio-oils can be used for transportation purposes as a substitute for petrol or diesel. The enrichment of the bio-oil produced by HTL may be achieved either by physical separation (such as extraction or solvent addition) or by chemical or catalytic processes. One of the bio-oil improvement methods is hydrodeoxygenation which is conducted in the presence of a suitable heterogeneous catalyst at around 350–600°C under a high-pressure H_2 environment. The C–C bonds present in the bio-oil are saturated by the reaction of oxygen with the added H_2 to form water. This results in higher stability, and the increase in the H/C ratio of the fuel leads toward higher energy content (Huber et al., 2006).

Hydrothermal gasification of biomass

It is a thermochemical conversion method of biomass at elevated temperature under pressurized water into gaseous products (Elliott, 2011). In this process, water acts both as a solvent and as a reactant at near-supercritical conditions. The decomposition of biomass in the presence of supercritical water causes the dissipation of reactive species, thereby ensuring a higher gas (CH_4 and H_2) yield (Savage et al., 2010). The process can be broadly classified into the following categories based on the target product:

- i Aqueous phase refining: production of gas at a subcritical temperature using a catalyst.
- ii Catalytic gasification in a near-critical state: production of CH_4-rich gas at a temperature between critical point and 500°C using a catalyst
- iii Supercritical water gasification (SCWG): production of H_2-rich gas at a temperature above 500°C without a catalyst or non-metallic catalyst

In the HTG process, very wet biomass can be utilized as a feedstock as the presence of water is desirable during the processing. It contributes toward the selection of a large variety of feedstocks for thermochemical processing and is economical due to energy savings in the drying operation of biomass. The advantages of using wet biomass and bio-waste in the HTG process over anaerobic digestion (AD) include rapid reaction period, use of selective catalysts, ease of deployment and overall flexibility regarding process design for specific products (Elliott, 2011). HTG of biomass is carried out at both the subcritical and supercritical regions of water. However, HTG under supercritical water conditions enhances the H_2 production. The gaseous products obtained from HTG include CO, H_2, CO_2 and CH_4, with trace amount of C_2H_4 and C_2H_6, which can be easily handled due to the less complex structure compared to liquid (oil) and solid products (ash, char) obtained from HTL, HTC and pyrolysis.

HTG can be performed in both batch and continuous modes of operation. In the batch process, experiments were carried out with different catalysts and concentrations, whereas the continuous process helps study the reaction kinetics. At a lower temperature in the range of 215–265°C, aqueous phase refining takes place in the presence of a heterogeneous catalyst to yield H_2 and CO_2. This process is not very much desirable unless in situ H_2 is utilized for biomass hydrogenation. At the near-critical state of water (350–500°C), biomass gasification leads to the production of CH_4 and CO_2 in the presence of a heterogeneous catalyst, whereas CO produced during the process undergoes further hydrogenation to produce CH_4. Lastly, SCWG occurs in between 600°C and 700°C to produce CO_2 and H_2 with or without a catalyst. It is primarily desired for biomass feedstocks with higher moisture content (above 30%) (Elliott et al., 1993).

In HTG technology, lignin present in biomass is broken down in hydrothermal media into low molecular weight phenolic compounds (Savage et al., 2010). Formaldehyde formed during elimination reactions undergoes condensation reactions with alkyl phenol molecules. Subsequently, higher molecular weight compounds result from polymerization of alkyl phenols where formaldehyde works as a binding agent. The presence of lignin, therefore, reduces gas production and escalates the yield of biochar. However, under the supercritical condition and in the presence of ruthenium catalysts, lignin along with cellulose can be gasified entirely to yield CO_2 and CH_4.

Figure 7.3 provides a schematic of a commercially viable HTG process with a continuous reactor. In a continuous gasification system, biomass is fed as an aqueous solution or slurry using high-pressure pumps. The additional heat energy generated is utilized in preheating the feedstock and for power generation. A pressure transducer is installed for monitoring effluent pressure before sending to a gas–liquid separator. H_2 gas is separated from the effluent which is then sent for upgrading to pressure swing adsorption (PSA) unit for removal of CO_2. At the same time, a membrane separation process, such as a polymer electrolyte membrane (PEM) electrolysis, is used to remove

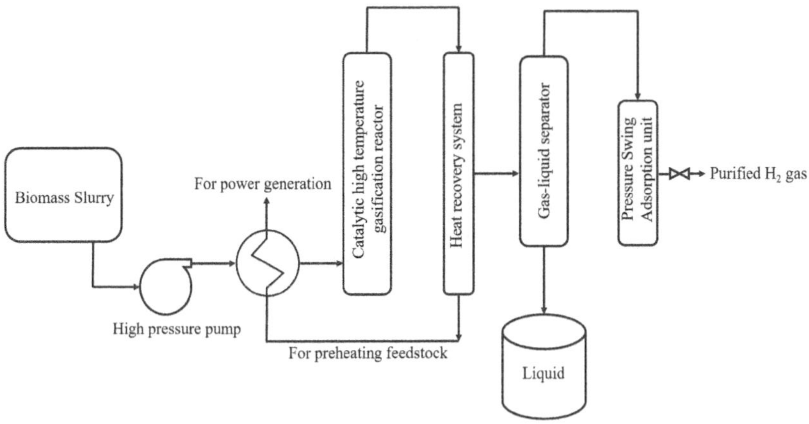

Figure 7.3 Schematic of continuous processing of biomass under HTG for hydrogen production. Reproduced with permission from Onwudili (2014).

CH_4 and other undesirable gases. In the HTG process, catalysts play a vital role which includes product selectivity and reduces the severity of the process conditions.

Production of H_2 via HTG involves a series of complex chemical reactions, as shown in Figure 7.4. Under the supercritical conditions, reforming of biomass includes hydrolysis followed by degradation to produce glucose and other monomers. The monomers undergo further reforming, which gives simple organic acids. These organic acids undergo tertiary reforming producing gaseous products such as CH_4, CO, H_2 and CO_2. To increase the yield of H_2, CO and CH_4 are separated from the gas mixture. Water–gas shift reaction converts the separated CO to H_2, whereas methane reforming also leads to the production of hydrogen. Studies reveal that important intermediates that are responsible for gas formation consist of aldehydes and small organic acids such as formic acid, acetic acid, levulinic acid, lactic acid, formic aldehyde and acetic aldehyde (Onwudili, 2014).

There are different categories of catalysts utilized for HTG, namely metallic, alkaline and carbonaceous catalysts. At subcritical water conditions, it is necessary to use a suitable catalyst such as nickel, rubidium, rhodium and alkaline salts for HTG. Elliott and his team have suggested nickel (Ni) and ruthenium (Ru) as the most effective metallic catalysts for this process (Elliott et al., 1993; Elliott, 2011). Ni was prominently used over Ru due to its lower cost, although Ru has higher stability under hydrothermal conditions. The gasification process is enhanced by the use of Ni as catalyst, but the catalyst often gets deactivated. A research team at Kyoto University in Japan developed a catalyst on carbonaceous material with Ni in the dispersed phase to achieve higher efficiency at moderately low temperatures (around 200°C) (Matsumura, 2015).

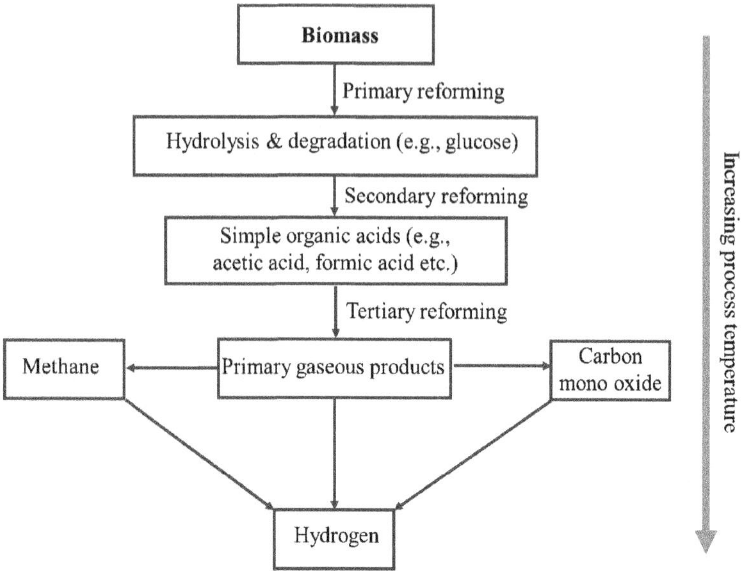

Figure 7.4 Schematic of the intermediate stages of hydrogen production from biomass using HTG. Reproduced with permission from Onwudili (2014).

Alkali catalysts also enhance gasification efficiency and lower char production. However, the yield of CO_2 during the process results in the deposition of carbonate salt within the apparatus. The difficulty of recovery of the alkaline catalyst leads toward the use of ceramics, a metal oxide solid catalyst. Metal catalysts are normally utilized around 400°C due to their deactivation problem at high temperatures, whereas the high stability of alkali catalysts makes them to be used up to 600°C. Sometimes, quartz capillary reactors are employed to prevent the metal reactor walls from the effect of catalyst. Later, it has been witnessed that activated carbon can be used in supercritical conditions for HTG. Researchers have observed suspended activated carbon particles as an effective catalyst present in the feedstock slurry. They have treated various feedstocks with this catalyst and suggested that its effectiveness varied with feedstock type.

At supercritical water conditions (around 600°C), H_2-rich gas is achieved by complete conversion of biomass. The effect of temperature on product gas yield and composition is quite significant at supercritical water conditions, while the effect of pressure is negligible. The efficiency of this process is, however, lower than that of conventional gasification processes such as steam reforming, but it is a carbon-neutral process. The high temperature in HTG favors the H_2 production and is carried out usually at 400–600°C. However, for the sustainable conversion of biomass, optimum process conditions should be chosen between H_2 yield and production cost.

HTG has remarkable benefits over conventional gasification processes which yield low-quality syngas, and the presence of impurities such as tar or char leads to clogging problems. This technology, therefore, excludes the cost of purification of low-quality syngas obtained from the traditional processes (Kirkels and Verbong, 2011). Moreover, it has several other advantages, such as the utilization of wet biomass as feedstock with suitable thermal efficiency, H_2-rich gas yields and reduced tar and soot formation (Schmieder et al., 2000). The products of HTG have widespread applications, such as power generation, chemical synthesis and fuel cells. Furthermore, the efficiency and overall cost of the HTG process can be improved soon with the help of rigorous research.

Hydrothermal carbonization of biomass

It is a comparatively new technology that primarily converts biomass into stable biochar and other valuable products under saturated pressure of 2–10 MPa and at a temperature of 180–250°C (Funke et al., 2013). The product obtained has carbon content in the lignite range, and the mass yield is about 35% to 60% depending upon process conditions (Hoekman et al., 2011; Xiao et al., 2012). Besides the solids, the aqueous phase obtained during the process contains mostly dissolved organics with a minimum amount of gas. The process largely depends on the nature of feedstock, solid loading and operating conditions. Compared to the conventional carbonization technique, the HTC process requires lesser energy and time for a higher yield.

A lower carbonization temperature and liquid-phase synthesis of abundantly available biomass and use of other chemicals employing catalysts or nanoparticles make the process very popular in recent years. HTC improves calorific value and dewatering capacity, thereby enhancing the solid biochar yield and process efficiency. However, to have a higher yield of solids such as char, the operating conditions such as temperature and pressure must be adequately controlled. In the HTC process, almost complete conversion of carbon takes place from the biomass feedstock to the biochar with the carbon conversion efficiency of approximately 100%. The application of HTC has increased from primary fuel source to the manufacturing of supercapacitors, fuel cells to the fields such as material science and nanotechnology.

Many researchers have performed HTC processing of biomass to convert cellulose, hemicellulose and carbohydrates into carbon-rich materials (Titirici et al., 2008). In the HTC process, carbohydrates present in the feedstocks are generally dehydrated into 5-hydroxy methyl furfural or furfural, followed by polymerization of furfurals to polyfurans. Finally, the intermediate products such as polyfurans are further dehydrated at the intermolecular stage for carbonization.

During the HTC process, hydrolysis of oligosaccharides such as sucrose and starch results in the formation of respective monosaccharides, i.e., glucose and fructose (Oomori et al., 2004). The monosaccharides formed undergo further reactions including C–C bond breaking in the ring structure, followed by dehydration to form furfural. Further decomposition of these compounds generates organic acids having lower molecular weight such as aldehydes, acids and phenols, resulting in decreased pH (Sinag et al., 2003). Subsequently, the decomposed compounds and monomers undergo condensation and polymerization reactions which are enhanced by aldol condensation or intermolecular dehydration and polymer aromatization. With an increase in concentration toward the critical supersaturation point, chemical species diffuses at the surface after nucleation occurs. Finally, the process completes with a thick suspension of powdered coal and water formed in the reactor. The bio-coal thus formed has a crystal structure. Some additives such as bases or acids can be added to the HTC process, which may catalyze the hydrolysis process leading toward higher yield and better product composition (Krylova and Zaitchenko, 2018).

Spherical carbonaceous particles in the micrometer range are obtained from HTC of carbohydrates which are hydrophilic and readily disperse in water. The size of the product particles depends on residence time, solid loading and precursor concentration. Besides, product distribution is affected by solid loading. HTC is primarily used for the production of char due to its higher energy content, good hydrophobicity and better grindability. The char obtained from this process is less dusty, which enhances pelletization characteristics.

Hydrochar and pelletization

The hydrochar obtained by HTC can act as an effective binder for other biomass feedstocks due to the presence of furan and phenolic resins. The quality of the pellets is determined by mechanical durability, which is the capability to remain intact in the course of storage and transportation. High carbonization temperature enhances the durability of the pellets. However, at a higher temperature, it becomes more brittle. On the other hand, the pellets formed at a temperature lower than 200°C are less durable and usually swell when submerged in water. The study by Liu et al. (2014) suggests that blended hydrochar/lignite pellets with more than 50% hydrochar fraction improve tensile strength, durability and combustion characteristics. However, the crumbling of pellets leads to reduced combustion efficiency and increased emissions.

The commercial-scale HTC process has suffered from higher operational costs as the process requires high temperature and pressure. Recirculation of spent liquor enhances the energy, and mass yield of hydrochar, thereby increasing the process economy. Shen and his team performed a pilot-scale study, and they have suggested that nearly 80% of the energy is recovered by the char produced from the original feedstock (Shen et al., 2017).

Hydrothermal gasification of algal biomass

Hydrothermal processing is a very inspiring technology for the utilization of algae in wet condition without incorporating energy-intensive pre-drying operation. Many researchers have demonstrated HTC of algal biomass yielding biochar at a pressure of less than 2 MPa and temperature of about 200°C having a residence time shorter than 1 h, and the produced char has properties similar to bituminous coal. Microalgae have also been utilized for the generation of bio-oils (such as bio-crude) by the process of HTL. Biller and Ross (2012) have investigated the effect of alumina-supported metal catalysts in the bio-oil yield (40 wt.%) from microalgae at a temperature of 300–350°C having a fuel value of 39 MJ/kg. Generation of bio-oil from microalgae accelerates with an increase in lipid content; however, the yield is much lower for macroalgae due to higher carbohydrate and mineral matter contents. Anastasakis and Ross (2011) have reported that catalytic HTL in the presence of KOH of brown macroalgae produces bio-oil (19.3 wt.%) with a calorific value of 36.5 MJ/kg. Similarly, HTG can develop as a low-cost process of H_2 generation in comparison with electrolysis as well as conventional gasification. In addition, the opportunity of CO_2 trapping makes the process more effective, and the overall efficiency is also higher.

Researchers have found that the efficiency of HTG of microalgae is generally higher than that of HTL as high protein is utilized and it produces nitrogen-free fuel. The HTG reduces organics in the aqueous phase, which increases the gasification efficiency of the carbonaceous material. Metallic catalysts such as Ni and Ru have been used for the HTG of algae. However, marine microalgae and macroalgae offer a lower conversion efficiency in comparison with freshwater algae. The variation in conversion efficiencies highlights the significance of segregating algal biomass for hydrothermal technologies into liquid fuels, char or gaseous fuels such as hydrogen or syngas (Onwudili et al., 2013).

HTG of microalgal biomass is an eco-friendly conversion involving partial oxidation in steam yielding a mixture of combustible gases, at high temperature. Onwudili et al. (2013) have evaluated reaction products of various algal biomass feedstocks from catalytic HTG to estimate their suitability for power generation as well as nutrient recycling. The gaseous product constitutes H_2, CO_2, CH_4 and N_2, which can be used to generate power using a diesel engine or gas turbine (Kumar et al., 2017).

The microalgal cellular monomers such as carbohydrates (e.g., glucose or xylose) decompose and under ionic conditions isomerize to yield fructose (Kabyemela et al., 1999; Matsumura et al., 2006). Further decomposition at a temperature of 300–400°C under free-radical conditions yields intermediates as smaller organic acids and furfurals. Dehydration and decarboxylation at temperatures of 450–650°C reduce smaller organic acids to gaseous products. The furfurals then undergo polymerization reaction to form char under subcritical conditions. Amino acids, on the other hand, undergo hydrolysis in both subcritical and supercritical states into smaller amino compounds besides acid intermediates releasing NH_3 gas. The intermediates then decompose further to yield smaller intermediates and gases. At a temperature of 500°C, a black solid product is also observed along with liquids and gases.

Various products such as solid char, bio-oils and gases are obtained during the hydrothermal treatment process based on the operating parameters. It is, therefore, essential to extract the products as well as analyze them. The liquid samples obtained during the hydrothermal process are required to be separated from the char. After the separation, the reactor is rinsed with deionized water followed by a known quantity of dichloromethane (DCM) to dissolve the obtained char. The mixture of aqueous residual and DCM is then filtered to segregate solid residuum from liquid filtrates under vacuum. The filter medium is cleaned with additional DCM to ensure complete separation of tar into the liquid phase. DCM is then separated from the aqueous layer and is dried using anhydrous sodium sulfate. The tar produced is further dissolved in DCM for analysis in gas chromatography (GC). The key constituents present in tar are identified using GC/MS by comparing the mass spectral data (Onwudili et al., 2013).

The carbon balance after hydrothermal reactions is usually carried out by elemental analysis of the solid residuum, tar or oil obtained and water-soluble products. Water-soluble products are the solid residues settled after evaporation of the liquid phase comprising both inorganic and organic residuals. The analysis of the gaseous products is achieved using offline gas chromatography. Various constituents such as H_2, O_2, CO, N_2 and CO_2 are analyzed using GC fixed with thermal conductivity detectors (GC/TCD). However, C_1–C_4 gases are analyzed using GC connected with a flame ionization detector (GC/FID). The gas chromatographs give the mole percent of individual gases in the product gas (Onwudili et al., 2013). The char obtained during the hydrothermal process is analyzed elementally using a CHNSO analyzer. The morphological study of the surface of the raw feedstock and char is conducted using a field emission scanning electron microscope (FESEM). Surface area and total pore volume of the char produced are determined for a better understanding of the products (Safari et al., 2016).

Conclusions

This chapter discusses the potential of various hydrothermal processing technologies for the generation of fuel and valuable chemicals. A large number of biomass feedstocks have been tested by researchers to obtain the desired product. This process directly utilizes high-moisture biomass feedstock, thereby excluding the energy-intensive drying operation making it more efficient. Hydrothermal processing of biomass is usually performed by either liquefaction, gasification or carbonization. Quality and yield of the hydrothermal process depend upon various factors such as feedstock type, solvent condition, catalyst, residence time and operating conditions. Therefore, it is crucial to select an appropriate process based on the desired final product with suitable operating conditions, proper selection of catalyst and water condition for the process. Like the other thermochemical processes, the hydrothermal process also requires separation, extraction and analysis of products. But, this technology provides a prominent way for the utilization of various types of biomass materials and its capability to operate at moderate pressure and temperature than the conventional processes.

The production of biofuels from microalgal species under hydrothermal conditions has received significant attention contributing to the fast growth rate of microalgae, CO_2 capture and little competition with food crops. Likewise, algal-based H_2 production and other value-added chemicals play a significant role in the energy industry, and it has shifted the industry toward large-scale production and the concept of biorefinery. However, the hydrothermal route of conversion has few disadvantages, such as the inability to handle a huge amount of water which requires complex and expensive reactors. Hence, the subject requires further research and experimentation for efficient and successful commercial applications.

References

Akhtar, J. and Amin, N.A.S. (2011). A review on process conditions for optimum bio-oil yield in hydrothermal liquefaction of biomass. *Renewable and Sustainable Energy Reviews* 15:1615–1624.

Anastasakis, K. and Ross, A.B. (2011). Hydrothermal liquefaction of the brown macro-alga *Laminaria saccharina*: Effect of reaction conditions on product distribution and composition. *Bioresource Technology* 102:4876–4883.

Biller, P. and Ross, A.B. (2012). Hydrothermal processing of algal biomass for the production of biofuels and chemicals. *Biofuels* 3:603–623.

Bobleter, O. (1994). Hydrothermal degradation of polymers derived from plants. *Progress in Polymer Science* 19:797–841.

Briens, C., Piskorz, J. and Berruti, F. (2008). Biomass valorization for fuel and chemicals production – A review. *International Journal of Chemical Reactor Engineering* 6(1). Doi: 10.2202/1542-6580.1674

Dimitriadis, A. and Bezergianni, S. (2017). Hydrothermal liquefaction of various biomass and waste feedstocks for biocrude production: A state of the art review. *Renewable and Sustainable Energy Reviews* 68(1):113–125.

Elliott, D.C. (1980). Process development for biomass liquefaction. *American Chemical Society, Division of Fuel Chemistry. Las Vegas, USA* 25(4):257–263.

Elliott, D.C. (2011). Hydrothermal Processing. In: Brown, R.C. (ed.) *Thermochemical Processing of Biomass: Conversion into Fuels, Chemicals and Power.* John Wiley & Sons Ltd, pp. 200–231.

Elliott, D.C., Sealock, L.J. and Baker, E.G. (1993). Kinetics, catalysis, and reaction engineering. *Industrial & Engineering Chemistry Research* 32(8):1542–1548.

Erlach, B., Harder, B. and Tsatsaronis, G. (2012). Combined hydrothermal carbonization and gasification of biomass with carbon capture. *Energy* 45(1):329–338.

Funke, A., Reebs, F. and Kruse, A. (2013). Experimental comparison of hydrothermal and vapothermal carbonization. *Fuel Processing Technology* 115:261–269.

Guo, Y., Wang, S.Z., Xu, D.H., Gong, Y.M., Ma, H.H. and Tang, X.Y. (2010). Review of catalytic supercritical water gasification for hydrogen production from biomass. *Renewable and Sustainable Energy Reviews* 14(1):334–343.

Heidari, M., Dutta, A., Acharya, B. and Mahmud, S. (2019). A review of the current knowledge and challenges of hydrothermal carbonization for biomass conversion. *Journal of the Energy Institute* 92(6):1779–1799.

Hoekman, S.K., Broch, A. and Robbins, C. (2011). Hydrothermal carbonization (HTC) of lignocellulosic biomass. *Energy and Fuels* 25:1802–1810.

Huber, G.W., Iborra, S. and Corma, A. (2006). Synthesis of transportation fuels from biomass: Chemistry, catalysts, and engineering. *Chemical Reviews* 106:4044–4098.

Kabyemela, B.M., Adschiri, T., Malaluan, R.M. and Arai, K. (1999). Glucose and fructose decomposition in subcritical and supercritical water: Detailed reaction pathway, mechanisms, and kinetics. *Industrial & Engineering Chemistry Research* 38:2888–2895.

Kambo, H.S. and Dutta, A. (2015). A comparative review of biochar and hydrochar in terms of production, physico-chemical properties and applications. *Renewable and Sustainable Energy Reviews* 45:359–378.

Karagöz, S., Bhaskar, T., Muto, A., Sakata, Y. and Uddin, M.A. (2004). Low-temperature hydrothermal treatment of biomass: Effect of reaction parameters on products and boiling point distributions. *Energy and Fuels* 18:234–241.

Kirkels, A.F. and Verbong, G.P.J. (2011). Biomass gasification : Still promising ? A 30-year global overview. *Renewable and Sustainable Energy Reviews* 15(1):471–481.

Krylova, A.Y. and Zaitchenko, V.M. (2018). Hydrothermal carbonization of biomass: A review. *Solid Fuel Chemistry* 52:91–103.

Kumar, G., Shobana, S., Chen, W.H., Bach, Q.V., Kim, S.H., Atabani, A.E. and Chang, J.S. (2017). A review of thermochemical conversion of microalgal biomass for biofuels: Chemistry and processes. *Green Chemistry* 19(1):44–67.

Kumar, M., Olajire Oyedun, A. and Kumar, A. (2018). A review on the current status of various hydrothermal technologies on biomass feedstock. *Renewable and Sustainable Energy Reviews* 81(2):1742–1770.

Liu, Z., Quek, A. and Balasubramanian, R. (2014). Preparation and characterization of fuel pellets from woody biomass, agro-residues and their corresponding hydrochars. *Applied Energy* 113:1315–1322.

Maddi, B., Panisko, E., Wietsma, T., Lemmon, T., Swita, M., Albrecht, K. and Howe, D. (2017). Quantitative characterization of aqueous byproducts from hydrothermal liquefaction of municipal wastes, food industry wastes, and biomass grown on waste. *ACS Sustainable Chemistry and Engineering* 5(3):2205–2214.

Matsumura, Y. (2015). Hydrothermal gasification of biomass. In: *Recent Advances in Thermochemical Conversion of Biomass*. Elsevier, pp. 251–267.

Matsumura, Y., Yanachi, S. and Yoshida, T. (2006). Glucose decomposition kinetics in water at 25 MPa in the temperature range of 448–673 K. *Industrial & Engineering Chemistry Research* 45(6):1875–1879.

McKendry, P. (2002). Energy production from biomass (part 1): Overview of biomass. *Bioresource Technology* 83(1):37–46.

Nielsen, R.P., Olofsson, G. and Søgaard, E.G. (2012). CatLiq - High pressure and temperature catalytic conversion of biomass: The CatLiq technology in relation to other thermochemical conversion technologies. *Biomass and Bioenergy* 39:399–402.

Nonaka, M., Hirajima, T. and Sasaki, K. (2011). Upgrading of low rank coal and woody biomass mixture by hydrothermal treatment. *Fuel* 90(8): 2578–2584.

Onwudili, J.A. (2014). Hydrothermal gasification of biomass for hydrogen production. In: Jin, F. (ed.) *Application of Hydrothermal Reactions to Biomass Conversion*. Green Chemistry and Sustainable Technology. Springer-Verlag Berlin Heidelberg, pp. 219–246.

Onwudili, J.A., Lea-Langton, A.R., Ross, A.B. and Williams, P.T. (2013). Catalytic hydrothermal gasification of algae for hydrogen production: Composition of reaction products and potential for nutrient recycling. *Bioresource Technology* 127:72–80.

Oomori, T., Khajavi, S.H., Kimura, Y., Adachi, S. and Matsuno, R. (2004). Hydrolysis of disaccharides containing glucose residue in subcritical water. *Biochemical Engineering Journal* 18:143–147.

Peterson, A.A., Vogel, F., Lachance, R.P., Fröling, M., Antal, M.J. and Tester, J.W. (2008). Thermochemical biofuel production in hydrothermal media: A review of sub- and supercritical water technologies. *Energy & Environmental Science* 1:32–65.

Safari, F., Norouzi, O. and Tavasoli, A. (2016). Hydrothermal gasification of *Cladophora glomerata* macroalgae over its hydrochar as a catalyst for hydrogen-rich gas production. *Bioresource Technology* 222:232–241.

Saisu, M., Sato, T., Watanabe, M., Adschiri, T. and Arai, K. (2003). Conversion of lignin with supercritical water-phenol mixtures. *Energy and Fuels* 17(4):922–928.

Sasaki, M., Furukawa, M., Minami, K., Adschiri, T. and Arai, K. (2002). Kinetics and mechanism of cellobiose hydrolysis and retro-aldol condensation in subcritical and supercritical water. *Industrial & Engineering Chemistry Research* 41(26): 6642–6649.

Savage, P., Levine, R. and Huelsman, C. (2010). Hydrothermal processing of biomass. In: Crocker, M. (ed.) *Thermochemical Conversion of Biomass to Liquid Fuels and Chemicals*. RSC Publishing, pp. 192–215.

Schaleger, L.L., Figueroa, C. and Davis, H.G. (1982). Direct liquefaction of biomass: Results from operation of continuous bench scale unit in liquefaction of water slurries of Douglas fir wood. *Symposium on Biotechnology in Energy Production and Conservation,* Gatlinburg, Tennessee, USA.

Schmieder, H., Abeln, J., Boukis, N., Dinjus, E., Kruse, A., Kluth, M., Petrich, G., Sadri, E. and Schacht, M. (2000). Hydrothermal gasification of biomass and organic wastes. *The Journal of Supercritical Fluids* 17(2):145–153.

Sealock, L.J., Elliott, D.C., Baker, E.G. and Butner, R.S. (1993). Chemical Processing in high-pressure aqueous environments. 1. Historical perspective and continuing developments. *Industrial & Engineering Chemistry Research* 32:1535–1541.

Shen, Y., Yu, S., Ge, S., Chen, X., Ge, X. and Chen, M. (2017). Hydrothermal carbonization of medical wastes and lignocellulosic biomass for solid fuel production from lab-scale to pilot-scale. *Energy* 118(1):312–323.

Sinag, A., Kruse, A. and Schwarzkopf, V. (2003). Formation and degradation pathways of intermediate products formed during the hydropyrolysis of glucose as a model substance for wet biomass in a tubular reactor. *Engineering in Life Sciences* 3(12):469–473.

Tekin, K. and Karagöz, S. (2013). Non-catalytic and catalytic hydrothermal liquefaction of biomass. *Research on Chemical Intermediates* 39:485–498.

Tekin, K., Karagöz, S. and Bektaş, S. (2014). A review of hydrothermal biomass processing. *Renewable and Sustainable Energy Reviews* 40:673–687.

Titirici, M.M., Antonietti, M. and Baccile, N. (2008). Hydrothermal carbon from biomass: A comparison of the local structure from poly- to monosaccharides and pentoses/hexoses. *Green Chemistry* 10:1204–1212.

Toor, S.S., Rosendahl, L. and Rudolf, A. (2011). Hydrothermal liquefaction of biomass: A review of subcritical water technologies. *Energy* 36(5):2328–2342.

Tran, K.Q. (2016). Fast hydrothermal liquefaction for production of chemicals and biofuels from wet biomass – The need to develop a plug-flow reactor. *Bioresource Technology* 213:327–332.

Xiao, L.P., Shi, Z.J., Xu, F. and Sun, R.C. (2012). Hydrothermal carbonization of lignocellulosic biomass. *Bioresource Technology* 118:619–623.

Xu, X., Matsumura, Y., Stenberg, J. and Antal, M.J. (1996). Carbon-catalyzed gasification of organic feedstocks in supercritical Water. *Industrial & Engineering Chemistry Research* 35(8):2522–2530.

Yu, D., Aihara, M. and Antal, M.J. (1993). Hydrogen production by steam reforming glucose in supercritical water. *Energy and Fuels* 7(5):574–577.

Zhang, L., Xu, C.C. and Champagne, P. (2010). Overview of recent advances in thermo-chemical conversion of biomass. *Energy Conversion and Management* 51(5):969–982.

Zhang, Y. (2010). Hydrothermal liquefaction to convert biomass into crude oil. In: Blaschek, H.P., Ezeji, T.C., Scheffran, J. (eds.) *Biofuels from Agricultural Wastes and Byproducts*. Blackwell Publishing, pp. 201–232.

8 Production of bio-jet and synthetic fuels via catalytic and supercritical fluid upgrading of algal bio-crude oil

Vikranth Volli, Anjani Ravi Kiran Gollakota, Naga Chaitanya Kavuri, and Chi-Min Shu

Introduction

The dependence on conventional fossil fuels due to the rapid increase in population, industrialization and exigent energy demand has driven the growing interest in the utilization of bio-based non-fossil fuels for energy. The combustion of fossil fuels results in the emission of greenhouse gases (GHGs), including chlorofluorocarbons, CO, CO_2 and NO_x that pose a significant threat to the environment. This led to an alarming level of energy consumption from sources such as coal, petroleum and natural gas, which grew at a rate of 2.9% in 2019, a twofold increase (10-year average of 1.5%/year) since 2010. More than two-thirds of the global energy demand was accounted for by China, the USA and India together. The consumption of coal, petroleum, natural gas, electricity and renewables increased by 1.4% (162 mtoe), 1.5% (1.4 million barrels/d), 5.3% (195 billion cubic meters), 3.7% and 14.5% (71 mtoe), respectively, by 2018 (Eurostat, 2015).

The steady demand for crude oil (2.2 million barrels/day) accounting for half of the overall growth by 2020 was driven by the increase in importance for petroleum products (fuel oil, motor gasoline, liquefied petroleum gas, kerosene-type jet fuel, gas oil and diesel) and the average refinery utilization scaled to its highest since 2007. By 2030, particularly in the transport sector, the annual growth rate is expected to increase by 1.3%. The principal consumers of petroleum products within the transport sector were road transport, non-energy purposes (e.g., lubricants, heating oils, bitumen for road surfaces, etc.), shipping and aviation.

The sustained increase of air transportation has assimilated a substantial growth in the modern world, improving business and trade worldwide. Approximately 142 million metric tons of jet fuel was used in 2010, and with an increase in air traffic (factor of 4.2), this consumption is expected to increase by 2.8 times by 2040, expanding the jet fuel cost by \$2.82/gallon (ICAO, 2017). The intent of the international aviation sector (39th session of the ICAO) was to maintain net carbon emissions from 2020 at the same level and improve fuel efficiency by 2% per annum. ExxonMobil, Shell, Chevron,

British Petroleum, Valero and Marathon Petroleum Group produce roughly 50% of total US jet fuel production. The typical prerequisites of aviation drop-in fuel as per ASTM standards were density at 15°C of 775–840 g/L, flash point (minimum temperatures at which vapors ignite in the presence of flame source) of minimum 38°C, freezing point (at lower temperatures, wax crystals are formed and completely disappear when the fuel is rewarmed) maximum of −47°C, aromatic content (smoke and soot formation) maximum of 25 vol%, sulfur content maximum of 0.3 wt.% and heating value of 42.8 MJ/kg (Prakash and Davies, 2020). However, factors such as increased jet fuel costs, environmental effects (noise and GHG emissions), safe working of the aircraft engine, consistency and higher efficiency have inspired the scientific community in the development of alternatives to conventional aviation fuels.

The use of aviation biofuels as an alternative to conventional fossil stock could be a feasible solution as biofuels are renewable and have low carbon and GHGs, their overall cost benefits are higher, and they are flexible to current engine designs and can fulfill the essential properties of conventional aviation fuels. The different sources of raw materials used for biofuel production include oil, sugar, starch and lignocellulosic and algae-based feedstocks, which are mostly biomass. They are socially and environmentally viable sources that offer monetary returns in developing countries (availability of cultivable land). It can be beneficial for biodiversity where abandoned and moderately degraded land can be systematically rehabilitated to produce algal biomass. The suitable zoning of urban and agro-industrial residues, infrastructure, legal and regulatory requirements can help address the problems related to waste management in urban areas. However, feedstock segregation, manufacturing technologies and the distinctive characterization of the produced fuel are exceedingly complex.

In this chapter, a brief explanation of different routes for the production of biofuels from various algal raw materials is presented. The role of macroalgae and microalgae, in particular as feedstock for the production of bio-crude; their properties; extraction techniques; and their limitations are discussed in detail. In addition, the upgrading of algal bio-crude via catalytic cracking, hydroprocessing and supercritical systems, and their reaction mechanism are assessed. Furthermore, opportunities, challenges and the prospects of macroalgae- and microalgae-based upgraded bio-jet fuels are also included.

Production of biofuels and bio-crudes from micro- and macroalgae

Conventional and unconventional sources

Most of the aviation fuel used today (80% of energy consumption) is produced from conventional fossil crude (deposits of microscopic marine organisms transformed at high temperature and pressure located at the bottom of the oceans) via traditional petroleum refining techniques. In this process, crude oil is heated in fractionation columns at different temperatures and condensed back to various fractions such as naphtha (motor gasoline), middle distillate (diesel, kerosene, heating oil, special solvents) and bottom/residual fraction (boiler fuel). Jet fuels from crude oil are separated as a petroleum fraction in the temperature range of 205–260°C that is rich in paraffin and aromatics (20–40% of n/iso-paraffins, naphthenes and aromatics). However, lower aromatic content would result in leakage of the fuel in aircraft engines (Liu et al., 2013). The research and development on synthetic paraffinic kerosene (SPK) and synthetic iso-paraffins (SIP) liquids from coal (CTL), natural gas (GTL) and biomass (BTL) using the Fischer–Tropsch (FT) process have been extensively studied. The conversion of coal to liquids is based on liquefaction that mainly comprises three stages, including the production of syngas by gasification, FT reaction followed by upgrading of hydrocarbons, and separation of beneficial chemicals in the temperature range of 500–1,200°C and 30–50 bar pressure. However, the generation of CO_2 and SO_2 as the undesired FT gaseous products could significantly influence the hydrocarbon yields (De Toni et al., 2014).

The conversion of natural gas in the presence of cobalt and iron as a catalyst via FT-based GTL synthesis on producing long-chain hydrocarbons in the range of transportation fuels, middle distillates and lubricants was commercialized in the year 1993 (Jiang et al., 2016). Technologies such as autothermal reforming (ATR), partial oxidation (POX) and catalytic steam methane reforming (SMR) were used for the production of syngas at 500–950°C at a pressure of 2–4 MPa, followed by FT reaction, and upgrading resulting in the formation of liquefied petroleum gas, naphtha, gasoline, diesel and high-quality petrochemicals. Biomass can be converted to bio-crude via physical (e.g., extraction, transesterification and emulsification), thermal (e.g., combustion, pyrolysis, gasification and liquefaction) and biochemical (e.g., fermentation and anaerobic digestion) conversion (Volli et al., 2020). However, gasification/pyrolysis, followed by the FT process, is best suited for the production of bio-jet fuels. The operating temperatures and pressures were in the range of 900°C and 21 bar, respectively. However, more recent studies were focused on the production of bio-crude from biomass to liquids and algae (e.g., microalgae and macroalgae) that are discussed in detail in the subsequent sections.

The primary unconventional petroleum sources include Canadian oil sands, Venezuelan VHOs and oil shale (kerogen) that are bitumen (a mixture of hydrocarbons) in sand/porous rock located at Alberta (1.6 trillion barrels), Venezuela (1.2 trillion barrels) and Colorado (1 trillion barrels), respectively, in large concentrations. Surface mining, in situ extraction and vacuum pyrolysis, followed by conventional petroleum refining, are used for the production of jet fuel compatible with current aviation systems (Davidson et al., 2014). However, the use of nonpolar solvents, water pollution, increased GHG emissions (>10–25% emissions of crude oil), supercritical conditions and their technical and economic viability represent a severe environmental concern.

Microalgae- and macroalgae-based bio-crude oil

The dwindling crude oil reserves, questionable ecology and environmental impacts of increased CO_2 and GHG emissions have drawn substantial interest in the utilization of sustainable (solar, wind, hydro and geothermal) and economically feasible alternative sources as feedstock for bioenergy. Research on algal biomass for the production of bio-crude for aviation fuels is one among the top eyed researches nowadays. Algae are non-vascular photosynthetic organisms that store energy in the form of lipids, carbohydrates and proteins living in salt- and freshwaters. The potential of algae as a feedstock for biofuel production can be attributed to the following factors (Chisti, 2019):

- i Algae have simple cellular structure
- ii Algae can be harvested throughout the year
- iii Algae have higher lipid content, the production of which can be adjusted by changing the growth medium
- iv Algae can be cultivated in salt- and wastewaters to achieve higher growth rates
- v Algae consumes CO_2 for growth helping in carbon capture
- vi Algal biofuels are non-toxic and have negligible sulfur content
- vii Algal production systems are adaptable to current technological applications

Algal biofuels are considered third-generation fuels, are responsible for a significant photosynthetic activity on earth and are capable of producing high-commercial-value by-products. Based on cell structure and pigmentation, they were classified into two groups, macroalgae (*Phaeophyceae, Rhodophyceae* and *Chlorophyceae*) and microalgae (*Bacillariophyceae, Chlorophyceae, Chrysophyceae* and *Cyanophyceae*), with annual production capacities of 20 million tons in 2010 (Stephens et al., 2013). Macroalgae are often cultured in oceans and can reach up to 60 m in length, having a thallus, stem and foot. They produce a small number of lipids (0.3–6%) with medium yields and energy density, whereas microalgae (1–50 μm) are cultured in photobioreactors and open ponds with high yields of lipid content (>70%) and energy density.

The different processes for the production of algae-derived fuels are categorized into cultivation (tubular/flat plate/dome reactors, open ponds and sea); harvesting (flocculation, gravity settling, centrifugation and filtration); post-harvest treatment including pretreatment (drying, size reduction, sonication, bead beating, preservation and storage); extraction (solvent, chemical and supercritical CO_2); and high-temperature process (combustion, gasification, pyrolysis and liquefaction) (Wang et al., 2016). The possible routes for the production of aviation fuels using different feedstocks are represented in Figure 8.1. The algal-based conversion technologies are categorized as lipid, thermochemical and biochemical conversion. They are mostly influenced by the quantity and feedstock type, economy, desired end form of products and energy.

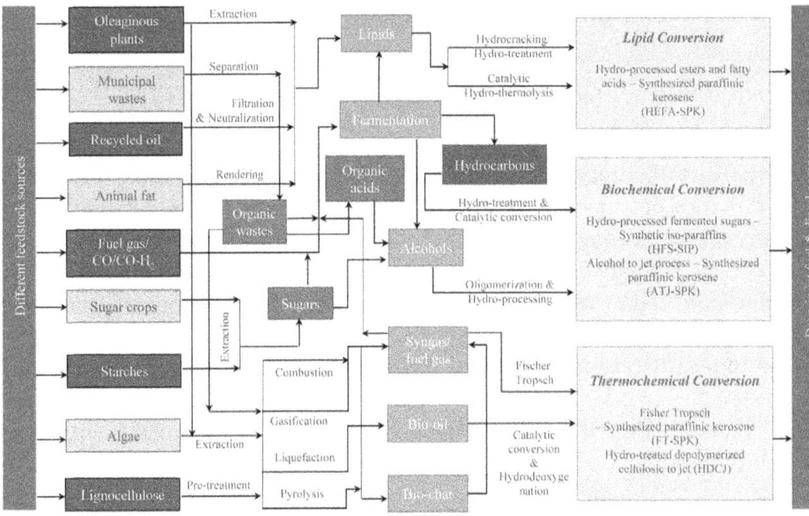

Figure 8.1 Block diagram for the possible routes in the production of jet fuels (reference: ICAO, 2017).

Lipid conversion (oil to jet, OTJ)

The primary composition of oil-bearing plants, recycled oils, animal fat and oleaginous residues is lipids. Hydrotreating is currently the best-known technology for the production of bio-crude and petroleum products that have similar characteristics to those of conventional aviation fuels. The reactions are classified into hydrocracking (e.g., destructive hydrogenation, isomerization and cracking of C–C bonds higher molecular weight components to lighter products) and hydrotreatment (e.g., nondestructive hydrogenation to improve the quality of petroleum distillates) (Sotelo-Boyas et al., 2012). The conventional hydrotreating process is used to refine sulfur, nitrogen and metals in petroleum products in the form of ammonia, H_2S and water. Hydrotreating of esters and fatty acid/triglycerides (HEFA) to hydrocarbon middle distillates (e.g., synthetic paraffinic kerosene, or SPK) can be used for the production of aviation fuels.

Conversion of lipids to bio-crudes via transesterification/alcoholysis is a reversible reaction similar to hydrolysis where different solvents (usually 6:1 ratio) (e.g., methanol, ethanol, butanol and propanol) and catalysts (3–10 wt.% of lipid) (e.g., homogeneous–heterogeneous acid, alkali, alkaline and enzymatic) at temperatures close to the boiling point of solvent were utilized. Hydrotreating of triglycerides offers flexibility in the use of low-quality feedstock and easier separation of by-products and catalyst. The key product of hydrotreating includes propane, n-aliphatic and paraffinic hydrocarbons (Raut et al., 2015). Distillation of the product yields naphtha, jet fuel and diesel with a maximum blend ratio of 50% from the HEFA-SPK process. However, the presence of contaminants (trace metals) in water and carbon oxide significantly affects the lifetime and recyclability of hydrotreating catalysts.

Biochemical methods (alcohol to jet, ATJ; and sugar to jet, STJ)

The biochemical conversion of algae typically utilizes enzymes as a catalyst to hydrolyze carbohydrates into starch/sugars that are subsequently fermented and distilled to produce aviation fuels and is currently in the initial stages of commercialization. Biochemical methods are classified as anaerobic digestion (bio-gasification) and alcoholic fermentation (Özçimen and Inan, 2015). The anaerobic conversion of organic matter to biogas via hydrolysis, fermentation, acetogenesis and methanogenesis reactions is called anaerobic digestion or bio-gasification. Alcoholic fermentation is the conversion of algal polysaccharides (e.g., *Ascophyllum*, *Porphyra* and *Palmaria* in macroalgae; and pectin, agar and alginate in microalgae) such as starch, sugar and cellulose to alcohols (ethanol or isobutanol), acids and a mixture of CO and H_2S gases in the absence of oxygen. Two routes were proposed in this process (Milledge and Harvey, 2016). They are hydroprocessed fermented sugars–synthetic iso-paraffins (HFS-SIP) and alcohol-to-jet synthetic paraffinic kerosene (ATJ-SPK). The HFS-SIP process involves direct fermentation of sugars to hydrocarbons from hydroprocessed esters and fatty acids.

The microorganisms produce substances such as farnesene, which is a SIP that can be converted into aviation fuel. A maximum blend of 10% can be used in aviation fuels from the HFS-SIP process. The ATJ-SPK process involves dehydration (removal of water), oligomerization (conversion of small molecules in more complex ones) and hydrogenation (addition of hydrogen) stages in the conversion of alcohols, such as isobutanol, into hydrocarbon. A maximum blending ratio of 50% can be used in aviation fuels from the ATJ-SPK process (Carvalho et al., 2011). However, preprocessing (mechanical comminution, acid, alkaline, organosolv, ammonia fiber explosion methods) and downstream processing involved can be energy-intensive. The process has limited productivity, requires higher reaction time and generates secondary wastes and sludge.

Thermochemical methods (gas to jet, GTJ; and pyrolysis to jet, PTJ)

The production of molecular precursors from algae for bio-crude is classified into four conventional methods. They are direct combustion, gasification, liquefaction and pyrolysis. They are mainly affected by moisture content and are dependent on specific feedstock characteristics. The combustion and gasification produce synthesis gas (CO, CH_4 and H_2), whereas pyrolysis and liquefaction produce bio-crude that can be further processed into bio-jet fuels (Yang et al., 2016).

Direct combustion is one of the conventional methods where dry biomass (moisture <50%) is used as a source of heat and electricity for household and industrial applications. In the case of dry algae, the heating values generated during combustion are in the range of 14–16 MJ/kg and were not extensively studied. Direct combustion of algae avoids the production of liquid fuels that often reduces the efficiency of the boilers. The ash and alkali content in algae is in the range of 3.5–46% that causes fouling issues during combustion. The direct combustion of algae (<0.18 mm) in fluidized-bed boilers could increase the heating value by 20% (Milledge and Harvey, 2016). The co-combustion of algae in coal-fired boilers significantly ameliorates the co-generation of heat and electricity. However, direct combustion of algae appears uneconomical due to lower heating values, solid inorganic residues and higher moisture content.

Gasification is a thermochemical conversion of carbonaceous materials in the temperature range of 800–1,000°C by POX in sub-stoichiometric concentration to produce combustible syngas mixture (H_2, CO, CO_2, CH_3, C_2H_4, N_2 and water vapor) with heating values in the range of 4–6 MJ/kg that can be catalytically reformed to long-chain hydrocarbons and liquid fuel via the FT process (Milledge et al., 2014). Conventional FT is categorized into low (210–260°C)- and high (310–340°C)-temperature processes where cobalt and iron are used as a catalyst at pressures of 2–3 MPa that supports the formation of linear hydrocarbons and alkenes/oxygenates, respectively. The water–gas shift reaction with H_2:CO ratio 2:1 is used for the formation of syngas that can be further upgraded to produce synthetic diesel and jet fuel fractions via hydrocracking (FT-SPK).

Various industrial gasification configurations include fixed-, fluidized- and moving-bed reactors with drying, devolatilization/pyrolysis and oxidation as different reaction stages. Literature studies suggest the availability of very few data on the gasification of algae. Gasification of *Laminaria digitata* and *Fucus serratus* (mixed macroalgae) resulted in syngas with low heating values. However, catalytic supercritical water gasification (400°C and 250 bar) of *Ulva lactuca* and *Saccharina latissima* produced syngas with higher methane content (Saber et al., 2016). The presence of NaOH, Ni, olivine and dolomite decreases C_2–C_4 yields increasing H_2 production. An increase in gasification efficiency by 58–92% was observed when Al_2O_3-supported Ru catalyst was used for *Saccharina latissima* feedstock. However, the use of a regenerated catalyst decreased H_2 and CH_4 yields. Co-gasification of algae with biomass-based feedstock could be a suitable alternative (Suganya et al., 2016). The complexity of the reactors, very high operating costs, the formation of unwanted by-products (e.g., ash, char and tars), the use of high-moisture feedstock, inefficient mixing and heat distribution are the problems associated with the effective implementation of gasification.

The conversion of biomass to stable liquid hydrocarbons at low temperatures (250–400°C) and high pressures (40–200 bar) in the presence of catalyst and hydrogen is called hydrothermal liquefaction (HTL). The products of liquefaction include bio-oil (10–73%), gas (8–20%) and char (0.2–0.5%) by mass. The steps involved in a batch-scale liquefaction process include feedstock preparation (slurry, catalyst and solvent), headspace reactor pressurization (H_2, N_2 and Ar), mixing and holding at selected residence time, cooling and product recovery (separation and solvent recovery) (Haider et al., 2018).

Feedstock pretreatment is not essential, and the use of sub-/supercritical waters demonstrated a significant increase in product yields reducing the operating costs. The reaction steps in liquefaction are as follows: depolymerization of biomass to monomers; bond cleavage, dehydration, decarboxylation, deamination of monomers to form new unstable fragments; polymerization, condensation and cyclization of unstable fragments to new compounds; and depolymerization and condensation with the change in temperature and time for stable product formation. HTL of algae has attracted much interest and is advantageous as high-moisture-content feedstock can be processed to produce bio-crude (O'Connell et al., 2019). HTL of macroalgae such as *Laminaria saccharina, Enteromorpha prolifera, Oedogonium, Cladophora* and *Derbesia* resulted in a bio-crude yield of 20–26% with energy recovery in the range of 56–79%, respectively. On the other hand, bio-oil yields from a few microalgal biomass are as follows: *Spirulina* (41%), *Scenedesmus* (24–45%), *Dunaliella* (37%) and *Desmodesmus* (49%) (Gollakota et al., 2018). However, co-liquefaction of microalgae and macroalgae increased the conversion by 50%, forming a synergistic effect. The optimization of process parameters such as feedstock composition, solvent, temperature, pressure, the atmosphere in the reactor and residence time is obligatory to evaluate the suitability of algal biomass for commercialization.

Thermal pyrolysis is a thermolytic technique where dry biomass is decomposed into organic components in the absence of oxygen. It is classified into slow (200–500°C and 0.01–2°C/min), fast (600–900°C and 10–200°C/min) and flash (>1,000°C and >200°C/min) pyrolysis with the primary component as syngas, bio-oil and solid char residue (Gollakota et al., 2016). Thermal pyrolysis of *Enteromorpha prolifera* at 300°C produced hydrocarbons and oxygenates in the presence of vacuum gas oil and ethanol. Thermal pyrolysis of dry chlorella resulted in bio-oil yield in the range of 50.8–57.9% with a heating value of 23.6 MJ/kg.

Fast pyrolysis of *Chlorella protothecoides* achieved bio-oil yields of 18–24 wt.% with a heating value of 30 MJ/kg. Thermal pyrolysis of *Laminaria digitata* and *Enteromorpha prolifera* in a free-fall pyrolysis reactor converted macroalgae into fuels and value-added chemicals. The process parameters that influence liquid product yields are temperature, residence time, particle size, heating rate and reactor configuration (fixed/fluidized/spouted fluid/circulating fluidized bed, rotating cone, vacuum and ablative) (Zhou and Hu, 2020). The conversion of pyrolysis oils via hydrotreated depolymerized cellulosic to jet (HDCJ) fuels involves the gasification of pyrolysis products with an oxidizing agent to produce syngas that is further processed to yield liquid hydrocarbon with varying chain length. These hydrocarbons are refined and separated to produce the desired jet fuels. The bio-oils produced during thermal pyrolysis have a high moisture content, high corrosiveness and a lower aromatic content (Taimbú et al., 2019). However, certain aromatic content is required in aviation fuels to avoid sealing problems in jet engines. Several research and development initiatives in developing pyrolysis oil to jet processing are still in the demonstration phase, and the ongoing research aims at maximizing energy potential from algal biomass to reach the stage of commercial development.

Bio-crude and need for upgrading

The bio-crude oils are dark brown, highly viscous, chemically complex liquids with a very strong odor and are significantly different from conventional petroleum fuels. The physicochemical properties depend on the type of feedstock and processing conditions (temperature, pressure, residence time, solvent and reactor configurations), are difficult to separate due to the presence of various higher molecular weight compounds (phenols, esters, ketones, alcohols, pyrazines, pyrimidines, pyrrolidines, ethers, furans, acids, aldehydes and aromatics) (both algal and lignocellulosic feedstocks) and are responsible for higher carbon and NO_x emissions when used directly in IC engines (Kolosz et al., 2020). Different engines, such as spark-ignition (gasoline with minimum octane number), compression-ignition (diesel with higher cetane number) and jet engines (fuels with superior cold flow properties), require various types of fuels for operations. Table 8.1 represents the comparison of the physicochemical properties of conventional fuels with vegetable oil, bio-crude obtained from algae, and pyrolysis of lignocellulosic biomass.

The essential physical properties of fuels include viscosity, heating value, moisture and oxygen content. The viscosities of algal bio-crude and pyrolysis oils (4.5–49.3 cSt) were much higher when compared to conventional petroleum products (1.5–2.7 cSt), which could significantly affect the cold flow properties of fuel at low temperatures (Volli et al., 2020). The higher oxygen content of bio-crude (15–41%) supports complete combustion but lowers the heating values. Higher octane/cetane numbers decrease the ignition delay in engines reducing the hydrocarbon emissions. The presence of phenols influences the shelf life of bio-crude as the long-term storage results in difficulties during handling, storage and transport due to an increase in viscosity (loss of volatiles/formation of an organic compound) (Beal et al., 2012). However, to overcome these disadvantages, upgrading techniques such as solvent and chemical extraction, emulsification, alcoholysis/esterification, supercritical fluids (SCFs), hydroprocessing (hydrotreating, hydro/catalytic cracking) and steam reforming were developed. Several efforts were made to lessen the undesirable properties (high oxygen, water, viscosity, corrosiveness, nitrogen content, thermal and chemical instability) by upgrading to enhance the quality of bio-oil and their subsequent use as a transportation fuel as explained in the following sections.

Table 8.1 Comparison of physicochemical properties of conventional fuels with vegetable oil and bio-crude obtained from algae and pyrolysis of lignocellulosic biomass

Characteristic	*Gasoline*	*Diesel*	*Kerosene*	*Jet fuel*	*Biodiesel*	*Vegetable oil*	*Algal bio-crude*	*Pyrolysis oils*
Water content (%)	<0.01	<0.05	<0.01	<0.05	0.05	<0.08	>4.72	>7.5
Specific gravity	0.7–0.8	0.84	0.79–0.81	0.77–0.84	0.8–0.9	0.91–0.98	0.86	1.02
Viscosity at 40°C	0.8	2.7	1.9	1.5	1.9–6.0	28.5–55.3	4.21	30–49.3
Flash point (°C)	−46	60	45.6	38	130	246–334	160	44–182
Pour point (°C)	−40	−35	−49	−47	−15	−8–10	−12.6	2–16
HHV (MJ/kg)	44–46	42–45	45.6	42.8	39.3	20.9–24.5	29–45	21
Initial boiling point (°C)	40	172	153	205	150	174	190	197
Final boiling point (°C)	204	341	245	300	354	322	500	538
Carbon (wt.%)	87	85.72	84.5	86.5	77	48.4–61.9	26.5–62.1	55–73
Hydrogen (wt.%)	13	13.2	15.5	13.5	12	3.5–6.1	4.1–93	5.5–8.7
Nitrogen (wt.%)	–	0.18	–	–	–	2.5–6.2	2.9–9.7	1.8–5.8
Sulfur (wt.%)	–	0.05	0.04	<0.3	0.05	0.2–2.5	0.6–2.8	0.1–0.5
Oxygen (wt.%)	–	0.85	–	–	11	24.5–31.7	19.5–41.1	15.5–28.9
Chain length	C_4–C_{12}	C_8–C_{25}	C_{10}–C_{16}	C_4–C_{16}	C_8–C_{22}	C_{10}–C_{22}	C_{14}–C_{20}	C_{10}–C_{27}

Note: Reproduced with permission from Volli et al. (2020).

Upgrading of biofuels and bio-crudes

The physicochemical properties of algal bio-crude indicate that its applications as a transportation fuel are minimal because of the presence of many undesirable components. The necessity of upgrading algal bio-crude is vital to make it useful as jet/liquid fuel. A variety of both physical and chemical methods/technologies are available for improving algal bio-crude. Figure 8.2 shows the classification of these techniques based on their process conditions and their requirements. These techniques can be classified in different ways and perspectives. Some of these techniques/processes will be operating at mild conditions, whereas others require high pressures and temperatures to upgrade the algal bio-crude.

Steam reforming

In steam reforming, synthesis gas, i.e., a mixture of hydrogen and carbon monoxide from hydrocarbon fuels, is generated. Fossil fuels, when reacting with steam at high temperatures (700–1,000°C) in the presence of a catalyst, produce syngas. Most industries use nickel as a catalyst for the steam reforming process, which is a well-established technology that is now being extended for reforming bio-oils as well (Huber et al., 2006). The syngas produced by steam reforming of bio-oils can be converted into a range of fuels, including jet fuels as well. Different types of reactors, such as fixed-bed reactors and fluidized-bed reactors, were explored for the steam reforming process of bio-oils (Czernik et al., 2007). The most influential factors in steam reforming are found to be temperature, catalyst loading and steam/carbon ratio. As in the case of fossil fuels, Ni-based catalysts are widely used for steam reforming of the bio-oils (Vagia and Lemonidou, 2008). However,

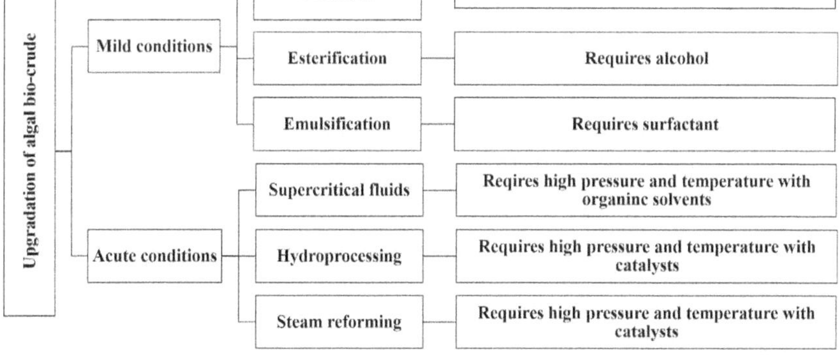

Figure 8.2 Classification of algal bio-crude upgrading techniques based on process conditions.

deactivation of the catalyst due to coking is one of the major disadvantages of steam reforming.

Solvent, chemical and SCF extraction

The most common practice in the upgrading of bio-oil from algae and other biomass is solvent extraction. Solvent addition is based on efficiency, selectivity and desired class of lipids which upgrades the bio-oils by increasing heating value and reducing viscosity (the ability of the solvent to prevent lipid degradation). The use of polar solvents ($C_2H_5COOCH_3$, CH_3COCH_3, $CH_3CHOHCH_3$, CH_3OH and C_2H_5OH) enhances cracking of hydrogen and covalent bond linkages between the lipids and organelles (van der Waals interactions) increasing the heating value of the bio-oils (Diebold and Czernik, 1997; Boucher et al., 2000). It can be because the heating values of polar solvents that are being added themselves have higher heating values. The different steps involved in the algal solvent extraction are penetration of organic solvent through the cell wall, interaction with the selective class of lipids forming a solvent–lipid complex, and diffusion of the complex into the bulk (concentration gradient) depending upon the dielectric constant and biological matrix of the cell wall (Halim et al., 2012).

To improve the extraction kinetics, solvent extraction methods were modified as accelerated solvent extraction (ASE), microwave-assisted extraction (MAE) and pressurized solvent extraction (PSE). The dipole rotation of solvent molecules helps the faster disruption of weak hydrogen bonds, higher temperature accelerates the extraction kinetics, and high pressure maintains the phase equilibrium allowing the solvent to pass through the pore matrix. Solvent addition prevents aging (interaction of solvent and bio-oil components), reducing the viscosity due to physical dilution (Liu et al., 2014). Bio-oils contain a wide variety of organic compounds that have commercial and industrial value (chemicals such as phenol, organic acids and alkanes). Several separation techniques, such as adsorption, distillation and fractionation (by phase separation or aqueous extraction), were developed. Vacuum distillation was employed for the upgrading of bio-oil from microalgae, which resulted in increased heating value and a decrease in oxygen content and boiling point ranges in distilled bio-oil.

The development of cost-effective and efficient refining techniques can help the commercialization of these technologies (Zhang et al., 2007). The conventional lipid extraction techniques of algal biomass involve organic solvents that have high selectivity and solubility. However, large-scale usage of such processes results in poor solvent recovery/separation and release of the high toxicity of organic solvents to the environment.

Alternatively, solvent-free methods, such as SCF extraction, were developed. The term supercritical fluid (SCF) refers to a fluid at a temperature and pressure which are above its critical point. SCFs are difficult to distinguish between gas and liquid phases, thus leading to higher rates of heat and mass transfer (Xu and Etcheverry, 2008). Commercial applications of SCFs date back to the early 1990s, and both inorganic SCFs (e.g., CO_2, N_2O, NH_3, SF_6 and H_2O) and organic (e.g., C_2H_6, C_3H_8, C_5H_{12}, CHF_3 and Freon-22) SCFs were tested for their compatibility and efficiency. However, SC-CO_2 has gained considerable attention due to its lower critical temperature (31.1°C) and pressure (7.3 MPa). Very few studies were focused on the influence of solvating power (selectivity) during the supercritical extraction from macroalgae. The SC-CO_2 extraction of ω-3 fatty acids and fucoxanthin form *Hypnea charoides* and *Undaria pinnatifida* was performed at 25–60°C, 200–400 bar and CO_2 flow rate of 1–4 mL/min. It was observed that 80% of extraction was achieved within 3 h of operation (Cheung, 1999; Quitain et al., 2013).

Table 8.2 represents the comparison of various literatures using algal SC-CO_2 extraction. Recently, SCFs were used in HTL of biomass for the production of bio-oil as they can improve the quality and increase the yield. Besides, SCFs can also be used in the upgrading process of bio-oils. Investigations have been conducted on several solvents such as ethanol, methanol and water (Michalak et al., 2015) as SCFs. The process can be seen as the esterification of any alcohol with an acid catalyst in supercritical conditions. Some of the notable research outcomes of applying supercritical conditions for the upgrading of bio-oils have resulted in better fuel properties when compared to the product stream of solvent extraction processes (Xu et al., 2014). When Pt/C catalyst in combination with supercritical water was used for the upgrading of bio-oil produced by HTL, it has increased the heating value and decreased oxygen, nitrogen viscosity and acid number (Duan and Savage, 2011a). It has also shown similar properties to those of petroleum-based fuels. However, these extraction techniques require high-end equipment with high power consumption.

Esterification and emulsification

Esterification is one of the renowned techniques used in the production of biodiesel, in which alcohol is added to the bio-oil to convert fatty acids into esters. In this process, organic acids (-RCOOH) react with an alcohol (-ROH) to form esters and water. The term alcoholysis is generally used instead of esterification because the alcohol added to the bio-crude can also react with aldehydes and can form acetals (Figure 8.3). Esterification helps to reduce the adhesion and dimerization, improving the volatility, sample separation and detectability. Esterification is based on the lower water solubility of esters when compared to their corresponding acids represented as:

$$\text{Biooil} + \text{R} - \text{OH} \overset{\text{Acid Catalyst}}{\Leftrightarrow} \text{Upgraded biooil} + H_2O \tag{8.1}$$

The temperatures are usually lower than the boiling points of the solvents used, and for methanol, esterification is performed at 60°C. The catalysts used in this process can be both homogeneous and heterogeneous. However,

Table 8.2 Comparison of various literatures on algal SC-CO_2 extraction

Algal feedstock	*Supercritical fluid (SCF)*	*Temperature (°C)*	*Pressure (bar)*	*Apparatus type*	*Description*	*References*
Arthrospira maxima	SC-CO_2 + ethanol	50–60	250	Flow	γ-Linolenic acid and lipids with γ-linolenic acid of 45%	Mendes et al. (2005)
Arthrospira platensis	SC-CO_2	48	200	–	For 4 h of reaction time, the maximum predicted yield of 10.2 g/kg was obtained. Components responsible for antioxidant activity were flavonoids (85 g/kg), β-carotene (78 g/kg), vitamin A (113 g/kg) and α-tocopherol (3.4 g/kg)	Wang et al. (2007)
Botryococcus braunii	SC-CO_2	50–85	200–250	Flow	The higher fatty acid content of ω-6 (C14:0, C16:0 and C22:6) was observed	Santana et al. (2012)
Botryococcus braunii, Chlorella vulgaris, Dunaliella salina and *Arthrospira maxima*	Hexane, ethanol, acetone, SC-CO_2 + 10 mol% ethanol	40–60	125–300	Flow	Alkadienes, carotenoids, β-carotene and ω-6 (C18:3) were extracted with total lipid (28–73%) and γ-linolenic acid (4–72%)	Mendes et al. (2005)
Chlorella vulgaris	SC-CO_2	55	350	Flow	With an increase in pressure from 15 to 35 MPa, the fractional yield increased from 11.3 to 171.1 mg/100 g (oil)	Mendes et al. (1995)
Haematococcus pluvialis	SC-CO_2	70	500	–	In a reaction time of 4 h, a maximum of 23 mg/g astaxanthin was extracted	Thana et al. (2008)
Haematococcus pluvialis and *Arthrospira maxima*	SC-CO_2 + 9.4 mol% ethanol	60	300	Column flow	Astaxanthin (1.7 wt.%) and phycocyanin (1.1 wt.%) with >97% of maximum recovery	Valderrama et al. (2003)
Synechococcus	SC-CO_2	50–60	300–500	Flow	Carotenoids (1.5 μg/mg dry weight of microalga); chlorophyll a (0.71 μg/mg dry weight of microalga)	Macias-Sanchez et al. (2007)

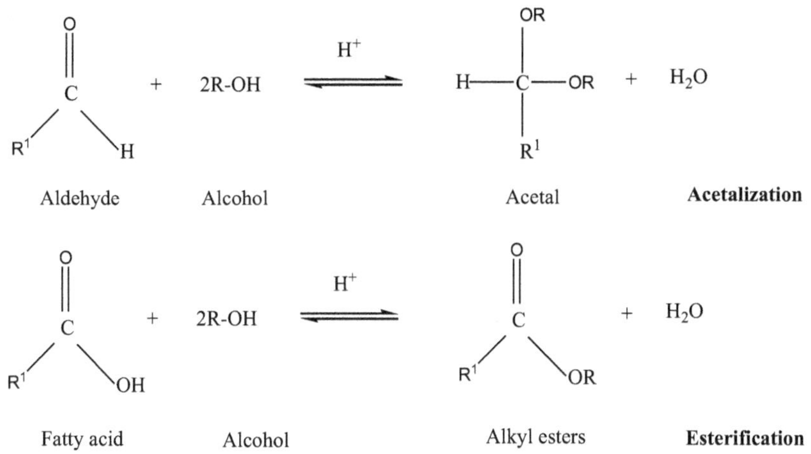

Figure 8.3 Acetalization and esterification reactions involved in bio-oil upgrading using alcoholysis. Reproduced with permission from Saber et al. (2016).

the heterogeneous process is preferred for bio-oil upgrading due to the ease in the separation process (Peng et al., 2009). When compared to the sub- and supercritical ethanol for catalytic upgrading, the supercritical upgrading process has shown more promise than the others. Further, the esterification reaction is reversible, enhancing the separation of by-products, and can be used as reactive distillation to increase the ester yield (Mahfud et al., 2007).

The presence of fatty acids in bio-oils naturally created an interest among the researchers to employ esterification for the upgrading process. It has been observed that the upgrading of algal bio-oil (extracted from *Spirulina*) by catalytic esterification has exhibited some better tribological features such as improved friction resistance, and anti-wear properties. The proposed reaction mechanism was as follows: ester formation by the adsorption of organic acids and catalytic conversion, the addition of alcohol, phase separation followed by chemical esterification, and enzymatic esterification in dilute solutions by suitable microorganisms (Xu et al., 2014).

It has been clear that there is an improvement in bio-oil performance such as the reduction of acidity, viscosity, density, aging rate, oxygen and water content, and an increase in heating value due to esterification and acetalization. One of the major drawbacks of this approach is the inability to remove nitrogen content in bio-oils significantly. Still, this approach seems to be prominent because of its simplicity and economic viability. Further, emulsifying the bio-oil with conventional petroleum products (gasoline, diesel, kerosene, etc.) helps to reduce the viscosity and increase the calorific value and cetane/octane number as liquid fuel (Jiang and Ellis, 2010). It has been observed that the miscibility of pyrolysis-derived bio-oils is lower with petroleum-derived fuels, but the miscibility enhances when mixed with surfactants. The emulsification process is relatively simple in implementation, but the addition of surfactants results in higher costs. The energy requirements for the emulsification process are also high. Bio-oil upgraded by emulsification process have desired ignition properties, but can still be relatively corrosive (Garcia-Perez et al., 2007).

Hydroprocessing

Hydroprocessing is a traditional petrochemical refinery process in which the removal of heteroatoms (oxygen, nitrogen and sulfur) from feedstock is achieved in the presence of pressurized hydrogen and selective catalyst for further processing and final blending. Hydroprocessing of bio-crude increases hydrogen content (H/C ratio), which significantly influences the heating value, thermal/oxidative stability and combustion properties affecting the performance of petroleum products. Hydroprocessing is classified into hydrocracking, hydrogenation and hydrotreating based on process conditions and desired end-use. Different reactions involved in hydroprocessing are shown in Figure 8.4. Cracking (thermal and catalytic) involves the production of short-chain hydrocarbons by breaking complex carbon–carbon molecules, whereas hydrocracking saturates free radicals occurring after the bond cleavage. Thermal cracking promotes coke formation due to the presence of oxygenates in bio-crude, which is an undesirable reaction.

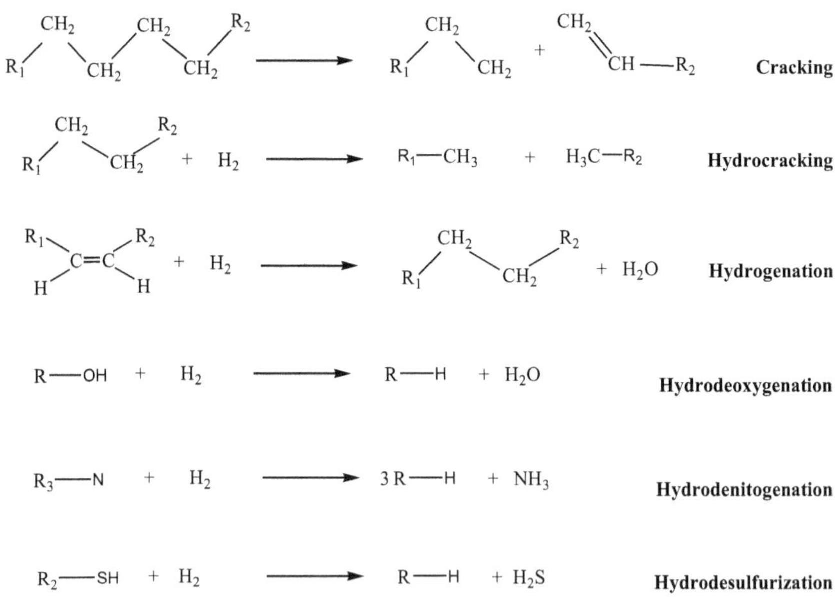

Figure 8.4 Different reactions occurring in hydroprocessing. Reproduced with permission from Saber et al. (2016).

The typical requirements of this process are high pressure (up to 20 MPa) and moderate temperatures (300–450°C) with a supply of hydrogen (Grange et al., 1996; Elliott, 2007).

Although hydrotreating aims to remove oxygen, nitrogen and sulfur in bio-oil, the cracking of heavy molecules also happens partially due to the presence of high-pressure hydrogen in this process. The presence of significant amounts of oxygen-rich compounds such as esters, ketones, phenols, acids and aldehydes has intrigued several researchers to use hydrotreating for the upgrading of bio-oils (Elliott and Hart, 2009). It requires moderate temperatures (350°C) and high pressure (2,000 psi) and a suitable catalyst. At the end of the hydrocracking process, the outcome consists of an aqueous phase, an organic liquid phase, coke and some gases. In comparison with hydrotreating, the hydrogen in hydrocracking is predominantly used in breaking C-C bonds, while in hydrotreating, hydrogen breaks other bonds such as C-N, C-O and C-S. One of the consequences of hydrotreating is hydrogenation, and some investigators also use the combination of both hydrotreating and hydrocracking (Zhu et al., 2013).

The approach was first to upgrade the bio-oil by hydrotreating and then to separate the heavy molecules from it for them to subject to hydrocracking. It can be easily said that hydrocracking is the best way to break heavy molecules and convert the bio-oils into efficient fuels. Still, the only concern is its demand for high temperatures and pressures, which results in substantial costs. Therefore, hydrotreating catalysts, such as Co/Mo-based and Ni/Mo-based catalysts are widely used in increasing the selectivity and reducing the formation of undesirable side products (Saber et al., 2016).

The cracking process can be achieved in the absence of hydrogen, but with the help of microporous alumino-silicate materials called zeolites. They are popular in industries as catalysts and adsorbents due to their microporous structure and availability of large surface areas. This process requires temperatures of 350–500°C at which deoxygenation happens along with cracking at atmospheric pressure. At the end of this process, products such as water, organics (hydrophobic/hydrophilic), coke and some gases (CO_2, CO and light alkanes) will be generated. Due to the absence of hydrogen, the zeolite cracking process results in a low H/C ratio in the upgraded bio-oil (Mortensen et al., 2011).

Even though hydrotreating is a well-established process in upgrading crude oils, certain disadvantages need to be solved by further research. The need for a high-pressure hydrogen source demands massive costs in designing high-pressure equipment. The average catalyst lifetime in this process is very low (200 h) in many of the cases (Mortensen et al., 2011). The deactivation of the catalyst due to coking is another concern that needs to be addressed. Despite these disadvantages, hydrotreating can be a promising process in upgrading bio-oils because of their ability to remove N, O and S. Further research is required in developing a sustainable hydrogen source and for increasing catalyst lifetime. The process of removing oxygen, nitrogen and sulfur from bio-crude is represented as hydrodeoxygenation (HDO), hydrodenitrogenation (HDN) and hydrodesulfurization (HDS), respectively, in the form of water, ammonia and hydrogen sulfide. The primary reaction during hydroprocessing is HDO, and as per stoichiometry, 1.5 mol of monoatomic hydrogen is required to upgrade each carbon atom in bio-crude as given in Eq. (8.2):

$$CH_{1.33}O_{0.43} + 0.77H_2O \rightarrow CH_2 + 0.43H_2O \tag{8.2}$$

The HDO process is not often selective and alters the formation of carbon in the liquid phase due to secondary reactions (polymerization and condensation of carbon to tar and coke and gasification to form methane or carbon oxides). The reactivity of oxygenates (alcohols, ketones, carboxylic acids and esters) is simple, whereas compounds such as phenols and furans require acute operating conditions (600°C and higher hydrogen content). The details of the reaction mechanism were explained in our previous studies (Volli et al., 2020). The presence of nitro compounds in petroleum residues and algal bio-crude is classified into non-heterocyclic (anilines, aliphatic amines and nitriles) and heterocyclic (indoles, pyrroles, pyrans and amines) compounds that are formed due to the depolymerization of polypeptides and proteins.

The hydrodenitrogenation reaction mechanism involves three stages: hydrogenation of the ring containing the nitrogen atoms, hydrogenolysis for bond cleavage between nitrogen and carbon bond, and the removal of nitrogen in the form of ammonia. The marginal levels of sulfur in algal bio-crude do not mandate HDS during its upgrading. However, industrial HDS is performed in fixed-bed reactors at 300–400°C between 30 and 130 atm in the presence of Co/Mo-impregnated alumina catalyst. Numerous studies were reported on the upgrading of bio-crude from different algal biomass feedstocks. *Chlorella* was used as a feedstock in HTL at 350–405°C and 60–66 bar with Co/Mo catalyst for a residence time of 2 h. A considerable reduction of oxygen (85%) and nitrogen (65%) was observed along with coke and gas yields (Biller et al., 2015). The two-step hydrotreatment (i.e., non-catalytic at 350°C and upgrading at 400°C) for bio-crude from *Chlorella pyrenoidosa* using different catalysts (Ru/C and Ni) and solvent (n-hexane) at 60 bar and 4 h was performed. The influence of reaction time (1–8 h) and catalytic loading (5–80%) on *Nannochloropsis* sp. bio-crude over 5% Pd/C catalyst in supercritical water was studied (Duan and Savage, 2011b).

The use of MoS_2/Co-impregnated AlF_3 support catalyst for hydrotreating bio-crude obtained from HTL of algae at 405°C and 136 bar pressure using continuous flow system resulted in an almost oxygen-free product (Elliott et al., 2013). The comparison of feedstock properties to the upgraded algal bio-crude is shown in Table 8.3. Though hydroprocessing of bio-crude is widely accepted, other factors, such as water formation, exothermic heat release, coke formation, low sulfur content and hydrogen consumption, significantly influence the process efficiency. Moreover, all the aforementioned

Table 8.3 Physicochemical properties of feedstock and upgraded product via hydroprocessing of different algal biomass sources

Algae feedstock	*Feedstock (wt.%)*								*Product (wt.%)*								*References*
	C	*H*	*N*	*S*	*O*	*A*	*M*	*HHV*	*C*	*H*	*N*	*S*	*O*	*A*	*M*	*HHV*	
Seaweed meal	39.4	5.1	2.9	0.6	52.1	16.5	-	14.4	75.5	9.1	3.65	0.6	11.6	-	-	35.97	Bach et al. (2014)
Nannochloropsis	56.8	9.3	10.1	0.3	19.9	3.42	68.8	24.1	79.3	11.9	5.2	0.2	3.1	0.2	7.1	34.8	Shakya et al. (2015)
Pavlova	54.3	8.6	8.67	0.8	24.1	3.4	17.7	22.6	78.3	10.1	4.7	0.2	6.3	0.4	5.6	32.2	
Chaetomorpha	26.5	4.1	3.4	2.1	31.1	36.6	5.1	10.3	70.9	7.7	6.8	0.1	11.4	-	-	32.5	Neveux et al. (2014)
Cladophora	30.9	5.1	5.2	2.3	34.9	25.5	6.7	12.7	71.6	8.1	7.1	0.9	10.6	-	-	33.3	
Oedogonium	36.6	5.7	4.8	0.4	30.9	20.6	6.5	15.8	72.1	8.1	6.3	0.8	10.4	-	-	33.7	
Derbesia	29.2	4.8	4.5	2.8	27.4	34.7	6.4	12.4	73.1	7.5	6.5	0.7	10.6	-	-	33.2	
Ulva	27.7	5.5	3.5	5.1	41.1	30.7	7.2	11.7	72.6	8.2	5.8	0.4	11.1	-	-	33.8	
Scenedesmus	38.1	5.6	5.5	0.5	30.4	20.1	-	16.8	74.9	9.1	5.9	0.7	9.6	-	-	36.2	López et al. (2015)
Nannochloropsis gaditana	47.6	7.5	6.9	0.5	25.1	12.4	-	23.1	76.1	10.3	4.5	0.4	8.8	-	-	38.1	

Reference: Volli et al. (2020). C: carbon; H: hydrogen; N: nitrogen; S: sulfur; O: oxygen; A: ash; M: moisture; HHV: higher heating value.

reactions may occur simultaneously and are kinetics controlled, making mathematical modeling extremely difficult. The use of different catalysts favors one reaction path over another, resulting in various secondary reactions. Furthermore, inappropriate process design and catalyst choice may result in low selectivity, catalyst poisoning and corrosive effects.

Current status, future prospects and recommendations

The first flight trial with bio-jet fuel that was carried out by Virgin Atlantic Airways in 2008 was successful in producing bio-jet fuel as per ASTM standards by 2011, and tested the jet fuel with Pratt and Whitney engines in Canada in 2013. Similarly, a blend of 25% bio-jet fuel from Jatropha seeds and 75% air turbine fuel (ATF) was used for a 43-min flight by SpiceJet in 2018, reducing the fuel cost by ~20–30%. The lack of mature technologies necessitates exhaustive research and development (quality, quantity, lifecycle assessment (LCA) and cost of production) in the production of bio-jet fuel (Kolosz et al., 2020). HEFA-SPK and FT-SPK are the sole mature processes for the production of bio-jet fuel.

The absence of significant amounts of naphthenic and aromatic fractions limits the commercial utilization of synthetic hydrocarbon fuels to 50% blends with conventional fuels. The cost of the oleo-chemical feedstock is higher, which makes HEFA an undesirable process for bio-jet fuel production. However, the use of alternative feedstock, such as non-edible crops and seed oil cakes, was also studied. Biodiesel could be competitive for the bio-jet fuel production process, and it can be circumvented if any of the pathways to convert lignocellulose to bio-jet fuel get commercialized (Yang et al., 2016). The widespread availability and lower costs of the renewable lignocellulosic and algal biomass feedstocks provide a virtuous alternative to bio-jet fuel production.

The emerging production and upgrading (e.g., hydroprocessing, alcoholysis, catalytic cracking, hydrodeoxygenation and oligomerization) strategies of macroalgal and microalgal bio-crude represent considerable potential for process integration and commercialization. The two main widely accepted upgrading processes include HTL and fast/flash pyrolysis. Numerous representative companies, such as Chevron (USA), Algenol (USA), Isuzu Motors (Japan), Algae Link (Netherlands), AlgaFuel (Portugal), Pond Biofuels (Canada), Varicon Aqua Solutions (UK), Shell (Brazil) and Solazyme (USA), have been pioneers and are approaching commercial reality (Saber et al., 2016). Although the energy return on investment is significantly high, advanced feedstock production strategies, integration of bioproduct outputs in biorefinery models, and economic sustainability remain the primary challenge. The uncertainties concerning the scale-up, lack of clear evidence of ecological sustainability, and understanding the complex conversion mechanisms need further detailed study (Beal et al., 2012). Assessing the optimal reaction conditions, improvements in the biorefinery process, catalyst reusability and characterization would provide valuable insight into the design of efficient reactor configurations. The absence of significant amounts of naphthenic and aromatic fractions limits the commercial utilization of synthetic hydrocarbon fuels to 50% blends with conventional fuels.

The need for feedstock supply chain management and government policies could make bio-jet fuel economically attractive. The USA supports the bio-jet fuel by categorizing it by renewable identification numbers (RINs) under the US Renewable Fuel Standard under different categories, namely D4, D5 and D7. Under this policy, a company receives a bio-ticket for the delivery of a volume of bio-jet fuel. Companies can trade this ticket with each other for fulfilling their EU Renewable Energy Directive (RED) or Fuel Quality Directive (FQD) obligations. Furthermore, bio-jet fuel requires extra processing to meet the standards of aviation fuel that adds to its cost. Additionally, the products that are blended with the bio-jet fuel should also be incentivized. In this regard, numerous airlines and aircraft manufacturers, along with various aviation industry associations, are working to tackle these aspects of bio-jet fuel.

There is a strong role of industry and customers to grow the bio-jet fuel industry. Some of the examples are KLM's corporate program and the Fly Green Fund. These programs provide a way for customers to cover up the price premium of using bio-jet fuel. The discussion states that there is a strong need to develop technologies to use low-cost feedstock effectively and efficiently, promising supply chains to provide feedstock and bio-jet fuel in a timely and efficient manner, drafting of policies and mandates to support bio-jet fuel, and gathering industry and customer support for the exponential growth of the bio-jet fuel industry.

Conclusions

The depleting reserves of fossil fuels, increasing cost and emissions of aviation fuels have driven the scientific community to reconnoiter the use of sustainable biomass-derived bio-jet fuels as alternative renewable energy resources. The feedstock is considered the most critical factor for the commercial feasibility of bio-jet fuel, and '*algae*,' a third-generation feedstock, have shown promising results in the generation of valuable products including biodiesel, bioethanol, biogas, pharmaceuticals and nutraceuticals. The algal-based bio-crude production including lipid (OTJ), biochemical (ATJ and STJ) and thermochemical (GTJ and PTJ) conversion methods was developed based on feedstock composition, production, upgrading processes and environmental sustainability. All these methods have their pros and cons. The catalytic longevity and presence of trace metal contamination downstream were the significant difficulties in lipid conversion. The biochemical techniques are in the early stages to be commercialized, have limited productivity and contain secondary waste. The separation of secondary molecules from bio-crude generated during the thermochemical process requires higher operating costs and energy. Further, bio-crude is highly viscous and contains higher O, N and S contents that significantly affect the heating value and combustion properties.

Several efforts were made to upgrade the algal bio-crude via catalytic cracking, hydroprocessing and SCF extraction, and their reaction mechanism was assessed to reduce the undesirable properties. Most of these upgrading processes operate at relatively high temperatures and pressures that influence the upgrading cost. The chemical processes such as catalytic cracking seem to be less effective in removing nitrogen content when compared to other processes. Hydroprocessing (HTL) methodologies are found to be effective in upgrading bio-oils, and they are the nearest to be commercialized. They suffer from anomalies such as coking and catalyst deactivation. However, the commercialization of supercritical systems for bio-oil upgrading is overpriced. The lack of comprehensive economic evaluation for the production and upgrading techniques for bio-jet fuels remains elusive. Research on large-scale production and optimization of process parameters using the latest innovations is necessary for the future. The economic evaluation of these techniques with optimized scale-up possibilities can be a superior prospect for future research.

Acknowledgments

The authors greatly appreciate the PS&DPL group, National Yunlin University of Science and Technology, Taiwan, China, for their cooperation and support.

References

Bach, Q. V., Sillero, M. V., Tran, K. Q., Skjermo, J. (2014). Fast hydrothermal liquefaction of a Norwegian macro-alga: Screening tests. *Algal Research* 6:271–276.

Beal, C. M., Hebner, R. E., Webber, M. E., Ruoff, R. S., Seibert, A. F. (2012). The energy return on investment for algal bio-crude: Results for a research production facility. *Bioenergy Research* 5:341–362.

Biller, P., Sharma, B. K., Kunwar, B., Ross, A. B. (2015). Hydroprocessing of bio-crude from continuous hydrothermal liquefaction of microalgae. *Fuel* 159:197–205.

Boucher, M. E., Chaala, A., Roy, C. (2000). Bio-oils obtained by vacuum pyrolysis of softwood bark as a liquid fuel for gas turbines. Part I: properties of bio-oil and its blends with methanol and a pyrolytic aqueous phase. *Biomass and Bioenergy* 19:337–350.

Carvalho, J., Ribeiro, A., Castro, J., Vilarinho, C., Castro, F. (2011). Biodiesel production by microalgae and macroalgae from north littoral portuguese coast. In *Wastes: Solutions, Treatments and Opportunities*, pp. 5–10. https://core.ac.uk/download/pdf/55615137.pdf.

Cheung, P. C. K. (1999). Temperature and pressure effects on supercritical carbon dioxide extraction of n-3 fatty acids from red seaweed. *Food Chemistry* 65:399–403.

Chisti, Y. (2019). Introduction to algal fuels. In Pandey, A., Chang, J.-S., Soccol, C. R., Lee, D.-J., Chisti, Y. (Second Eds.), *Biomass, Biofuels, Biochemicals*. Elsevier, pp. 1–31.

Czernik, S., Evans, R., French, R. (2007). Hydrogen from biomass-production by steam reforming of biomass pyrolysis oil. *Catalysis Today* 129:265–268.

Davidson, C., Newes, E., Schwab, A., Vimmerstedt, L. (2014). An overview of aviation fuel markets for biofuels stakeholders. *NREL*, pp. 1–52. https://doi.org/10.2172/1148623.

De Toni, A. R., Cancino, L. R., Oliveira, A. A. M., Oliveira, E., Rocha, M. I. (2014). Conventional and alternative aviation fuels: An overview on composition, properties and aircraft engine test. *15th Brazilian Congress of Thermal Sciences and Engineering*, pp. 1–8.

Diebold, J. P., Czernik, C. (1997). Additives to lower and stabilize the viscosity of pyrolysis oils during storage. *Energy and Fuels* 11:1081–1091.

Duan, P., Savage, P. E. (2011a). Upgrading of crude algal bio-oil in supercritical water. *Bioresource Technology* 102:1899–1906.

Duan, P., Savage, P. E. (2011b). Catalytic hydrotreatment of crude algal bio-oil in supercritical water. *Applied Catalysis B: Environmental* 104:136–143.

Elliott, D. C. (2007). Historical developments in hydroprocessing bio-oils. *Energy and Fuels* 21:1792–1815.

Elliott, D. C., Hart, T. R. (2009). Catalytic hydroprocessing of chemical models for bio-oil. *Energy and Fuels* 23:631–637.

Elliott, D. C., Hart, T. R., Schmidt, A. J., Neuenschwander, G. G., Rotness, L. J., Olarte, M. V., Zacher, A. H., Albrecht, K. O., Hallen, R. T., Holladay, J. E. (2013). Process development for hydrothermal liquefaction of algae feedstocks in a continuous-flow reactor. *Algal Research* 2:445–454.

Eurostat. (2015). Oil and petroleum products - A statistical overview. Eurostat UN-ECE. https://doi.org/10.3112/SPEED.V3I3.1258.

Garcia-Perez, M., Chaala, A., Pakdel, H., Kretschmer, D., Roy, C. (2007). Characterization of bio-oils in chemical families. *Biomass and Bioenergy* 31:222–242.

Gollakota, A. R. K., Kishore, N., Gu, S. (2018). A review on hydrothermal liquefaction of biomass. *Renewable and Sustainable Energy Reviews* 81:1387–1392.

Gollakota, A. R. K., Reddy, M., Subramanyam, M. D., Kishore, N. (2016). A review on the upgrading techniques of pyrolysis oil. *Renewable and Sustainable Energy Reviews* 58:1543–1568.

Grange, P., Laurent, E., Maggi, R., Centeno, A., Delmon, B. (1996). Hydrotreatment of pyrolysis oils from biomass: Reactivity of the various categories of oxygenated compounds and preliminary techno-economical study. *Catalysis Today* 29:297–301.

Haider, M. S., Castello, D., Michalski, K. M., Pedersen, T. H., Rosendahl, L. A. (2018). Catalytic hydrotreatment of microalgae bio-crude from continuous hydrothermal liquefaction: Heteroatom removal and their distribution in distillation cuts. *Energies* 11:1–14.

Halim, R., Danquah M. K., Webley, P. A. (2012). Extraction of oil from microalgae for biodiesel production: A review. *Biotechnology Advances* 30:709–732.

Huber, G. W., Iborra, S., Corma, A. (2006). Synthesis of transportation fuels from biomass: Chemistry, catalysts, and engineering. *Chemical Reviews* 106:4044–4098.

ICAO global framework for aviation alternative fuels (GFAAF). Montreal: International Civil Aviation Organization. (2017). www.icao.int/environmentalprotection/GFAAF/Pages/default.aspx.

Jiang, R., Ingle, K. N., Golberg, A. (2016). Macroalgae (seaweed) for liquid transportation biofuel production: What is next? *Algal Research* 14:48–57.

Kolosz, B. W., Luo, Y., Xu, B., Maroto-Valer, M. M., Andresen, J. M. (2020). Life cycle environmental analysis of "drop in" alternative aviation fuels: A review. *Sustainable Energy and Fuels* 4:3229–3263.

Liu, G., Yan, B., Chen, G. (2013). Technical review on jet fuel production. *Renewable and Sustainable Energy Reviews* 25:59–70.

Liu, R., Wenting, F., Chenjie, S. (2014). Influence of acetone addition on the physicochemical properties of bio-oils. *Journal of the Energy Institute* 87:127–133.

López, B. D., Riede, S., Hornung, U., Kruse, A., Prins, W. (2015). Hydrothermal liquefaction of microalgae: Effect on the product yields of the addition of an organic solvent to separate the aqueous phase and the bio-crude oil. *Algal Research* 12:206–212.

Macias-Sanchez, M. D., Mantell, C., Rodriguez, M., de la Ossa, E. M., Lubian, L. M., Montero, O. (2007). Supercritical fluid extraction of carotenoids and chlorophyll a from Synechococcus sp. *The Journal of Supercritical Fluids* 39:323–329.

Mahfud, F. H., Melián-Cabrera, Manurung, R., Heeres, H. J. (2007). Biomass to fuels upgrading of flash pyrolysis oil by reactive distillation using a high boiling alcohol and acid catalysts. *Process Safety and Environmental Protection* 85:466–472.

Mendes, R. L., Fernandes, H. L., Coelho, J. P., Elsa, C., Reis, E. C., Cabral, J. M. S., Novais, J. M. Palavra, A. F. (1995). Supercritical CO_2 extraction of carotenoids and other lipids from Chlorella vulgaris. *Food Chemistry* 53:99–103.

Mendes, R. L., Nobre, B. P., Cardoso, M. T., Pereira, A. P., Palavra, A. F. (2003) Supercritical carbon dioxide extraction of compounds with pharmaceutical importance from microalgae. *Inorganica Chimica Acta* 356:328–334.

Mendes, R. L., Reis, A. D. A., Pereira, P., Cardoso, M. T., Palavra, A. F., Coelho, J. P. (2005). Supercritical CO_2 extraction of γ-linolenic acid (GLA) from the cyanobacterium *Arthrospira (Spirulina) maxima*: Experiments and modeling. *Chemical Engineering Journal* 105:147–152.

Michalak, I., Dmytryk, A., Wieczorek, P. P., Rój, E., Łęska, B., Górka, B., Messyasz, B., Lipok, J., Mikulewicz, M., Wilk, R., Schroeder, G., Chojnacka, K. (2015). Supercritical algal extracts: A source of biologically active compounds from nature. *Journal of Chemistry* 2015:597140:14.

Milledge, J. J., Harvey, P. J. (2016). Potential process 'hurdles' in the use of macroalgae as feedstock for biofuel production in the British Isles. *Journal of Chemical Technology and Biotechnology* 91:2221–2234.

Milledge, J. J., Smith, B., Dyer, P. W., Harvey, P. (2014). Macroalgae-derived biofuel: A review of methods of energy extraction from seaweed biomass. *Energies* 7:7194–7222.

Mortensen, P. M., Grunwaldt, J. D., Jensen, P. A., Knudsen, K. G., Jensen, A. D. (2011). A review of catalytic upgrading of bio-oil to engine fuels. *Applied Catalysis A: General* 407:1–19.

Neveux, N., Yuen, A. K. L, Jazrawi, C., Magnusson, M., Haynes, B. S., Masters, A. F., Montoya, A., Paul, M. A., Maschmeyer, T., de Nys, R. (2014). Bio-crude yield and productivity from the hydrothermal liquefaction of marine and freshwater green macroalgae. *Bioresource Technology* 155:334–341.

O'Connell, A., Kousoulidou, M., Lonza, L., Weindorf, W. (2019). Considerations on GHG emissions and energy balances of promising aviation biofuel pathways. *Renewable and Sustainable Energy Reviews* 101:504–515.

Özçimen, D., Inan, B. (2015). An overview of bioethanol production from algae. In Biernat, K. (Ed.), *Biofuels - Status and Perspective*. IntechOpen, pp. 141–162.

Peng, J., Ping, C., Hui, L., Xiaoming, Z. (2009). Catalytic upgrading of bio-oil by HZSM-5 in sub- and supercritical ethanol. *Bioresource Technology* 100:3415–3418.

Prakash, A., Davies, J. (2020). An investigation into the feasibility of alternative aviation fuels. *AIAA SciTech Forum*, pp. 1–19. https://doi.org/10.2514/6.2020-0642.

Quitain, A. T., Kai, T., Sasaki, M., Goto, M. (2013). Supercritical carbon dioxide extraction of fucoxanthin from *Undaria pinnatifida*. *Journal of Agricultural and Food Chemistry* 61:5792–5797.

Raut, N., Al-Balushi, T., Panwar, S., Vaidya, R. S., Shinde, G. B. (2015). Microalgal biofuel. In Biernat, K. (Ed.), *Biofuels - Status and Perspective*. IntechOpen, pp. 101–140.

Saber, M., Nakhshiniev, B., Yoshikawa, K. (2016). A review of production and upgrading of algal bio-oil. *Renewable and Sustainable Energy Reviews* 58:918–930.

Santana, A., Jesus, S., Larrayoz, M. A., Filho, R. M. (2012). Supercritical carbon dioxide extraction of algal lipids for the biodiesel production. *Procedia Engineering* 42:1755–1761.

Shakya, R., Whelen, J, Adhikari, S., Mahadevan, R., Neupane, S. (2015). Effect of temperature and Na_2CO_3 catalyst on hydrothermal liquefaction of algae. *Algal Research* 12:80–90.

Sotelo-Boyas, R., Trejo-Zarraga, F., Jesus Hernandez-Loyo, F. de. (2012). Hydroconversion of triglycerides into green liquid fuels. In Karamé, I. (Ed.), *Hydroconversion of Triglycerides into Green Liquid Fuels, Hydrogenation*. IntechOpen, pp. 187–216.

Stephens, E., de Nys, R., Ross, I. L., Hankamer, B. (2013). Algae fuels as an alternative to petroleum. *Journal of Petroleum & Environmental Biotechnology* 4:1000148:7.

Suganya, T., Varman, M., Masjuki, H. H., Renganathan, S. (2016). Macroalgae and microalgae as a potential source for commercial applications along with biofuels production: A biorefinery approach. *Renewable and Sustainable Energy Reviews* 55:909–941.

Taimbú, C. A., Martín, M., Grossmann, I. E. (2019). Process optimization for the hydrothermal production of algae fuels. *Industrial and Engineering Chemistry Research* 58:23276–23283.

Thana, P., Machmudah, S., Goto, M., Sasaki, M., Pavasant, P., Shotipruk, A. (2008). Response surface methodology to supercritical carbon dioxide extraction of astaxanthin from *Haematococcus pluvialis*. *Bioresource Technology* 99:3110–3115.

United Nations, department of economic and social affairs, population division (2019). World population prospects 2019: Highlights. ST/ESA/SER.A/423.

Vagia, E. C., Lemonidou, A. A. (2008). Hydrogen production via steam reforming of bio-oil components over calcium aluminate supported nickel and noble metal catalysts. *Applied Catalysis A: General* 351:111–121.

Valderrama, J. O., Perrut, M., Majewski, W. (2003). Extraction of astaxantine and phycocyanine from microalgae with supercritical carbon dioxide. *Journal of Chemical and Engineering Data* 48:827–830.

Volli, V., Gollakota, A. R. K., Purkait, M. K., Shu, C.-M. (2020). Conversion of waste biomass to bio-oils and upgrading by hydrothermal liquefaction, gasification, and hydrodeoxygenation. In Nanda, S., Vo, D.-V. N., Sarangi, P. K. (Eds.), *Biorefinery of Alternative Resources: Targeting Green Fuels and Platform Chemicals*. Springer, pp. 285–315.

Wang, L., Pan, B., Sheng, J., Xu, J., Hu, Q. (2007). Antioxidant activity of Spirulina platensis extracts by supercritical carbon dioxide extraction. *Food Chemistry* 105:36–34.

Wang, W.-C., Tao, L., Markham, J., Zhang, Y., Tan, E., Batan, L., Warner, E., Biddy, M. (2016). Review of biojet fuel conversion technologies. *National Renewable Energy Laboratory (NREL)*, pp. 1–106. https://www.nrel.gov/docs/fy16osti/66291.pdf.

Wang, Z., Adhikari, S., Valdez, P., Shakya, R., Laird, C. (2016). Upgrading of hydrothermal liquefaction bio-crude from algae grown in municipal wastewater. *Fuel Processing Technology* 142:147–156.

Xu, C., Etcheverry, T. (2008). Hydro-liquefaction of woody biomass in sub- and supercritical ethanol with iron-based catalysts. *Fuel* 87:335–345.

Xu, X., Zhang, C., Zhai, Y., Liu, Y., Zhang, R., Tang, W. (2014). Upgrading of bio-oil using supercritical 1-butanol over a Ru/c heterogeneous catalyst: Role of the solvent. *Energy and Fuels* 28:4611–4621.

Yang, X., Guo, F., Xue, S., Wang, X. (2016). Carbon distribution of algae-based alternative aviation fuel obtained by different pathways. *Renewable and Sustainable Energy Reviews* 54:1129–1147.

Zhang, Q., Jie, C., Tiejun, W., Ying, X. (2007). Review of biomass pyrolysis oil properties and upgrading research. *Energy Conversion and Management* 48:87–92.

Zhou, Y., Hu, C. (2020). Catalytic thermochemical conversion of algae and upgrading of algal oil for the production of high-grade liquid fuel: A review. *Catalysts* 10:1–24.

Zhu, Y., Albrecht, K. O., Elliott, D. C., Hallen, R. T., Susanne, B., Jones, S. B. (2013). Development of hydrothermal liquefaction and upgrading technologies for lipid-extracted algae conversion to liquid fuels. *Algal Research* 2:455–464.

9 Algal-derived physiologically active nutraceuticals

Dietary supplements, vitamins, carotenoids, fatty acids and other novel products

Nidhi Hans, Falguni Pattnaik, Anushree Malik, Sonil Nanda, Vivek Kumar, Ajay K. Dalai, and Satyanarayan Naik

Introduction

Identification of food as medicine to prevent various diseases in humans was significant until the emergence of new drug therapy in the 19th century. In the early 20th century, the scientific focus was nutrition science that helps in body growth and their role in preventing many dietary deficiency diseases (Admassu et al., 2015). Desire to improve health and achieve sustainability, and consumer's expectations toward suitable food increased exponentially. This led to the establishment of food-based dietary guidelines in more than 100 countries worldwide for public food and nutrition to promote healthy eating practices and lifestyles. Healthy food prevents many chronic diseases, such as obesity, heart disease, cancer, diabetes, stroke and osteoporosis (Ganesan et al., 2019). Many studies have demonstrated that natural products or nutraceuticals can reduce the risk of these diseases.

According to the US Institute of Medicine, nutraceutical has been defined as 'any food or part of food which provides medical or health benefits including the prevention and treatment of disease, beyond the traditional nutrients it contains' (Burja and Radianingtyas, 2008). In addition, nutraceuticals are bioactive compounds isolated or purified from foods to be utilized in the medicinal form. This may include the fortification of conventional food to increase nutritional or health benefit. A functional food can be (Howlett, 2008):

- i unmodified or natural food
- ii food in which one of the constituents has been intensified using biotechnological means
- iii food in which any component is added to provide benefit
- iv food from which one particular part is removed technologically so that food provides benefits not otherwise available

v food where another replaces one component with beneficial properties
vi food in which a part is modified enzymatically or chemically to provide benefit
vii food in which bioavailability of ingredients is changed

Recently, marine sources have gained significant attention worldwide as a nutraceutical in the food and food supplement industries. Among marine organisms, marine algae are rich sources of structurally diverse bioactive compounds with valuable nutraceutical properties. Algae are the most promising and suitable organism for developing functional foods as they have the potential of providing necessary bioactive compounds for human nutrition with medicinal and pharmaceutical values (Domínguez, 2013).

Algae are exposed to adverse environments such as light, rapid variation in temperature and high oxygen concentration that lead to the formation of free radicals. However, they can protect themselves from external stress without any damage by generating bioactive compounds (Lin et al., 2016). Microalgae and macroalgae are both utilized as sources of food or food ingredients in the nutraceutical industries. Macroalgae or seaweeds are the most abundant source of vitamins A, B_1, B_{12}, C, D and E, riboflavin, niacin, pantothenic acid and folic acid in addition to minerals such as calcium, phosphorus, sodium, potassium and iodine (Wells et al., 2017). They are also a great source of antioxidants and antimicrobial molecules (Baker et al., 2017). Polysaccharides from seaweeds possess many biological activities for use as a dietary fiber and prebiotic, which help promote the growth of intestinal microorganisms (Lopez-Santamarina et al., 2020). In addition, polysaccharides such as carrageenan and alginates extracted from seaweeds have been reported as sources of edible packaging material in the food industry (Song et al., 2011). Macroalgae are also a great source of natural metabolites such as proteins, essential fatty acids, pigments and phlorotannins with unique nutritional and therapeutic properties (Cikoš et al., 2021). Microalgae are also a rich source of various physiologically active molecules such as β-carotene, astaxanthin, polyunsaturated fatty acids (PUFAs), vitamins A, B_1, B_2, B_6, B_{12}, C, E and H and polysaccharides which can be used as dietary supplements (Fu et al., 2017).

The development of functional food requires physical and chemical methods that minimize the degradation of bioactive compounds. For the extraction and fractionation of selective bioactive molecules from marine algae, organic solvents such as ethanol, subcritical water and supercritical CO_2 are encouraged (Hans et al., 2019). According to the current food regulations, functional food must be safe for its intended use. It should be widely available and accessible in the market, fulfilling ever-increasing consumer expectations toward 'green' products.

This chapter focuses on the review and discussion on the extraction techniques implemented for the isolation of the bioactive compounds, different applications in various fields and a cursory description of the additional components such as the bioavailability of nutraceuticals and metal toxicity (biomagnification) in algal species. The entire layout of the chapter is depicted in Figure 9.1.

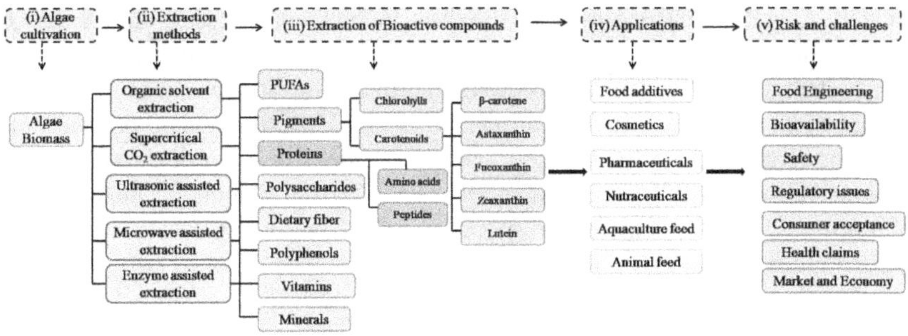

Figure 9.1 Marine algal-based bioactive compounds and their applications.

Algal-based nutraceuticals in day-to-day life

The market value of any products, especially foods and food supplements such as nutraceuticals, is significant for the promotion of the products and is directly dependent on the efficacy of the products floating in the market space. The inclusion and growth of nutraceuticals in the food business have become a necessity due to the various therapeutic effects. According to a study, the global nutraceutical market is forecasted to grow from US$209 billion to US$373 billion by 2025 with a significant growth rate (GlobeNewswire, 2019). In the nutraceutical market, algal-derived nutraceuticals share a substantial part with a global value of US$9.9 billion in 2018, which is anticipated to reach US$15 billion by 2024 (Businesswire, 2019).

A local, national or international market depends directly on several factors, which include the major factor as regulations and policies imposed on the products to run through a market. The consumer use of algal nutraceuticals and algal-based food or food supplements is dependent upon the regulations and policies raised by the federal or provincial bodies to make these market-available. These regulations built for the entry of the nutraceuticals in a consumer market space should be functionalized through balancing the benefits, safety and consumer access (Shao, 2017). Under the regulations and policymaking, several food regulatory organizations, such as Therapeutic Goods Administration (TGA), European Commission and US Federal Food and Drug Administration (FDA), maintain the allowable and restricted ingredients or nutraceuticals list to be present in the food and food supplements. A healthy market space always matters in the promotion of functional food products on both federal and provincial levels.

In recent years, the modern lifestyle, improper nutrition and physical inactivity adopted by people have led to an increase in chronic and non-communicable diseases. Today, obesity is considered a worldwide problem. In developed countries, heart diseases are a significant cause of death followed by cancer, osteoporosis, arthritis, diabetes and many others. The global demand for microalgal and microalgal nutraceuticals has great opportunities for food-processing industries to design novel products, which will potentially prevent an individual from these diseases. A diet constituting antioxidants, PUFAs, proteins, vitamins, minerals, etc., is required for a healthy lifestyle.

Even today, consumption of microalgal biomass is restricted to a few genera, such as *Spirulina*, *Chlorella*, *Dunaliella*, *Nostoc* and *Aphanizomenon*, due to their high nutrient content (Soletto et al., 2005). *Chlorella* and *Spirulina* have gained popularity in the global microalgal market. *Chlorella* species is considered as a functional food as it exhibits numerous health benefits. It helps lower blood sugar concentration and cholesterol and increases hemoglobin concentration in case of malnutrition (Caporgno and Mathys, 2018). Due to its increasing demand globally, the world market of US$49 million was produced in 2017 and it is estimated to reach US$68 million by 2024 (Koyande et al., 2019). According to the World Health Organization, *Spirulina* is known as a 'superfood,' as it is rich in protein and other bioactive molecules that humans cannot synthesize. Therefore, it helps reduce cholesterol levels and blood pressure, maintains blood sugar and increases hemoglobin levels in older people, thus improving their immune system (Selmi et al., 2011).

Seaweeds or their extracts are incorporated into food as functional ingredients to enhance nutritional properties and increase the content of dietary fibers and trace elements. They can absorb and retain water, thus being used in the production of low-calorie food. It is recommended to use seaweeds as dietary supplements daily to detoxify the body and reduce the harmful effect of environmental pollution on human health (Peixoto et al., 2016). Pigments present in seaweeds can influence food product color. Seaweeds are available in a diversified range of food products such as bread and confectionery, condiments, drinks, supplements, salads, soups and snacks (Bouga and Combet, 2015). Seaweed hydrocolloids such as carrageenan, agar and alginate are used as a thickening, texturing and stabilizing agent in food such as meat, dairy, liquid infant formula and toothpaste (Bixler, 2017). Seaweeds are indirectly incorporated in animal (e.g., chicken, egg, milk) and plant (e.g., vegetables, fruits, cereals) products to enhance their nutritional properties. Therefore, algal-based diets are crucial for human health. For their more full availability all over the world, it is essential to provide relevant information about which foods provide the best opportunities for health, well-being and reduced risk of disease.

Algal-derived bioactive components

Proteins, peptides and amino acids

The microalgae constitute up to 50% of protein, based on the dry weight. Protein content ranges from 3% to 47% in red macroalgae, from 2 to 32% in green macroalgae and from 1 to 24% in brown macroalgae (Cherry et al., 2019). Most of the microalgal proteins are rich in essential amino acids such as histidine, phenylalanine, threonine, tryptophan, valine, isoleucine, methionine and lysine, making them highly appropriate for use as direct supplements (Christaki et al., 2011). Proteins extracted from microalgae have recently gained attention due to their functional properties and bioactive components.

Phycobiliprotein, a water-soluble protein complex majorly present in blue-green and red algae (as phycoerythrin), is involved in photosynthesis. Purified phycobiliproteins can be used as fluorescent labels in immunological techniques and can be used as cosmetics and colorants in foods (Kamble et al., 2018). They are shown to have antioxidant, anti-inflammatory and hepatoprotective activities beneficial in treating cancer and neurodegenerative diseases (Chandra et al., 2017). Another diverse group of carbohydrate-binding proteins is lectins, which are purified from green marine algae. They play an essential role in many biological processes such as in interactions between host and pathogen, cell–cell communication, inducing apoptosis, agglutination of blood cells and showing the cytotoxic effect on colon and cervical cancer (Hori et al., 2000).

Proteins are hydrolyzed using enzyme, acid or alkaline hydrolysis to produce bioactive peptides of 2–20 amino acid residues, which may have great potential as nutraceuticals. Algal peptides exert a hormonal effect in various physiological stages of human development (Himaya et al., 2012). A sulfur-rich oligopeptide, phytochelatin, is involved in homeostasis and detoxification of metals in the cells (Pawlik-Skowrońska et al., 2007). Peptides isolated from *Chlorella* species have gastrointestinal enzyme resistance and help prevent DNA from oxidative damage of cells (Sheih et al., 2009a). Peptides possess angiotensin-converting enzyme (ACE) inhibitory activity, cholesterol-reducing activity, antiproliferative activity and antioxidant activity (Sheih et al., 2009b).

All the amino acids are produced in marine algae, out of which approximately 35% comprise essential amino acids. A higher amount of amino acids is present in green macroalgae than in red and brown macroalgae. In amino acid composition, phenylalanine represents the most elevated amino acids, followed by glycine, alanine, arginine, proline, glutamic acid and aspartic acid, whereas cysteine, tyrosine and methionine are present in lower amounts (Matanjun et al., 2009). In addition, red marine algae are featured as a rich source of the non-proteinaceous amino acid taurine. It is a part of bile juices, facilitating the reduction in the concentration of cholesterol in the bloodstream of humans (Murakami, 2017). Mycosporine-like amino acids (MAAs), laminine and kainoids are some other bioactive amino acids synthesized by marine algae. MAAs protect cells from UV-induced damage, laminine depresses the contractions of smooth muscles, and kainoids help in preventing neurophysiological disorders (Smit, 2004).

Various methods are employed for the extraction, purification and characterization of proteins, peptides and amino acids, such as ion-exchange chromatography, ultrafiltration and sodium dodecyl sulfate–polyacrylamide gel electrophoresis (SDS-PAGE) (Boobathy et al., 2010). Total protein content can be determined by spectrophotometric methods such as the Lowry method and the Bradford method. The amino acid compositions can be determined by the high-performance capillary electrophoresis (HPCE) technique or by high-performance liquid chromatography (HPLC) (Chen et al., 2005a). Amino acid compound-specific isotope analysis (AA-CSIA) is used to evaluate essential amino acids used by consumers to trace dietary protein in food sources.

Phenolic compounds

Phenolic compounds or phenolics are one of the major constituents of algal biomasses, which are considered as the secondary metabolites because of their exclusive involvement in the defensive activities toward several threats (Lin et al., 2016). Phenolic compounds are included in the most vital class of antioxidants derived from natural sources. Like in plants, algal phenolics are involved in the defensive action to scavenge the reactive oxygen species (ROS) such as hydrogen peroxide (H_2O_2), superoxide radicals ($^{\cdot}O_2^{-}$) and hydroxyl radical ($^{\cdot}OH$), which are generated in the mitochondria as the by-products in the energy production process or by the ultraviolet light exposure (Machu et al., 2015). The scavenging of these ROS reduces the oxidative stress caused by these oxygen species and protects the internal tissues and other functional compounds (e.g., fatty acids, proteins) from damage (Freile-Pelegrín and Robledo, 2013). In human bodies, these phenolic compounds can act similarly as above, and as an essential nutraceutical, these are being popularized for the prevention of several chronic diseases such as cancer, heart diseases, diabetes, hypertension and Alzheimer's disease (Machu et al., 2015).

Chemically, the phenolic compounds are the class of bioactive molecules with one or more aromatic rings with one or several numbers of hydroxyl groups (-OH), which make the phenolic compounds polar and always need a polar medium for their efficient extraction. Moreover, the molecules with larger numbers of phenolic rings (i.e., aromatic ring and hydroxyl groups) are structurally designated as polyphenols (Fomenko et al., 2019). Among the various classes of phenolic compounds, usually, the algal bodies constitute several polyphenols such as hydroxybenzoic acid, hydroxycinnamic acid, vanillic acid, lignans, bromophenols (e.g., colpol), MAAs, tichocarpols, various flavonoids, tannins and phlorotannins (Freile-Pelegrín and Robledo, 2013; Machu et al., 2015). According to several studies, among the different types of algal biomasses such as freshwater microalgae and marine algae (prominently macroalgae or seaweeds) such as red, brown and green algae, brown algae constitute a larger content of these phenolic compounds and show a greater degree of antioxidant activity (Mekinić et al., 2019). In the context of the extraction and isolation of the phenolics, there exist various techniques such as solvent extraction (e.g., polar solvents such as ethanol and methanol), supercritical fluid extraction and microwave- and ultrasound-assisted solvent extraction.

Yuan et al. (2018) performed a microwave-assisted extraction (with 70% methanol as the extraction medium) of phenolic compounds from four brown macroalgal species, estimated the phenolic contents and compared the result with the conventional solvent extraction method. The researchers concluded with the higher yield of phenolic compounds from the species *Ascophyllum nodosum* by microwave-assisted extraction with the total phenolics yield of 140 gallic acid equivalent (GAE) mg from 100 g of dry seaweed, which included phenolics such as hydroxybenzoic acid derivative, *p*-coumaric acid derivative, apigenin, medioresinol, gallocatechin and phlorotannins with their dimeric and hexameric forms (Yuan et al., 2018).

Another study on the extraction of phenolics from a brown algal species (*Sargassum vestitum*) and estimation of antioxidant activity has been done by Dang et al. (2018), which included the ultrasound-assisted technique with the conventional solvent extraction and microwave-assisted techniques. Their findings reveal that microwave-assisted extraction was found to be the efficient method of extracting the phenolic compounds because of the higher yield at a short extraction time of 75 s in comparison with the conventional solvent extraction (extraction time of 12 h) and ultrasound-assisted extraction technique (extraction time of 60 min). The phenolic extracts obtained from the microwave extraction showed the highest antioxidant activities in comparison with the other implemented techniques. The results conclude the retaining of the phenolic compounds from the degradation or other conversions due to the shorter extraction time (Dang et al., 2018).

Supercritical CO_2 ($SCCO_2$) extraction is considered a greener approach to the extraction of nutraceuticals, and the nonpolar nature of the extraction medium makes its implementation majorly restricted upon the extraction of nonpolar compounds such as fatty acids from the feedstock. Supercritical CO_2 occurs when the gaseous CO_2 exceeds its critical temperature (31°C) and critical pressure (7.3 MPa). However, the greater polarity of the polyphenols or simple phenolic compounds hampers the implementation of this extraction technique. Therefore, several researchers have tried the co-solvent-assisted $SCCO_2$ extraction method for the extraction of phenolic moieties from the various feedstocks such as algal biomasses.

Shipeng et al. (2015) performed the extraction of phenolics and flavonoids from the brown algae *Sargassum horneri* using both conventional solvent extraction and ethanol-assisted $SCCO_2$ extraction and estimated the highest yields of phenolic compounds and flavonoids by the latter method (0.64 mg/g and 5.7 mg/g, respectively). The extract also showed an elevated level of antioxidant activity than that of the extracts obtained from the solvent extraction.

The minimal temperature in the $SCCO_2$ extraction technique is the crucial parameter for retaining the antioxidant activity of the phenolic extractives. Besides brown algae, several red algal species are taken through various extraction methods to isolate different phenolic compounds. In the work done by Topuz et al. (2016), ultrasound-assisted ethanol extraction of red algae *Laurencia obtusa* was performed with a set of different temperatures, time periods and solvent-to-biomass ratios. From this study, the optimal yield was found to be 26.2 mg/g, obtained at a temperature of 45.3°C, solvent/biomass ratio of 24.3:1 and reaction time of 58 min, with the significant antioxidant activity (Topuz et al., 2016). The basic theory behind the excretion of the phenolic molecules relies on the stress level imposed on the algal biomasses. Phenolic compounds have also been extracted from the algal biomasses cultured in harsh conditions such as polluted water in the presence of heavy metals (Safafar et al., 2015).

Safafar et al. (2015) estimated the antioxidant activities of various microalgal species cultured in the media containing industrial effluents. Here, the researchers extracted phenolics along with other bioactive compounds from the five different microalgal species (i.e., *Phaeodactylum, Nannochloropsis, Chlorella, Dunaliella* and *Desmodesmus* spp.) grown on the industrial wastewater through the conventional solvent extraction method using methanol as the extraction medium. From this study, they found *Desmodesmus* as the highest contributor to the phenolic compounds, which showed an efficient and significant antioxidant activity (Safafar et al., 2015).

There are several more instances of the works, including the extraction of phenolic compounds and estimation of antioxidant activity, and some of them are presented in Table 9.1 with the corresponding extraction techniques implemented. Among the various phenolic compounds isolated from the algal biomasses, phlorotannins are the majorly found polyphenols in marine brown algae, which are the polymeric molecules of 1,3,5-trihydroxybenzene and its derivatives (Rastogi et al., 2017). Brown algal species such as *Fucus vesiculosus, Ascophyllum nodosum* and *Ecklonia cava* possess a significant amount of various phlorotannins (e.g., phloroglucinol, eckol, phlorofucofuroeckol A, phlorofucofuroeckol B, dioxynodehydroeckol, triphloroethol B, 2-phloroeckol, 7-phloroeckol, diphloroethol and fucofuroeckol A), which pharmacologically behave as the anti-inflammatory, antioxidant, antidiabetic, antibacterial and antiviral agents. These compounds are implemented in different fields such as cosmetics, food, pharmaceuticals and nutraceuticals (Sajna et al., 2015).

Lipids and polyunsaturated fatty acids

Microalgae possess 15–60% of total lipids on a dry-cell basis, and macroalgae contain 0.60–4.14% of total lipids, making them potential food supplements for humans (Rodrigues et al., 2015). Lipids are soluble in organic solvents such as chloroform, ether, hexane, acetone and benzene and insoluble in water. Lipids in the biological system include fats, waxes, sterols, fat-soluble vitamins, phospholipids, and monoglycerides, diglycerides and triglycerides. Algae are broadly explored for fatty acids, especially PUFAs, which represent 10–70% of the total fatty acids (Kumari et al., 2010).

Humans are incapable of synthesizing PUFAs, so PUFAs should be provided in a daily diet (Moi et al., 2018). Marine algae are an essential source

Table 9.1 Summary of the extraction processes of phenolic compounds from algal biomasses

Algal species	*Experimental methods*	*Key findings*	*References*
Ascophyllum nodosum, *Bifurcaria bifurcata* and *Fucus vesiculosus* (brown seaweeds)	• Ultrasound-assisted extraction with water as the extraction medium	• Eckol, tetrafuhalol, acacetin, hispidulin, quinic acid, gallocatechin, rosmarinic acid	Agregan et al. (2017)
Sargassum fusiforme (brown seaweed)	• Solvent extraction • Antioxidant activity assay by DPPH and FRAP	• Phlorotannin yield (including fuhalol, carmalol, eckol, phlorethols and fucophlorethols): 88.5/100 mg extract	Li et al. (2017)
Himanthalia elongata (brown seaweed)	• Solvent extraction with 60% methanol and phenolics purified by column chromatography • Antioxidant assay by DPPH and FRAP	• Total phenolic content (crude): 193 mg GAE/g • Phenolic content (purified): 211 mg GAE/g • Antioxidant (scavenging) capacity: 98%	Rajauria et al. (2016)
Acanthophora spicifera (red seaweed)	• Solvent extraction with ethyl acetate • DPPH antioxidant activity	• Total phenolic content: 40.5 µg GAE/mg • Antioxidant capacity: 45.6%	Zakaria et al. (2011)
Nannochloropsis sp. (marine green microalgae)	• Ultrasound-assisted extraction (400 W and 15 min) with DMSO (25–30%)	• Total phenolic content: 0.33 µg GAE/mg	Parniakov et al. (2015)
Chara fragilis (freshwater green macroalgae)	• Supercritical CO_2 extraction with 11.4% ethanol as co-solvent. • Pressure: 30 MPa; temperature: 40°C; and reaction time: 2 h	• Total phenolic content: 30.2 mg GAE/g	Fabrowska et al. (2016)
Macrocystis pyrifera (brown seaweed)	• Pretreatment using hexane • Extraction with water • temperature: 55°C; reaction time: 4 h; and solid/water ratio: 1:15 • Antioxidant assay as Trolox equivalent (TA)	• Phlorotannin yield (which majorly includes phloroeckol and phloroglucinol): 200 mg GAE/100 g • Antioxidant activity: 38.4 mg TA/100 g	Leyton et al. (2016)

Note: DPPH: 2,2-diphenyl-1-picrylhydrazyl; FRAP: ferric reducing antioxidant power; DMSO: dimethyl sulfoxide.

of high-value PUFAs, especially omega-3-fatty acids such as eicosapentaenoic acid (EPA) and docosahexaenoic acid (DHA) and omega-6-fatty acids such as gamma-linolenic acids (GLAs) and arachidonic acid (AA) (Borowitzka, 2013). It has been observed that PUFAs are essential for the human body, and their proper functioning affects human health. PUFAs help in regulating blood clotting, blood pressure and functions of the brain and nervous system. They protect against the development of cardiovascular disease (CVD) and reduce the risk of chronic diseases such as arthritis, diabetes and obesity (Endo and Arita, 2016). They can be widely used as a functional food, nutritional and pharmaceutical supplement due to their antibacterial, antifungal and antioxidant properties. Algal-derived nutritional products with DHA are marketed as infant formulas, infant food, dietary supplements and food additives (Nethravathy et al., 2019). The ratio of omega-3 to omega-6 essential fatty acid in macroalgae is approximately 1. Therefore, they can be used as ingredients in functional food formulations to meet the World Health Organization recommendations (i.e., <10) (van Ginneken et al., 2011).

Phospholipids, glycolipids and triacylglycerols are the structural components of fatty acids. Phospholipids are amphipathic lipid molecules that are the main component of the cell membrane and consist of a phosphate group attached to glycerol molecule or amino acid sphingosine resulting in compounds with fat-soluble (i.e., hydrophobic) and water-soluble (i.e., hydrophilic) regions. They are extracted from red algae and vary from 10% to 21% of total lipids (Holdt and Kraan, 2011). Most of the phospholipid membrane contains phosphatidylglycerol, phosphatidic acid, phosphatidylcholine, phosphatidylethanolamine, phosphatidylinositol and phosphatidylserine as the characteristic groups in their backbone. It helps in transporting molecules across the cell membrane, lubricates lungs and joints and delivers drugs throughout the body. It acts as a natural emulsifier that aids in digestion and absorption of fatty acids and cholesterol (Hamed et al., 2015).

Glycolipids are lipids with a carbohydrate attached by glycosidic bonds. They can be extracted from red algae and green algae, in which their content varies from 16 to 32 μmol/g and 11–32 μmol/g, respectively (Holdt and Kraan, 2011). Monogalactosyldiacylglycerol, digalactosyldiacylglycerol (DGDG), sulfolipid and sulfoquinovosyldiacylglycerol (SQDG) are different forms of glycolipids present in red and green algae. They maintain the stability of the cell membrane and serve as markers for cellular recognition that allow cells to connect to form tissues (Banfalvi, 2016).

Lipids from marine algae are extracted from the conventional solvent extraction method using organic solvents such as hexane, ethanol, chloroform, methanol, dichloromethane, diethyl ether and combinations of any two of these (de Boer et al., 2012). In the past few years, novel green extraction techniques are employed, such as supercritical fluid extraction, microwave-assisted extraction, ultrasound-assisted extraction, pressurized fluid extraction and enzyme-assisted extraction, for the isolation of lipids from algae (Ibañez et al., 2012; Hans et al., 2019). Both the methods have their advantages and disadvantages.

Isolated lipids are separated by thin-layer chromatography (TLC), HPLC or liquid chromatography–mass spectrometry (LC-MS) and can be characterized by various spectrometric techniques such as Fourier transform infrared spectroscopy (FTIR), nuclear magnetic resonance (NMR) spectroscopy and gas chromatography–mass spectrometry (GC-MS).

Polysaccharides

Polysaccharides are polymers composed of monosaccharides joined together with glycosidic bonds. Marine algae contain a large amount of polysaccharide in the range of 4–76% of dry weight. The polysaccharides present in the cell wall of microalgae are mainly composed of 2–10% of dry weight cellulose and 6–9% of dry weight hemicellulose. Red algae contain agar, carrageenans, xylans, floridean starch, sulfated galactan and porphyran. Alginic acid, fucoidan, laminarin and sargassan are isolated from brown algae, whereas green algae are composed of sulfuric acid polysaccharides, sulfated galactans and xylans (Kumar et al., 2008). The quality and quantity of polysaccharides vary with seasonal variation. Agar, alginate and carrageenan are the most common polysaccharides or phycocolloids extracted from microalgae.

Agar is extracted from red algae and is composed of agarose and agaropectin. It is used in nutrient medium for bacterial growth, food products (e.g., jellies and candies) and medicinal capsules, tablets and laxatives (Rasmussen and Morrissey, 2007). It shows anti-aggregation of red blood cells and anticancer activity, absorbs UV rays and helps in reducing the glucose level of blood (Chen et al., 2005b). Carrageenan is a sulfated linear polysaccharide attached to the sugar unit. It has applications in the food industry such as in ice creams, canned foods, bakery fillings and desserts, and in the brewing industry such as in refining wine, beer and honey. It is also used in pharmaceutical industries with anticoagulant, antitumor and antiviral activities. It is used in medicinal creams and lotions acting against dislodging mucus (Morrissey et al., 2001).

Alginates are isolated from brown algae and are present in acidic form, linear polyuronic acid, commonly known as alginic acid, and in salt form, which is present in the cell wall, and constitutes 40–47% of dry weight of biomass (Rasmussen and Morrissey, 2007). It has antibacterial, anti-inflammatory, antihypertensive and anticancer activities. It also helps in effectively reducing cholesterol (Patil et al., 2018). In addition to the various algal-based polysaccharides, algal-derived dietary fibers also play a vital role in treating several chronic diseases.

Dietary fibers

Dietary fibers are mostly indigestible due to a lack of specific degradation enzymes in the human gastrointestinal tract. It is of two types, i.e., soluble fibers and insoluble fibers. The soluble fibers form a thick and viscous gel in the human intestinal tract such as agar, alginic acid, furonan, laminaran and porphyrin. The content of total dietary fiber in microalgae varies from 25% to 75% of dry weight of biomass (Jiménez and Sánchez, 2000). Water-soluble polysaccharides help in preventing the low level of cholesterol (hypocholesterolemic) and glucose level (hypoglycemic) in the blood (Holdt and Kraan, 2011). Algal fibers are proven to be beneficial for human consumption due to their anticoagulant, antiviral and antitumor properties, reducing obesity and preventing large intestine cancer and diabetes (Ghosh et al., 2009).

Chlorophyll

With the growing concern of environmental safety related to synthetic pigments, pigments from marine algae are produced industrially due to the increasing consumer demands for natural foods. Chlorophyll and carotenoids are the dominant pigments found in marine algae. These major groups are subcategorized into many pigments that have their specific activity and health effects on humans. These pigments can be used as nutraceuticals and are also used as natural colorants in foodstuffs and drinks (Kraan, 2013).

Chlorophyll is a green photosynthetic and lipid-soluble pigment synthesized in photosystems I and II of all marine algae. It represents 0.5–1.5% of total dry biomass. Consumption of chlorophyll transforms it into pheophytin and pheophorbide. These derivatives are reported to stimulate liver functions and aid the metabolism of lipids, carbohydrates and proteins in humans. They are also claimed to possess antimutagenic and antioxidant effects and play an essential role in preventing cancer (Bishop and Zubeck, 2012).

Humans can easily digest chlorophyllin (CHL), a semisynthetic form of chlorophyll. Numerous experiments demonstrated that dietary supplementation of CHL reduces the risk of cancer by inhibiting various pathways involved in the metabolism of carcinogens (Nagini et al., 2015). According to the European Food Safety Authority (EFSA) [No. 231/2012-E140 (i)], it can be used as a nutrient source and additive in foods such as ice cream, jellies and candies (EFSA, 2015). Chlorophylls are isolated by supercritical fluid extraction and by organic solvent extraction techniques using methanol, ethanol and acetone. The extract obtained from supercritical CO_2 is pure and can be used in various food and pharmaceutical industries.

Carotenoids

Carotenoids are lipid-soluble pigments naturally synthesized by marine algae. They are biosynthetically derived from five-carbon isoprene units to form a highly conjugated 40-carbon structure with a molecular formula of $C_{40}H_{64}$. Cyclization of β-ionone ring and substitution reactions such as C-3 hydroxy, C-4 keto and C5–6 epoxy leads to the formation of different xanthophylls (Spolaore et al., 2006). Carotenoids belong to a terpenoid class with a tetraterpene backbone (C_{40}) responsible for a range of colors. They are accessory pigments for light-harvesting processes during photosynthesis and act as structural stabilizers for protein assembly in photosystems (Moldenhauer et al., 2017). It constitutes two major classes of highly conjugated polyene chromophoric molecules. The first includes carotenes, which are strictly hydrocarbons such as α-carotene, β-carotene and lycopene, and the second includes xanthophylls which are similar to carotenes but contain one or more oxygen molecules such as lutein, zeaxanthin, violaxanthin, neoxanthin and fucoxanthin (Christaki et al., 2013).

In green algae, carotenoids are synthesized and accumulated within plastid, but secondary xanthophylls such as astaxanthin accumulate in the cytoplasm. They are used as food pigments, nutraceuticals (precursor of vitamin A), cosmeceuticals and feed supplements in aquaculture sectors. They possess antibacterial, antifungal, antiviral, anti-oxidative and anti-inflammatory properties with beneficial effects on human health in preventing certain diseases such as cancers, CVD and macular degeneration (Guedes et al., 2011).

β-Carotene is a precursor of pro-vitamin A and represents about 10–14% of the dry algal biomass. According to the European Union (E-160a(ii)), it can be used as a natural coloring agent in food and drink products (EFSA, 2016). It has efficient antioxidant properties that help to prevent oxidative stress through the scavenging of free radicals. It also prevents photo-oxidative damage of skin when exposed to UV radiation, thus protecting against sunburns and premature aging (Godic et al., 2014). Consumption of β-carotene can help improve visual and retinal functions in patients with degenerative diseases of the retina such as night blindness, cataract and tunnel vision (Sher et al., 2018). Various studies have reported that β-carotene modifies the oxidative surface of low-density lipoprotein, thus inhibiting lipid-induced atherosclerotic CVD (D'Odorico et al., 2000). Many experiments also suggested that β-carotene helps enhance immunity against various infectious diseases and prevent certain cancers such as lung cancer and colon cancer. However, high doses of β-carotene supplements have been shown to increase the risk of cancer in smokers and asbestos workers (Druesne-Pecollo et al., 2010).

Astaxanthin is the second most important natural carotenoid pigment, and it ranges from 4% to 5% of dry weight in marine algae. The US Food and Drug Administration and many European countries have accepted astaxanthin as a color additive and a dietary supplement (Food and Drug Administration, 2002). It possesses severalfold more potent antioxidant activity than β-carotene, which helps in scavenging free radicals in the human body (Park et al., 2010). It protects the lipid membrane from peroxidation and prevents lipid and tissues from UV light photo-oxidation. It reduces the risk of age-related macular degeneration and cataract (Naguib, 2000). It also helps protect skin from sunburn when excessively exposed to sunlight, thus reducing skin cancer (Rao et al., 2013). It has anti-inflammatory activity; as a result, it helps in reducing gastric inflammation and ulcer diseases (Kang and Kim, 2017). Astaxanthin supplementation protects low-density lipoprotein cholesterol from oxidation and increases the high-density lipoprotein cholesterol in the bloodstream, which decreases the risk of coronary heart disease (Riccioni et al., 2011). It inhibits the senescence of mitochondrial cells by preventing oxidative damage, thus helping in the anti-aging of cells (Kidd, 2011). Many studies have reported that dietary astaxanthin significantly lowers the risk of various cancers, such as colon cancer, mammary cancer and prostate cancer, by modulating the immune response of tumor cells (Zhang and Wang, 2015). Astaxanthin is also an active therapeutic agent for neurodegenerative diseases such as Parkinson's, Huntington's and Alzheimer's diseases (Grimmig et al., 2017).

Fucoxanthin is one of the abundant carotenoids in nature, and its content varies from 172 to 720 mg/kg of dry weight in brown algae. Dietary fucoxanthin is converted into its active form, fucoxanthinol, by digestive enzymes in the gastrointestinal tract, which exerts biological and physiological functions in the body (Sugawara et al., 2002). It helps to prevent obesity in white adipose tissue by inducing the protein and mRNA expression of uncoupling protein 1 (UCP1) in the mitochondrial inner membrane. The induction of UCP1 by fucoxanthin releases chemical energy as heat in fat tissue, thereby stimulating thermogenesis to limit weight gain (Gammone and D'Orazio, 2015). It helps regulate blood glucose level, suppresses angiogenesis, protects from harmful UV radiations and has strong anti-inflammatory and neuroprotective activities (Pangestuti and Kim, 2011). Fucoxanthin has a strong inhibitory effect on leukemia cells, colon cancer, breast cancer and prostate cancer due to its high apoptotic activity (Nakazawa et al., 2009). Studies reported that fucoxanthin enhances the amount of DHA and AA in the liver of mice by up-regulating the enzymes involved in fatty acid bioconversion (Tsukui et al., 2007).

Zeaxanthin is mainly present in green and red marine algae, and its content varies from 4.5 to 5.5 mg/g of dry weight. Lutein is a colorant and additive in food products and is used as a pigment in fish and poultry (Jin et al., 2003). Supplements of zeaxanthin and lutein prevent the risk of macular degeneration and cataract in older people. Their antioxidant property minimizes the production of free radicals in the eye when exposed to light (Mares-Perlman et al., 2002). They also reduce the lipid peroxidation, protein oxidation and DNA damage in human lens epithelial cells against oxidative stress, thus preventing age-related-cataract (Gao et al., 2011). Many studies have reported that zeaxanthin and lutein help protect atherosclerosis at the initial stage of its progression (Dwyer et al., 2004). They possess anticancer activity and prevent premenopausal breast cancer (Stengel et al., 2011). Dietary intake of zeaxanthin and lutein can avoid a variety of chronic diseases by modulating immune response (Kim et al., 2000).

Carotenoids are extracted using organic solvents such as hexane, chloroform, methanol, acetone, diethyl ether and ethanol. Mechanical and chemical pretreatment steps are involved in the breakdown of the cell wall of algae, which facilitates the entry of the solvent into the cell. It is reported that carotenoid yield has increased 8–10 times more with effective cell disruption techniques (Michelon et al., 2012). Due to the safety of the environment and health, green solvents are employed to extract carotenoids from marine algae. Techniques such as supercritical fluid extraction, subcritical water extraction, enzyme-assisted extraction, pressurized liquid extraction, microwave-assisted extraction and ultrasound-assisted extraction are used for the higher recovery of carotenoids (Gong and Bassi, 2016; Hans et al., 2019). Characterization and profiling of extracted carotenoids are done by NMR spectroscopy and reversed-phase HPLC and LC-MS methods.

Vitamins

Marine algae can synthesize almost all vitamins such as A, B_1, B_2, B_6, B_{12}, C, E, and their content varies with species, season, algal growth rate, geographical area and environmental parameters. Algae are a good source of pro-vitamin A (derived from carotenoid β-carotene) that can be converted to vitamin A (retinol) with the help of carotene dioxygenase (Von Lintig and Vogt, 2000).

Green and red algae contain a higher concentration of vitamin B_{12} (cobalamin), which assists in curing the effects of aging, sleep abnormalities and anemia. Vitamin B_1 (thiamine) ranges from 0.14 to 2.02 μg/g, vitamin B_2 (riboflavin) from 0.31 to 6.15 μg/g, and vitamin B_9 from 61.4 to 161.6 μg/100 g in different species of algae (Sánchez-Machado et al., 2004). Vitamin C (ascorbic acid) is present in all red, brown and green algae. Its content varies from 500 to 3,000 mg/kg of dry matter for green and brown algae and 100–800 mg/kg for red algae. It helps in building the immune system, scavenging free radicals and triggering the absorption of iron in the intestine (Burtin, 2003). Vitamin E is a mixture of tocopherols (α-, β- and γ-tocopherols). All species of red, green and brown algae contain α-tocopherol with an antioxidant activity, which helps protect membrane lipids from oxidative damage (Misurcova, 2011).

Minerals

Many algae are rich in essential minerals (e.g., Na, K, Ca and Mg) and trace elements (e.g., Fe, Zn, Mn and Cu) (10–100 times higher than traditional vegetables), which are beneficial for human consumption. The content of minerals varies with species, physiological parameters, season, location and environmental conditions. Algae are rich in Na and K and help reduce high blood pressure in blood vessels of the lungs (Taboada et al., 2010). Intake of algae can reduce the risk of Ca deficiency, especially in pregnant women, children of age below 18 and the elderly (Hamed et al., 2015). Brown algae are a good source of iodine (up to 1.2% of dry weight), which is useful in metabolic growth and regulation. Its intake helps in the formation of thyroid hormone, which reduces the risk of goiter and hypothyroidism (Müssig, 2009). Microalgae are a potentially rich source of iron for the human diet.

Consumption of a large number of minerals can limit their absorption by binding to specific polysaccharides, which can induce certain toxic effects.

Applications of algae in human nutrition, animal feed and aquaculture

Human nutrition

At present, the consumption of microalgae is limited to a few species, including *Spirulina*, *Chlorella*, *Dunaliella salina* and *Aphanizomenon* spp. They are commercially produced as health supplements in the form of powder, tablets, capsules and liquid extract. They are incorporated into foods such as bread, pasta, candies, gums, cookies, noodles, yogurt or beverages, which enhance their nutritional quality. They have beneficial health effects such as scavenging free radicals and improve the cardiovascular system, brain and bone health, lose weight, modulate the immune system, lower cholesterol and triglycerides from the blood, remove toxins and improve energy and focus.

Biologically active compounds in marine algae have a potential role in the pharmaceutical and cosmeceutical industries. Secondary metabolites derived from marine algae are reported to have a beneficial effect on the skin and hair. Extracts of *Isochrysis*, *Nannochloropsis* and cyanobacteria are used in sunscreen formulations to prevent UV transmissions, sunburn, wrinkle formation and photoaging. Proteins and hydrolysates of *Porphyra* sp., *Spirulina* sp. and *Chlorella* sp. are used in moisturizers, skin lotions, creams and shampoos and are absorbed by the skin and hair and keep them moisturized for a longer time. *Spirulina* and *Chlorella* facilitate tissue regeneration and reduce wrinkles. Zeaxanthin from *Nannochloropsis oculata* prevents skin pigmentation and stimulates its whitening. *Fucus vesiculosus* extracts help reduce dark circles under the eyes (Ariede et al., 2017).

Algae also have a potential application in the pharmaceutical sector, as they possess antibacterial, antifungal and anticancer properties. Many studies show that algae have an antiviral activity that inhibits herpes simplex virus (HSV I and II), human immunodeficiency virus (HIV), dengue virus (DENV) and human papillomavirus (HPV) (Ahmadi et al., 2015). Recombinant protein from *Chlamydomonas reinhardtii* has led to the development of anti-cancer immunotoxins, monoclonal antibodies and therapeutic proteins. The chloroplast of *C. reinhardtii* can produce a malaria transmission-blocking antibody, which is likely to play a critical role in curing malaria in the future (Jones et al., 2013).

Animal feed

The addition of marine algae in animal feed increases animal health and improves the nutritional value of animal products, which has a significant impact on consumer's health. Algal biomass is included in the diet of ruminants, poultry, pig, rabbits, etc. The addition of microalgae *Schizochytrium* sp. in the diet of dairy cow and dairy ewes was reported to increase DHA, EPA, conjugated linoleic acid and *trans*-vaccenic acid content in milk fat and reduce the concentration of total saturated fatty acids. Products such as yogurt and feta cheese from PUFA-enriched milk are acceptable for human consumption, especially for pregnant women and infants (Papadopoulos et al., 2002). It also improves the mineral content of animal milk, mainly the concentration of iodine and selenium, which is essential for humans.

The incorporation of dietary seaweeds and *Chlorella* in laying hen improved the fatty acid content of EPA and DHA and albumen height and enhanced the yolk color (Carrillo et al., 2008). Algal biomass and extracts are also used in pet food, promoting the health of pets by strengthening their immune response and fertility and controlling the weight. It also affects their external appearance, such as healthy skin and shiny fur (Spolaore et al., 2006). Feeding pigs with algal biomass has been reported to increase the nutritional quality of their meat products (He et al., 2002). A lower serum cholesterol level and elevated high-density lipoprotein cholesterol in the blood were found in hypercholesterolemia-induced rabbits supplemented with microalgae (Colla et al., 2008).

Aquaculture

Marine algae are used as a nutrient source in aquaculture production worldwide. They are used as a feed for larvae, juvenile fish, bivalves, crustaceans and zooplankton. It was reported that fish fed on *Spirulina* had improved growth, flavor and protein efficiency (Palmegiano et al., 2005). Astaxanthin produced by *Haematococcus* is used as a pigment in salmon, sea bream, ornamental fish and shrimp, deposited in their skin, muscles, exoskeleton or gonads. Astaxanthin is a nutrient for aquatic animals, which helps improve their growth, survival and stress tolerance rate (Niu et al., 2009). Marine algae used for the production of rotifers help to increase lipid content and fatty acid composition (especially EPA and DHA) of rotifers in a short interval.

Algae in the diet of fish were reported to increase the growth, feed efficiency and lipid metabolism and protein deposition in muscles. Carotenoids synthesized from marine algae help to enhance biological functions in fish, such as expediting sexual maturity, increasing egg survival, improving fertilization rate and developing a better embryo. Carotenoids recover liver functions, defend against oxidative stress, protect against diseases and promote immune response by inducing biodefense mechanisms in aquaculture species (Higuera-Ciapara et al., 2006). The whole applications of different types of algal bioactive molecules are presented in Table 9.2, which also summarizes the health benefits of them, the major sources associated with these nutraceuticals and the various industries dealing with the particular bioactive molecule.

Table 9.2 Algal-derived bioactive compounds and their sources, health benefits and applications

Bioactive compounds	*Major sources*	*Health benefits*	*Applications*	*References*
Pigments				
Astaxanthin	• *Dunaliella salina* • *Haematococcus pluvialis* • *Chlorella zofingiensis* • *Nannochloropsis* sp.	• Strong antioxidant property • Anti-inflammatory • Anticancer • Improve cardiovascular health • Prevent neurodegenerative diseases • Enhance immune functions	• Food color • Animal feed • Skin health • Pharmaceuticals	Park et al. (2010), Rao et al. (2013), Riccioni et al. (2011) and Grimmig et al. (2017)
Fucoxanthin	• *Fucus vesiculosus* • *Ascophyllum nodosum* • *Isochrysis* sp.	• Antioxidant • Anti-obesity • Antiviral • Antidiabetic • Anticancer • Anti-inflammatory • Cure skin burn from UV rays	• Skincare • Anti-wrinkle • Anti-aging • Weight reduction	Sugawara et al. (2002), Gammone and D'Orazio (2015) and Ariede et al. (2017)
Zeaxanthin and lutein	• *Chlorella* • *Chlamydomonas* • *Scenedesmus almeriensis* • *Ascophyllum nodosum*	• Antioxidant • Prevent age-related macular degeneration • Anticancer • Protect atherosclerosis	• Food and health supplements • Fish feed • Animal feed • Pharmaceuticals • Cosmeceuticals	Jin et al. (2003), Mares-Perlman et al. (2002), and Dwyer et al. (2004)
β-Carotene	• *Dunaliella salina* • *Haematococcus pluvalis* • *Spirulina platensis*	• Vitamin A precursor • Antioxidant • Prevent macular degeneration • Cure skin burn from UV rays • Prevent liver fibrosis • Anticancer	• Natural food colorant • Pharmaceuticals • Cosmeceuticals • Animal feed and pet food • Aquaculture	Sher et al. (2018) and Godic et al. (2014)

Polyunsaturated fatty acids				
Arachidonic acid (AA)	• *Porphyridium* sp. • *Parietochloris incisa* • *Nannochloropsis oculata*	• Improves growth • Visual and functional development in infants • Develop the nervous system • Improve the immune system	• Infant formula • Nutritional supplements • Pharmaceuticals	Cheng-Wu et al. (2002), Su et al. (2016) and Tallima and El Ridi (2018)
Docosahexaenoic acid (DHA)	• *Crypthecodinium cohnii* • *Pavlova salina* • *Isochrysis galbana* • *Schizochytrium* sp.	• Anti-inflammatory • Anti-cardiovascular • Improve the nervous system and brain function	• Infant formula • Nutritional supplements • Aquaculture • Pharmaceuticals	Chu (2012)
Eicosapentaenoic acid (EPA)	• *Nannochloropsis* sp. • *Phaeodactylum tricornutum* • *Pavlova* • *Mundus subterraneus*	• Anti-inflammatory • Prevent cardiovascular disease • Mental development • Protect against atherosclerosis	• Nutritional supplements • Aquaculture • Pharmaceuticals	Borowitzka (2013), and Chu (2012)
Gamma-linolenic acids (GLAs)	• *Spirulina platensis* • *Dunaliella salina* • *Chlorococcum* spp.	• Antioxidant • Improve cardiovascular health • Reduce low-density lipoprotein cholesterol • Reduce blood pressure	• Infant formula • Nutritional supplements • Anti-aging • Anti-wrinkle • Treat skin hyperplasias	Choopani et al. (2016) and Santhosh et al. (2016)
Proteins				
Mycosporine-like amino acids (MAAs)	• *Aphanizomenon flos-aquae* • *Porphyra umbilicalis*	• Antioxidant • Anti-inflammatory • Protect against UV rays	• Anti-aging • Anti-wrinkle • Nourish skin and hairs	Daniel et al. (2004)
Peptides/ oligopeptides	• *Chlorella pyrenoidosa* • *Nannochloropsis oculata* • *Spirulina maxima*	• Angiotensin-converting enzyme (ACE) inhibitory activity • Reduce cholesterol • Antiproliferative • Antioxidant	• Diagnosis disorders of brain and central nervous system	Sheih et al. (2009b)

(Continued)

Bioactive compounds	*Major sources*	*Health benefits*	*Applications*	*References*
Phycocyanin	• *Spirulina platensis* • *Phaeodactylum tricornutum* • *Galdieria sulphuraria*	• Antioxidant • Anti-inflammatory • Neuroprotective • Hepatoprotective	• Natural food colorant • Cosmetics • Pharmaceuticals	Elias et al. (2008) and Romay et al. (2003)
Phycoerythrin	• *Porphyridium cruentum*	• Antioxidant • Anti-inflammatory • Anticancer • Prevent neurodegenerative diseases	• Diagnostic fluorescent agent • Skin-soothing and tightening • Anti-wrinkle • Food color	Aneiros and Garateix (2004), Elias et al. (2008) and
Polysaccharides				
Insoluble dietary fiber	• *Laminaria* sp. • *Fucus spiralis*	• Decrease digestive tract transit time • Resistance to microbial and human enzymes	• Food products • Bulking capacity • Improve laxation	de Jesus Raposo et al. (2016) and Li and Komarek (2017)
Sulfated polysaccharides	• *Chlorella pyrenoidosa* • *Porphyridium* sp. • *Phaeodactylum tricornutum* • *Kappaphycus alvarezii* • *Gelidium* sp. • *Gracilaria* sp.	• Antioxidant • Anti-inflammatory • Antiviral • Antihypertensive • Anticancer • Reduce cholesterol and glucose level	• Food products • Pharmaceutical industry • Creams and lotions • Viscosifiers • Lubricants	Murata and Nakazoe (2001), Morrissey et al. (2001) and Holdt and Kraan (2011)
Polyphenols				
Phlorotannins	• *Sargassum fusiforme* • *Ascophyllum nodosum* • *Bifurcaria bifurcate* • *Fucus vesiculosus*	• Antioxidant • Antidiabetic • Anti-human immunodeficiency virus • Antihypertensive • Matrix metalloproteinase enzyme inhibition • Hyaluronidase enzyme inhibition • Radioprotective • Anti-aging	• Cosmeceuticals • Pharmaceuticals • Food	Agregan et al. (2017) and Li et al. (2017)

Vitamins				
Vitamin B_{12} (cobalamin)	• *Spirulina* • *Chlorella* • *Dunaliella tertiolecta*	• Anti-aging • Anti-anemia • Cure sleep abnormalities • Improve bone health	• Production of red blood cells • Boost energy, skin and hair health	Lordan et al. (2011)
Vitamin C (ascorbic acid)	• *Prototheca moriformis* • *Gelidiella acerosa* • *Ulva reticulata*	• Antioxidant • Enhance the immune system • Reduce blood pressure • Anticancer	• Food nutrition • Pharmaceutical • Strengthen bones and teeth	Burtin (2003)
Vitamin E (tocopherol)	• *Spirulina platensis* • *Porphyridium cruentum* • *Ascophyllum* • *Ulva* • *Euglena gracilis*	• Anticancer • Improve cardiovascular health • Improve vision • Prevent neurodegenerative diseases • Protect against UV rays	• Food nutrition • Pharmaceutical • Skin creams and lotions	Ibañez et al. (2012) and Misurcova (2011)
Minerals				
Calcium, zinc, phosphorus, potassium, iron and magnesium	• *Chlorella vulgaris* • *Dunaliella tertiolecta* • *Isochrysis galbana* • *Spirulina platensis*	• Reduce blood pressure • Reduce deficiency in pregnant women • Anti-osteoporosis • Anti-goiter • Anti-depressant • Promote growth • Strengthen bones and teeth	• Dietary supplements • Health and nutrition products • Pharmaceuticals	Taboada et al. (2010) and Hamed et al. (2015)

Digestion and bioavailability of nutraceuticals

Human beings, who are considered as the potential food consumers, prefer to uptake the food items both for the nutritional benefits and to utilize the biological functions of the foods or food supplements for the intensive development of the internal or external structures of the body (Jafari et al., 2019). The effectiveness of the biologically active components (or nutraceuticals) present in the food or food supplements is directly dependent upon their bioavailability, bioaccessibility, absorption and transformation (Dima et al., 2020). Bioaccessibility refers to the extraction of the bioactive molecules from the food matrix into the body fluids, and depending upon the polarity, the molecules get dissolved differently; for example, polar molecules (e.g., polyphenols, polysaccharides) dissolve in gastrointestinal fluid, whereas hydrophobic molecules (e.g., lipids, phospholipids) are extracted in the intestinal fluid by the help of several lipolytic enzymes (Katouzian et al., 2017).

The extraction and solubilization of the bioactive molecules from the food matrix (e.g., algal or non-algal) depend directly upon the physicochemical properties of the food matrix (e.g., morphological features, dimension, texture, electrical properties), the properties of the bioactive molecules such as polarity, pH and ionic strength, and the concentration of the different types of enzymes helping in the degradation of food materials in the body fluids (Rafiee and Jafari, 2019). In the second stage of the process, the extracted bioactive compounds get absorbed in the epithelial membrane possessing several types of mechanisms, and the major types of bioactive compounds get absorbed in the small intestine (Dima et al., 2020). The third and the most crucial stage of the bioavailability is the degradation or metabolism (transformation) of the absorbed biomolecules into specific molecular fragments, which serve several functions in the human body. The last stage of bioavailability can be related to the digestion of the bioactive compounds, and the process of digestion is mechanized in specific ways for different bioactive components. Depending upon the nature of the bioactive compounds, digestion or the degradation process commenced from the chewing process from the mouth.

The major algal nutraceuticals such as carotenoids isomerize to the inactive *trans*-form, and proteins or peptides are hydrolyzed into the building blocks, say amino acids, by the breaking action of mild acids or proteolytic enzymes (Mahmoud and Chien, 2018; Rostamabadi et al., 2019). Similarly, polysaccharides and fibrous components obtained from the algal biomasses are hydrolyzed in the acidic atmosphere or in the presence of glycosidic enzymes (α-amylase, β-glucosidase, etc.) into corresponding sugars or other compounds (Wells et al., 2017). Although there are very few studies available on the bioavailability of polyphenols, some studies reveal that the higher availability of polyphenols, such as catechol and phlorotannins (major compounds found in brown algae), may proceed to the lower bioavailability of proteins (Ganesan et al., 2019).

In the process of bioavailability, the stability of specific bioactive compounds is hampered by the early degradation reactions in the presence of various pro-oxidants or by the insolubility of the products in the body fluids (Edelman et al., 2019). To overcome this constraint, some recent research studies have focused their interest on the encapsulation or emulsification of several algal-derived nutraceuticals (e.g., astaxanthin, other carotenoids, phytosterols) using various nanocarriers such as protein-based, carbohydrate-based and other emulsifying agents (Edelman et al., 2019; Mohammadi et al., 2020; Nooshkam and Varidi, 2020). The nanoencapsulation of these unstable biomolecules by these emulsifying agents increases the stability of the molecules by 40–60%, enhances the antioxidant activities to a certain higher level and ultimately improvises the bioavailability of several algal or non-algal nutraceuticals (Edelman et al., 2019; Mohammadi et al., 2020).

Biomagnification

Biomagnification is the accumulation of toxic chemicals (toxins) in the tissues of resistant species at a successively higher stage in a food chain or food web. The accumulation of toxins occurs successively from the lower level of the food chain (algae) to the highest level, say human beings in multiple manners. There are certain specific toxic molecules such as methylmercury, dichlorodiphenyltrichloroethane (DDT), several heavy metals (e.g., As, Pb, Cr, Cd) and benzene derivatives, which do not have the scope to be degraded irrespective of the environmental conditions and become persistent in the tissues of algae or other organisms (McQueen et al., 2020).

The persistence and concentration of the toxins in an organism are directly dependent upon the tolerance levels of them to accommodate the foreign toxic molecules. According to the studies, algal biomasses are considered as an efficient candidate for the adsorption of various toxic metals such as Al, Zn, Pb, Cd and As and for the treatment of contaminated effluents by adsorbing the toxic materials (Li et al., 2020). These studies prove the higher acceptability of several toxins from the aquatic media, some of which persist in the algal tissues for a prolonged period due to the incredible stability of the toxic molecules toward all the environmental and internal circumstances. In this way, toxic molecules are intensified to the subsequent organisms of the food chain, depending upon the algae for their food.

Challenges for algal ingredients in nutraceuticals

Marine algal-based functional foods or their extracts in foods play an essential role in preventing many diseases or help in stimulating an immune response. While significant progress has been made in commercializing certain algal-derived products, many challenges are yet to be overcome regarding algae production and their use as nutraceuticals. The main problem is the lack of public awareness of food and feed products derived from algae. Differences in the climatic and environmental conditions led to the variation in the bioactive composition of algae. The strong fishy flavor, odor and taste of algae limit their use in food products. Currently, proteins and carbohydrates from algae are produced in low volume in industries and have high cost compared with agro-food products.

Production and commercialization of algal-based products need to meet many conditions such as food safety and health claims before launching them. Possible risks associated with algae include toxin compounds, allergens, contamination with pathogens or production of harmful secondary metabolites. There are reports of severe allergic reactions after consuming wild-harvested Spirulina tablets for the first time (Le et al., 2014). Higher doses of kainic acid, an amino acid present in some of the red algae, are reported to damage neurons (Mouritsen et al., 2013). Some sicknesses and deaths were reported after the intake of toxic species of *Gracilaria* and *Caulerpa* (Cheney, 2016). Contamination of the heavy metals such as Arsenic, Lead and Cadmium is seen in seaweed processed food and nutritional supplements (Almela et al., 2006). Prostaglandin, the algae toxin, has low toxicity and can cause food poisoning after ingestion (Hammann et al., 2016).

Various claims are made regarding the quality and contamination of algal-derived dietary supplements at the production and processing stages due to the lack of proper Good Manufacturing Practice (GMP) regulations. According to the European Union regulations on food safety (Regulation No. 178/2002) and novel food (Regulation No. 258/97), one has to provide scientific evidence that novel algal-derived food product is safe for human and animal consumption (Regulation, 1997; European Commission, 2002). Shortly, genetically modified algae will be produced, which will increase algal productivity and will decrease the production cost of algal-derived products. Therefore, to make algal-based functional food successful in the market, one should spread awareness and knowledge about the health benefits of algae-related functional ingredients. There should be developments in science and technology to create and optimize functional foods from algal biomass and monitor their production. Besides, scientific evidence linked with the consumption of bioactive compounds in marine algae to prevent certain diseases should be provided.

Conclusions

Seawater and freshwater algae are known for their potential applications in various fields. Apart from being a biofuel feedstock, algal direct foods, food additives and bioactive compounds have existed from ancient times. To obtain the functional food ingredients on a large scale and for a large group of people, algal biomasses are now widely grown to form the commercial formulation. The enormous cultivation of algal biomasses also helps in absorbing atmospheric CO_2, which creates environmental sustainability in both direct and indirect ways.

In the context of nutraceuticals or bioactive compounds, both microalgae and macroalgae constitute many vital components, including polyphenols such as phlorotannins, complexes of omega fatty acids, vitamins, proteins, several polysaccharides and various minerals. These bioactive components have proven their potentiality in the direction of the antiviral, antibacterial, antifungal, anti-aging and, most importantly, antioxidant activities. Therefore, along with the food supplements and other functional foods derived from algae, algal resources and bioactive compounds derived from the algal bioresources are popularly utilized for the numerous implementations in pharmaceuticals and cosmeceuticals for treating several chronic diseases (e.g., Alzheimer's, Parkinson's and cardiac diseases) and skin disorders. Considering the diminishing arable lands, pervasive contaminants in food crops and leafy vegetables, and lifestyle-related diseases, it is quite promising that the food ingredients extracted from algal biomass can act as nutritional and therapeutic compounds for humankind.

References

Admassu, H., Zhao, W., Yang, R., Gasmalla, M., Alsir, E. (2015). Development of functional foods: seaweeds (algae) untouched potential and alternative resource—a review. *International Journal of Scientific Technology Research* 4(1): 108–115.

Agregan, R., Munekata, P. E., Franco, D., Dominguez, R., Carballo, J., Lorenzo, J. M. (2017). Phenolic compounds from three brown seaweed species using LC-DAD–ESI-MS/MS. *Food Research International* 99: 979–985.

Ahmadi, A., Zorofchian Moghadamtousi, S., Abubakar, S., Zandi, K. (2015). Antiviral potential of algae polysaccharides isolated from marine sources: a review. *BioMed Research International* 2015: 825203.

Almela, C., Clemente, M. J., Vélez, D., Montoro, R. (2006). Total arsenic, inorganic arsenic, lead and cadmium contents in edible seaweed sold in Spain. *Food and Chemical Toxicology* 44: 1901–1908.

Aneiros, A., Garateix, A. (2004). Bioactive peptides from marine sources: pharmacological properties and isolation procedures. *Journal of Chromatography B* 803: 41–53.

Ariede, M. B., Candido, T. M., Jacome, A. L. M., Velasco, M. V. R., de Carvalho, J. C. M., Baby, A. R. (2017). Cosmetic attributes of algae-A review. *Algal Research* 25:483–487.

Baker, D., Basondwah, S., Jambi, E., Rahimuddin, S. A., Abuzaid, M., Aly, M., Navarro, D. M. A. F. (2017). An assessment of the antioxidant and antimicrobial activity of six species of edible Irish seaweeds. *Pakistan Journal of Biological Sciences* 22: 16–30.

Banfalvi, G. (ed.) (2016). *Permeability of Biological Membranes*. Berlin: Springer.

Bishop, W. M., Zubeck, H. M. (2012). Evaluation of microalgae for use as nutraceuticals and nutritional supplements. *Journal of Nutrition Food Sciences* 2: 1–6.

Bixler, H. J. (2017). The carrageenan controversy. *Journal of Applied Phycology* 29: 2201–2207.

Boobathy, S., Soundarapandian, P., Prithivraj, M., Gunasundari, V. (2010). Biochemical characterization of protein isolated from seaweed, *Gracilaria edulis*. *Current Research Journal of Biological Sciences* 2: 35–37.

Borowitzka, M. A. (2013). High-value products from microalgae—their development and commercialisation. *Journal of Applied Phycology* 25: 743–756.

Bouga, M., Combet, E. (2015). Emergence of seaweed and seaweed-containing foods in the UK: focus on labeling, iodine content, toxicity and nutrition. *Foods* 4: 240–253.

Burja, A. M., Radianingtyas, H. (2008). 'Nutraceuticals and functional foods from marine microbes: an introduction to a diverse group of natural products isolated from marine macroalgae, microalgae, bacteria, fungi, and cyanobacteria' in Barrow, C. and Shahidi, F. (eds.) *Marine Nutraceuticals and Functional Foods*. CRC Press, pp. 367–403.

Burtin, P. (2003). Nutritional value of seaweeds. *Electronic Journal of Environmental, Agricultural and Food Chemistry* 2: 498–503.

Businesswire. Global Algae Products Market Nears $15 Billion by 2024- Market Size of Pharmaceuticals Nutraceuticals, Food Beverages, Cosmetics, and Other Applications - ResearchAndMarkets.com. https://www.businesswire.com/news/home/20190916005637/en/Global-Algae-Products-Market-Nears-15Billion#:~:text=The%20global%20algae%20products%20market, as%20a%20promising%20substitute%20for (Accessed on June 8, 2020).

Caporgno, M. P., Mathys, A. (2018). Trends in microalgae incorporation into innovative food products with potential health benefits. *Frontiers in Nutrition* 5: 58.

Carrillo, S., López, E., Casas, M. M., Avila, E., Castillo, R. M., Carranco, M. E., Pérez-Gil, F. (2008). 'Potential use of seaweeds in the laying hen ration to improve the quality of n-3 fatty acid enriched eggs' in Borowitzka, M. A., Critchley, A. T., Kraan, S., Peters, A., Sjøtun, K., Notoya, M. (eds.) *Nineteenth International Seaweed Symposium*. Netherlands: Springer, pp. 271–278.

Chandra, R., Parra, R., MN Iqbal, H. (2017). Phycobiliproteins: a novel green tool from marine origin blue-green algae and red algae. *Protein and Peptide Letters* 24: 118–125.

Chen, F., Wang, S., Guo, W., Hu, M. (2005a). Determination of amino acids in *Sargassum fusiforme* by high performance capillary electrophoresis. *Talanta* 66: 755–761.

Chen, H. M., Zheng, L., Yan, X. J. (2005b). The preparation and bioactivity research of agaro-oligosaccharides. *Food Technology and Biotechnology* 43: 29–36.

Cheney, D. (2016). 'Toxic and harmful seaweeds' in Fleurence, J., Levine, I. (eds.) *Seaweed in Health and Disease Prevention*. Academic Press, pp. 407–421.

Cheng-Wu, Z., Cohen, Z., Khozin-Goldberg, I., Richmond, A. (2002). Characterization of growth and arachidonic acid production of *Parietochloris incisa* comb. nov (*Trebouxiophyceae, Chlorophyta*). *Journal of Applied Phycology* 14: 453–460.

Cherry, P., O'Hara, C., Magee, P. J., McSorley, E. M., Allsopp, P. J. (2019). Risks and benefits of consuming edible seaweeds. *Nutrition Reviews* 77: 307–329.

Choopani, A., Poorsoltan, M., Fazilati, M., Latifi, A. M., Salavati, H. (2016). *Spirulina*: a source of gamma-linoleic acid and its applications. *Journal of Applied Biotechnology Reports* 3: 483–488.

Christaki, E., Bonos, E., Giannenas, I., Florou-Paneri, P. (2013). Functional properties of carotenoids originating from algae. *Journal of the Science of Food and Agriculture* 93: 5–11.

Christaki, E., Florou-Paneri, P., Bonos, E. (2011). Microalgae: a novel ingredient in nutrition. *International Journal of Food Sciences and Nutrition* 62(8): 794–799.

Chu, W. L. (2012). Biotechnological applications of microalgae. *International e-Journal of Science, Medicine Education* 6: S24–S37.

Cikoš, A. M., Jerković, I., Molnar, M., Šubarić, D., Jokić, S. (2021). New trends for macroalgal natural products applications. *Natural Product Research* 35: 1180–1191.

Colla, L. M., Muccillo-Baisch, A. L., Costa, J. A. V. (2008). *Spirulina platensis* effects on the levels of total cholesterol, HDL and triacylglycerols in rabbits fed with a hypercholesterolemic diet. *Brazilian Archives of Biology and Technology* 51: 405–411.

Dang, T. T., Bowyer, M. C., Van Altena, I. A., Scarlett, C. J. (2018). Optimum conditions of microwave-assisted extraction for phenolic compounds and antioxidant capacity of the brown alga *Sargassum vestitum*. *Separation Science and Technology* 53(11): 1711–1723.

Daniel, S., Cornelia, S., Fred, Z. (2004). UV-A sunscreen from red algae for protection against premature skin aging. *Cosmet Toiletries Manufacture Worldwide* 2004: 139–143.

de Boer, K., Moheimani, N. R., Borowitzka, M. A., Bahri, P. A. (2012). Extraction and conversion pathways for microalgae to biodiesel: a review focused on energy consumption. *Journal of Applied Phycology* 24(6): 1681–1698.

de Jesus Raposo, M. F., De Morais, A. M. M. B., De Morais, R. M. S. C. (2016). Emergent sources of prebiotics: seaweeds and microalgae. *Marine Drugs* 14: 27.

Dima, C., Assadpour, E., Dima, S., Jafari, S. M. (2020). Bioavailability of nutraceuticals: role of the food matrix, processing conditions, the gastrointestinal tract, and nanodelivery systems. *Comprehensive Reviews in Food Science and Food Safety* 19(3): 954–994.

D'Odorico, A., Martines, D., Kiechl, S., Egger, G., Oberhollenzer, F., Bonvicini, P., Willeit, J. (2000). High plasma levels of α- and β-carotene are associated with a lower risk of atherosclerosis: results from the Bruneck study. *Atherosclerosis* 153(1): 231–239.

Domínguez, H. (2013). 'Algae as a source of biologically active ingredients for the formulation of functional foods and nutraceuticals' in Domínguez, H. (ed.) *Functional Ingredients from Algae for Foods and Nutraceuticals*. Cambridge: Woodhead Publishing, pp. 1–19.

Druesne-Pecollo, N., Latino-Martel, P., Norat, T., Barrandon, E., Bertrais, S., Galan, P., Hercberg, S. (2010). Beta-carotene supplementation and cancer risk: a systematic review and metaanalysis of randomized controlled trials. *International Journal of Cancer* 127(1): 172–184.

Dwyer, J. H., Paul-Labrador, M. J., Fan, J., Shircore, A. M., Merz, C. N. B., Dwyer, K. M. (2004). Progression of carotid intima-media thickness and plasma antioxidants: the Los Angeles Atherosclerosis Study. *Arteriosclerosis, Thrombosis, and Vascular Biology* 24(2): 313–319.

Edelman, R., Engelberg, S., Fahoum, L., Meyron-Holtz, E. G., Livney, Y. D. (2019). Potato protein-based carriers for enhancing bioavailability of astaxanthin. *Food Hydrocolloids* 96: 72–80.

EFSA Panel on Food Additives and Nutrient Sources added to Food (ANS) (2015). Scientific Opinion on the re-evaluation of chlorophylls (E 140 (i)) as food additives. *EFSA Journal* 13(5): 4089.

EFSA Panel on Food Additives and Nutrient Sources added to Food (ANS). (2016). Safety of the proposed extension of use of synthetic β-carotene [E 160a (ii)] in foods for special medical purposes in young children. *EFSA Journal* 14(3): 4434.

Elias, R. J., Kellerby, S. S., Decker, E. A. (2008). Antioxidant activity of proteins and peptides. *Critical Reviews in Food Science and Nutrition* 48(5): 430–441.

Endo, J., Arita, M. (2016). Cardioprotective mechanism of omega-3 polyunsaturated fatty acids. *Journal of Cardiology* 67(1): 22–27.

European Commission. (2002). Regulation (EC) No 178/2002 of the European Parliament and of the Council of 28 January 2002 laying down the general principles and requirements of food law, establishing the European Food Safety Authority and laying down procedures in matters of food safety. *Official Journal of the European Communities* 31(01/02/2002): 1–24.

Fabrowska, J., Ibañez, E., Łęska, B., Herrero, M. (2016). Supercritical fluid extraction as a tool to valorize underexploited freshwater green algae. *Algal Research* 19: 237–245.

Fomenko, S. E., Kushnerova, N. F., Sprygin, V. G., Drugova, E. S., Lesnikova, L. N., Merzlyakov, V. Y., Momot, T. V. (2019). Lipid composition, content of polyphenols, and antiradical activity in some representatives of marine algae. *Russian Journal of Plant Physiology* 66(6): 942–949.

Food and Drug Administration. (2002). Code of federal regulations, Title 21-food and drugs, chapter I-food and drug administration, department of health and human services, subchapter B-food for human consumption, part 101-food labeling, subpart A-general provisions, sec. 101.12 reference amounts customarily consumed per eating occasion. US Government Printing Office: Washington, DC.

Freile-Pelegrín, Y., Robledo, D. (2013). 'Bioactive phenolic compounds from algae' in Hernández-Ledesma, B., Herrero, M. (eds.) *Bioactive Compounds from Marine Foods: Plant and Animal Sources*. Chichester: John Wiley Sons Ltd., pp. 113–129.

Fu, W., Nelson, D. R., Yi, Z., Xu, M., Khraiwesh, B., Jijakli, K., Salehi-Ashtiani, K. (2017). 'Bioactive compounds from microalgae: current development and prospects' in ur-Rahman, A. (ed.) *Studies in Natural Products Chemistry (Vol. 54)*. Netherlands: Elsevier, pp. 199–225.

Gammone, M. A., D'Orazio, N. (2015). Anti-obesity activity of the marine carotenoid fucoxanthin. *Marine Drugs* 13(4): 2196–2214.

Ganesan, A. R., Tiwari, U., Rajauria, G. (2019). Seaweed nutraceuticals and their therapeutic role in disease prevention. *Food Science and Human Wellness* 8(3): 252–263.

Gao, S., Qin, T., Liu, Z., Caceres, M. A., Ronchi, C. F., Chen, C. Y. O., Yeum, K. J., Taylor, A., Blumberg, J. B., Liu, Y., Shang, F. (2011). Lutein and zeaxanthin supplementation reduces H_2O_2-induced oxidative damage in human lens epithelial cells. *Molecular Vision* 17: 3180.

Ghosh, T., Chattopadhyay, K., Marschall, M., Karmakar, P., Mandal, P., Ray, B. (2009). Focus on antivirally active sulfated polysaccharides: from structure–activity analysis to clinical evaluation. *Glycobiology* 19(1): 2–15.

GlobeNewswire. Global Nutraceuticals Market Assessment 2019: Market Overview, Share, Growth Drivers, and Key Trends. https://www.globenewswire.com/news-release/2019/08/20/1904073/0/en/Global-Nutraceuticals-Market-Assessment-2019-Market-Overview-Share-Growth-Drivers-and-Key-Trends.html (Accessed on June 8, 2020).

Godic, A., Poljšak, B., Adamic, M., Dahmane, R. (2014). The role of antioxidants in skin cancer prevention and treatment. *Oxidative Medicine and Cellular Longevity* 2014:860479.

Gong, M., Bassi, A. (2016). Carotenoids from microalgae: a review of recent developments. *Biotechnology Advances* 34: 1396–1412.

Grimmig, B., Kim, S. H., Nash, K., Bickford, P. C., Shytle, R. D. (2017). Neuroprotective mechanisms of astaxanthin: a potential therapeutic role in preserving cognitive function in age and neurodegeneration. *Geroscience* 39(1): 19–32.

Guedes, A. C., Amaro, H. M., Malcata, F. X. (2011). Microalgae as sources of carotenoids. *Marine Drugs* 9(4): 625–644.

Hamed, I., Özogul, F., Özogul, Y., Regenstein, J. M. (2015). Marine bioactive compounds and their health benefits: a review. *Comprehensive Reviews in Food Science and Food Safety* 14(4): 446–465.

Hammann, M., Rempt, M., Pohnert, G., Wang, G., Boo, S. M., Weinberger, F. (2016). Increased potential for wound activated production of Prostaglandin E2 and related toxic compounds in non-native populations of *Gracilaria vermiculophylla*. *Harmful Algae* 51: 81–88.

Hans, N., Naik, S. N., Malik, A. (2019). 'Platform molecules from algae by using supercritical CO_2 and subcritical water extraction' in Ravishankar, G. A., Ambati, R. R. (eds.) *Handbook of Algal Technologies and Phytochemicals*. CRC Press, pp. 229–243.

He, M. L., Hollwich, W., Rambeck, W. A. (2002). Supplementation of algae to the diet of pigs: a new possibility to improve the iodine content in the meat. *Journal of Animal Physiology and Animal Nutrition* 86(3–4): 97–104.

Higuera-Ciapara, I., Felix-Valenzuela, L., Goycoolea, F. M. (2006). Astaxanthin: a review of its chemistry and applications. *Critical Reviews in Food Science and Nutrition* 46(2): 185–196.

Himaya, S. W. A., Ngo, D. H., Ryu, B., Kim, S. K. (2012). An active peptide purified from gastrointestinal enzyme hydrolysate of Pacific cod skin gelatin attenuates angiotensin-1 converting enzyme (ACE) activity and cellular oxidative stress. *Food Chemistry* 132(4): 1872–1882.

Holdt, S. L., Kraan, S. (2011). Bioactive compounds in seaweed: functional food applications and legislation. *Journal of Applied Phycology* 23(3): 543–597.

Hori, K., Matsubara, K., Miyazawa, K. (2000). Primary structures of two hemagglutinins from the marine red alga, *Hypnea japonica*. *Biochimica et Biophysica Acta (BBA)-General Subjects* 1474(2): 226–236.

Howlett, J. (ed.) (2008). *Functional Foods: From Science to Health and Claims*. ILSI Europe.

Ibañez, E., Herrero, M., Mendiola, J. A., Castro-Puyana, M. (2012). 'Extraction and characterization of bioactive compounds with health benefits from marine resources: macro and micro algae, cyanobacteria, and invertebrates' in Maria, H. (ed.) *Marine Bioactive Compounds*. Springer, pp. 55–98.

Jafari, S. M., Vakili, S., Dehnad, D. (2019). Production of a functional yogurt powder fortified with nanoliposomal vitamin D through spray drying. *Food and Bioprocess Technology* 12(7): 1220–1231.

Jiménez-Escrig, A., Sánchez-Muniz, F. J. (2000). Dietary fibre from edible seaweeds: chemical structure, physicochemical properties and effects on cholesterol metabolism. *Nutrition Research* 20(4): 585–598.

Jin, E. O. N. S. E. O. N., Polle, J. E., Lee, H. K., Hyun, S. M., Chang, M. (2003). Xanthophylls in microalgae: from biosynthesis to biotechnological mass production and application. *Journal of Microbiology and Biotechnology* 13(2): 165–174.

Jones, C. S., Luong, T., Hannon, M., Tran, M., Gregory, J. A., Shen, Z., Mayfield, S. P. (2013). Heterologous expression of the C-terminal antigenic domain of the malaria vaccine candidate Pfs48/45 in the green algae *Chlamydomonas reinhardtii*. *Applied Microbiology and Biotechnology* 97(5): 1987–1995.

Kamble, S. P., Vikhe, G. P., Chamle, D. R. (2018). Extraction and purification of phycoerythrin—a natural colouring agent from *Spirulina platensis*. *Journal of Pharmaceutical, Chemical and Biological Sciences* 6(2): 78–84.

Kang, H., Kim, H. (2017). Astaxanthin and β-carotene in *Helicobacter pylori*-induced gastric inflammation: a mini-review on action mechanisms. *Journal of Cancer Prevention* 22(2): 57.

Katouzian, I., Esfanjani, A. F., Jafari, S. M., Akhavan, S. (2017). Formulation and application of a new generation of lipid nano-carriers for the food bioactive ingredients. *Trends in Food Science Technology* 68: 14–25.

Kidd, P. (2011). Astaxanthin, cell membrane nutrient with diverse clinical benefits and anti-aging potential. *Alternative Medicine Review* 16(4): 355–364.

Kim, H. W., Chew, B. P., Wong, T. S., Park, J. S., Weng, B. B., Byrne, K. M., Hayek, M. G., Reinhart, G. A. (2000). Modulation of humoral and cell-mediated immune responses by dietary lutein in cats. *Veterinary Immunology and Immunopathology* 73(3–4): 331–341.

Koyande, A. K., Chew, K. W., Rambabu, K., Tao, Y., Chu, D. T., Show, P. L. (2019). Microalgae: a potential alternative to health supplementation for humans. *Food Science and Human Wellness* 8(1): 16–24.

Kraan, S. (2013). 'Pigments and minor compounds in algae' in Dominguez, H. (ed.) *Functional Ingredients from Algae for Foods and Nutraceuticals.* Woodhead Publishing, pp. 205–251.

Kumar, C. S., Ganesan, P., Suresh, P. V., Bhaskar, N. (2008). Seaweeds as a source of nutritionally beneficial compounds-a review. *Journal of Food Science and Technology* 45(1): 1.

Kumari, P., Kumar, M., Gupta, V., Reddy, C. R. K., Jha, B. (2010). Tropical marine macroalgae as potential sources of nutritionally important PUFAs. *Food Chemistry* 120(3): 749–757.

Le, T. M., Knulst, A. C., Röckmann, H. (2014). Anaphylaxis to *Spirulina* confirmed by skin prick test with ingredients of *Spirulina* tablets. *Food and Chemical Toxicology* 74: 309–310.

Leyton, A., Pezoa-Conte, R., Barriga, A., Buschmann, A. H., Mäki-Arvela, P., Mikkola, J. P., Lienqueo, M. E. (2016). Identification and efficient extraction method of phlorotannins from the brown seaweed *Macrocystis pyrifera* using an orthogonal experimental design. *Algal Research* 16: 201–208.

Li, H., Watson, J., Zhang, Y., Lu, H., Liu, Z. (2020). Environment-enhancing process for algal wastewater treatment, heavy metal control and hydrothermal biofuel production: A critical review. *Bioresource Technology* 298: 122421.

Li, Y., Fu, X., Duan, D., Liu, X., Xu, J., Gao, X. (2017). Extraction and identification of phlorotannins from the brown alga, *Sargassum fusiforme* (Harvey) Setchell. *Marine Drugs* 15(2): 49.

Li, Y. O., Komarek, A. R. (2017). Dietary fibre basics: health, nutrition, analysis, and applications. *Food Quality and Safety* 1(1): 47–59.

Lin, D., Xiao, M., Zhao, J., Li, Z., Xing, B., Li, X., Chen, H. (2016). An overview of plant phenolic compounds and their importance in human nutrition and management of type 2 diabetes. *Molecules* 21(10): 1374.

Lopez-Santamarina, A., Miranda, J. M., Mondragon, A. D. C., Lamas, A., Cardelle-Cobas, A., Franco, C. M., Cepeda, A. (2020). Potential use of marine seaweeds as prebiotics: a review. *Molecules* 25(4): 1004.

Lordan, S., Ross, R. P., Stanton, C. (2011). Marine bioactives as functional food ingredients: potential to reduce the incidence of chronic diseases. *Marine Drugs* 9(6): 1056–1100.

Machu, L., Misurcova, L., Vavra Ambrozova, J., Orsavova, J., Mlcek, J., Sochor, J., Jurikova, T. (2015). Phenolic content and antioxidant capacity in algal food products. *Molecules* 20(1): 1118–1133.

Mahmoud, S. A., Chien, P. (2018). Regulated proteolysis in bacteria. *Annual Review of Biochemistry* 87: 677–696.

Mares-Perlman, J. A., Millen, A. E., Ficek, T. L., Hankinson, S. E. (2002). The body of evidence to support a protective role for lutein and zeaxanthin in delaying chronic disease. Overview. *The Journal of Nutrition* 132(3): 518S–524S.

Matanjun, P., Mohamed, S., Mustapha, N. M., Muhammad, K. (2009). Nutrient content of tropical edible seaweeds, *Eucheuma cottonii*, *Caulerpa lentillifera* and *Sargassum polycystum*. *Journal of Applied Phycology* 21(1): 75–80.

McQueen, A. D., Lotufo, G. R., Pickard, S. W., Lenox, A. M., Moore, D. W., von Stackelberg, K., Suedel, B. C. (2020). Evaluation of dredged sediment for aquatic placement: interpreting contaminant bioaccumulation. *Environmental Monitoring and Assessment* 192: 1–11.

Mekinić, I. G., Skroza, D., Šimat, V., Hamed, I., Čagalj, M., Perković, Z. P. (2019). Phenolic content of brown algae (*Pheophyceae*) species: extraction, identification, and quantification. *Biomolecules* 9(6): 244.

Michelon, M., de Borba, T. D. M., da Silva Rafael, R., Burkert, C. A. V., de Medeiros Burkert, J. F. (2012). Extraction of carotenoids from *Phaffia rhodozyma*: a comparison between different techniques of cell disruption. *Food Science and Biotechnology* 21(1): 1–8.

Misurcova, L. (2011). 'Chemical composition of seaweeds' in Kim, S. K. (ed.) *Handbook of Marine Macroalgae: Biotechnology and Applied Phycology*. John Wiley Sons, pp. 173–192.

Mohammadi, M., Jafari, S. M., Hamishehkar, H., Ghanbarzadeh, B. (2020). Phytosterols as the core or stabilizing agent in different nanocarriers. *Trends in Food Science Technology* 201: 73–88.

Moi, I. M., Leow, A. T. C., Ali, M. S. M., Rahman, R. N. Z. R. A., Salleh, A. B., Sabri, S. (2018). Polyunsaturated fatty acids in marine bacteria and strategies to enhance their production. *Applied Microbiology and Biotechnology* 102(14): 5811–5826.

Moldenhauer, M., Sluchanko, N. N., Buhrke, D., Zlenko, D. V., Tavraz, N. N., Schmitt, F. J., Friedrich, T. (2017). Assembly of photoactive orange carotenoid protein from its domains unravels a carotenoid shuttle mechanism. *Photosynthesis Research* 133(1–3): 327–341.

Morrissey, J., Kraan, S., Guiry, M. D. (eds.) (2001). *A Guide to Commercially Important Seaweeds on the Irish Coast*. Dublin: Irish Bord Iascaigh Mhara/Irish Sea Fisheries Board.

Mouritsen, O. G., Dawczynski, C., Duelund, L., Jahreis, G., Vetter, W., Schröder, M. (2013). On the human consumption of the red seaweed dulse (*Palmaria palmata* (*L.*) Weber Mohr). *Journal of Applied Phycology* 25(6): 1777–1791.

Murakami, S. (2017). The physiological and pathophysiological roles of taurine in adipose tissue in relation to obesity. *Life Sciences* 186: 80–86.

Murata, M., Nakazoe, J. I. (2001). Production and use of marine algae in Japan. *Japan Agricultural Research Quarterly: JARQ* 35(4): 281–290.

Müssig K. (2009). 'Iodine-induced toxic effects due to seaweed consumption' in Preedy, V. R., Burrow, G. N., Watson, R. R. (eds.) *Comprehensive Handbook of Iodine*. Academic Press, pp. 897–908.

Nagini, S., Palitti, F., Natarajan, A. T. (2015). Chemopreventive potential of chlorophyllin: a review of the mechanisms of action and molecular targets. *Nutrition and Cancer* 67(2): 203–211.

Naguib, Y. M. (2000). Antioxidant activities of astaxanthin and related carotenoids. *Journal of Agricultural and Food Chemistry* 48(4): 1150–1154.

Nakazawa, Y., Sashima, T., Hosokawa, M., Miyashita, K. (2009). Comparative evaluation of growth inhibitory effect of stereoisomers of fucoxanthin in human cancer cell lines. *Journal of Functional Foods* 1(1): 88–97.

Nethravathy, M. U. Mehar, J. G., Mudliar, S. N., Shekh, A. Y. (2019). Recent advances in microalgal bioactives for food, feed, and healthcare products: commercial potential, market space, and sustainability. *Comprehensive Reviews in Food Science and Food Safety* 18(6): 1882–1897.

Niu, J., Tian, L. X., Liu, Y. J., Yang, H. J., Ye, C. X., Gao, W., Mai, K. S. (2009). Effect of dietary astaxanthin on growth, survival, and stress tolerance of postlarval shrimp, *Litopenaeus vannamei*. *Journal of the World Aquaculture Society* 40(6): 795–802.

Nooshkam, M., Varidi, M. (2020). Maillard conjugate-based delivery systems for the encapsulation, protection, and controlled release of nutraceuticals and food bioactive ingredients: a review. *Food Hydrocolloids* 100: 105389.

Palmegiano, G. B., Agradi, E., Forneris, G., Gai, F., Gasco, L., Rigamonti, E., Zoccarato, I. (2005). *Spirulina* as a nutrient source in diets for growing sturgeon (*Acipenser baeri*). *Aquaculture Research* 36(2): 188–195.

Pangestuti, R., Kim, S. K. (2011). Biological activities and health benefit effects of natural pigments derived from marine algae. *Journal of Functional Foods* 3(4): 255–266.

Papadopoulos, G., Goulas, C., Apostolaki, E., Abril, R. (2002). Effects of dietary supplements of algae, containing polyunsaturated fatty acids, on milk yield and the composition of milk products in dairy ewes. *The Journal of Dairy Research* 69(3): 357.

Park, J. S., Chyun, J. H., Kim, Y. K., Line, L. L., Chew, B. P. (2010). Astaxanthin decreased oxidative stress and inflammation and enhanced immune response in humans. *Nutrition Metabolism* 7(1): 18.

Parniakov, O., Apicella, E., Koubaa, M., Barba, F. J., Grimi, N., Lebovka, N., Vorobiev, E. (2015). Ultrasound-assisted green solvent extraction of high-added value compounds from microalgae *Nannochloropsis* spp. *Bioresource Technology* 198: 262–267.

Patil, N. P., Le, V., Sligar, A. D., Mei, L., Chavarria, D., Yang, E. Y., Baker, A. B. (2018). Algal polysaccharides as therapeutic agents for atherosclerosis. *Frontiers in Cardiovascular Medicine* 5: 153.

Pawlik-Skowrońska, B., Pirszel, J., Brown, M. T. (2007). Concentrations of phytochelatins and glutathione found in natural assemblages of seaweeds depend on species and metal concentrations of the habitat. *Aquatic Toxicology* 83(3): 190–199.

Peixoto, M. J., Svendsen, J. C., Malte, H., Pereira, L. F., Carvalho, P., Pereira, R., Ozório, R. O. (2016). Diets supplemented with seaweed affect metabolic rate, innate immune, and antioxidant responses, but not individual growth rate in European seabass (*Dicentrarchus labrax*). *Journal of Applied Phycology* 28(3): 2061–2071.

Rafiee, Z., Jafari, S. M. (2019). 'Application of lipid nanocarriers for the food industry' in Merillon J. M., Ramawat, K. G. (eds.) *Bioactive Molecules in Food*. Springer, pp. 1–43.

Rajauria, G., Foley, B., Abu-Ghannam, N. (2016). Identification and characterization of phenolic antioxidant compounds from brown Irish seaweed *Himanthalia elongata* using LC-DAD–ESI-MS/MS. *Innovative Food Science Emerging Technologies* 37: 261–268.

Rao, A. R., Sindhuja, H. N., Dharmesh, S. M., Sankar, K. U., Sarada, R., Ravishankar, G. A. (2013). Effective inhibition of skin cancer, tyrosinase, and antioxidative properties by astaxanthin and astaxanthin esters from the green alga Haematococcus pluvialis. *Journal of Agricultural and Food Chemistry* 61(16): 3842–3851.

Rasmussen, R. S., Morrissey, M. T. (2007). Marine biotechnology for production of food ingredients. *Advances in Food and Nutrition Research* 52: 237–292.

Rastogi, R. P., Madamwar, D., Pandey, A. (eds.) (2017). *Algal Green Chemistry: Recent Progress in Biotechnology*. Netherlands: Elsevier.

Regulation, H. A. T. (1997). Regulation (EC) No 258/97 of the European Parliament and of the Council of 27 January 1997 concerning novel foods and novel food ingredients. *Official Journal of the European Communities* 40: 1–7.

Riccioni, G., D'Orazio, N., Franceschelli, S., Speranza, L. (2011). Marine carotenoids and cardiovascular risk markers. *Marine Drugs* 9(7): 1166–1175.

Rodrigues, D., Freitas, A. C., Pereira, L., Rocha-Santos, T. A., Vasconcelos, M. W., Roriz, M., Duarte, A. C. (2015). Chemical composition of red, brown and green macroalgae from Buarcos bay in Central West Coast of Portugal. *Food Chemistry* 183: 197–207.

Romay, C. H., Gonzalez, R., Ledon, N., Remirez, D., Rimbau, V. (2003). C-phycocyanin: a biliprotein with antioxidant, anti-inflammatory and neuroprotective effects. *Current Protein and Peptide Science* 4(3): 207–216.

Rostamabadi, H., Falsafi, S. R., Jafari, S. M. (2019). Nanoencapsulation of carotenoids within lipid-based nanocarriers. *Journal of Controlled Release* 298: 38–67.

Safafar, H., Van Wagenen, J., Møller, P., Jacobsen, C. (2015). Carotenoids, phenolic compounds and tocopherols contribute to the antioxidative properties of some microalgae species grown on industrial wastewater. *Marine Drugs* 13(12): 7339–7356.

Sajna, K. V., Gottumukkala, L. D., Sukumaran, R. K., Pandey, A. (2015). 'White biotechnology in cosmetics' in Pandey, A., Höfer, R., Taherzadeh, M., Nampoothiri, K. M., Larroche, C. (eds.) *Industrial Biorefineries White Biotechnology*. Elsevier, pp. 607–652.

Sánchez-Machado, D. I., López-Cervantes, J., López-Hernández, J., Paseiro-Losada, P. (2004). Simultaneous determination of thiamine and riboflavin in edible marine seaweeds by high-performance liquid chromatography. *Journal of Chromatographic Science* 42(3): 117–120.

Santhosh, S., Dhandapani, R., Hemalatha, N. (2016). Bioactive compounds from microalgae and its different applications-a review. *Advances in Applied Science Research* 7(4): 153–158.

Selmi, C., Leung, P. S., Fischer, L., German, B., Yang, C. Y., Kenny, T. P., Gershwin, M. E. (2011). The effects of *Spirulina* on anemia and immune function in senior citizens. *Cellular Molecular Immunology* 8(3): 248.

Shao, A. (2017). 'Global market entry regulations for nutraceuticals, functional foods, dietary/food/health supplements' in Bagchi, D., Nair, S. (eds.) *Developing New Functional Food and Nutraceutical Products*. Academic Press, pp. 279–290.

Sheih, I. C., Wu, T. K., Fang, T. J. (2009a). Antioxidant properties of a new antioxidative peptide from algae protein waste hydrolysate in different oxidation systems. *Bioresource Technology* 100(13): 3419–3425.

Sheih, I. C., Fang, T. J., Wu, T. K. (2009b). Isolation and characterisation of a novel angiotensin I-converting enzyme (ACE) inhibitory peptide from the algae protein waste. *Food Chemistry* 115(1): 279–284.

Sher, I., Tzameret, A., Peri-Chen, S., Edelshtain, V., Ioffe, M., Sayer, A., Rotenstreich, Y. (2018). Synthetic 9-cis-beta-carotene inhibits photoreceptor degeneration in cultures of eye cups from rpe65rd12 mouse model of retinoid cycle defect. *Scientific Reports* 8(1): 1–9.

Shipeng, Y., Woo, H. C., Choi, J. H., Park, Y. B., Chun, B. S. (2015). Measurement of antioxidant activities and phenolic and flavonoid contents of the brown seaweed *Sargassum horneri*: comparison of supercritical CO_2 and various solvent extractions. *Fisheries and Aquatic Sciences* 18: 123–130.

Smit, A. J. (2004). Medicinal and pharmaceutical uses of seaweed natural products: a review. *Journal of Applied Phycology* 16(4): 245–262.

Soletto, D., Binaghi, L., Lodi, A., Carvalho, J. C. M. D., Converti, A. (2005). Batch and fed-batch cultivations of *Spirulina platensis* using ammonium sulphate and urea as nitrogen sources. *Aquaculture* 243(1–4): 217–224.

Song, Y., Liu, L., Shen, H., You, J., Luo, Y. (2011). Effect of sodium alginate-based edible coating containing different antioxidants on quality and shelf life of refrigerated bream (*Megalobrama amblycephala*). *Food Control* 22(3–4): 608–615.

Spolaore, P., Joannis-Cassan, C., Duran, E., Isambert, A. (2006). Commercial applications of microalgae. *Journal of Bioscience and Bioengineering* 101(2): 87–96.

Stengel, D. B., Connan, S., Popper, Z. A. (2011). Algal chemodiversity and bioactivity: sources of natural variability and implications for commercial application. *Biotechnology Advances* 29(5): 483–501.

Su, G., Jiao, K., Chang, J., Li, Z., Guo, X., Sun, Y., Zeng, X., Lu, Y., Lin, L. (2016). Enhancing total fatty acids and arachidonic acid production by the red microalgae *Porphyridium purpureum*. *Bioresources and Bioprocessing* 3(1): 1–9.

Sugawara, T., Baskaran, V., Tsuzuki, W., Nagao, A. (2002). Brown algae fucoxanthin is hydrolyzed to fucoxanthinol during absorption by Caco-2 human intestinal cells and mice. *The Journal of Nutrition* 132(5): 946–951.

Taboada, C., Millán, R., Míguez, I. (2010). Composition, nutritional aspects and effect on serum parameters of marine algae *Ulva rigida*. *Journal of the Science of Food and Agriculture* 90(3): 445–449.

Tallima, H., El Ridi, R. (2018). Arachidonic acid: physiological roles and potential health benefits–a review. *Journal of Advanced Research* 11: 33–41.

Topuz, O. K., Gokoglu, N., Yerlikaya, P., Ucak, I., Gumus, B. (2016). Optimization of antioxidant activity and phenolic compound extraction conditions from red seaweed (*Laurencia obtuse*). *Journal of Aquatic Food Product Technology* 25(3): 414–422.

Tsukui, T., Konno, K., Hosokawa, M., Maeda, H., Sashima, T., Miyashita, K. (2007). Fucoxanthin and fucoxanthinol enhance the amount of docosahexaenoic acid in the liver of KKAy obese/diabetic mice. *Journal of Agricultural and Food Chemistry* 55(13): 5025–5029.

van Ginneken, V. J., Helsper, J. P., de Visser, W., van Keulen, H., Brandenburg, W. A. (2011). Polyunsaturated fatty acids in various macroalgal species from north Atlantic and tropical seas. *Lipids in Health and Disease* 10(1): 104.

Von Lintig, J., Vogt, K. (2000). Filling the gap in vitamin a research molecular identification of an enzyme cleaving β-carotene to retinal. *Journal of Biological Chemistry* 275(16): 11915–11920.

Wells, M. L., Potin, P., Craigie, J. S., Raven, J. A., Merchant, S. S., Helliwell, K. E., Smith, A. G., Camire, M. E., Brawley, S.H. (2017). Algae as nutritional and functional food sources: revisiting our understanding. *Journal of Applied Phycology* 29(2): 949–982.

Yuan, Y., Zhang, J., Fan, J., Clark, J., Shen, P., Li, Y., Zhang, C. (2018). Microwave assisted extraction of phenolic compounds from four economic brown macroalgae species and evaluation of their antioxidant activities and inhibitory effects on α-amylase, α-glucosidase, pancreatic lipase and tyrosinase. *Food Research International* 113: 288–297.

Zakaria, N. A., Ibrahim, D., Sulaiman, S. F., Supardy, A. (2011). Assessment of antioxidant activity, total phenolic content and in-vitro toxicity of Malaysian red seaweed, *Acanthophora spicifera*. *Journal of Chemical and Pharmaceutical Research* 3(3): 182–191.

Zhang, L., Wang, H. (2015). Multiple mechanisms of anti-cancer effects exerted by astaxanthin. *Marine Drugs* 13(7): 4310–4330.

10 Algal-derived pharmaceuticals

Antimicrobial, antiviral, antifungal, neuroprotective products, therapeutic proteins and drugs

Sharmila Narayanan and Santhi Raju Pilli

Introduction

There is a high demand for the production of chemicals, pharmaceutical compounds, food products and biofuels. These all can be produced from renewable feedstocks such as lignocellulosic materials based on forestry, agricultural or aquatic biomasses. In this view, biorefinery plays a very crucial role in combining essential steps to convert biomass into finished products (Cherubini, 2010). Flexible biorefinery plants are necessary to advance the development of bio-based products and processes with innovations. A lot of research is being carried out to scale up the biomass fractions of algae (hereafter referred to as algae or microalgae) to make large amounts of intermediate products that are necessary for the market growth for the introduction. Two primary possible operations in biorefinery are hydrothermal treatment and biochemical processes (Ruiz et al., 2017).

Biorefinery plants are used for different purposes such as verification of the process parameters obtained in the lab and pilot-scale data and preparation of design data for further upscaling. Feedstock fractions of materials made by lignocelluloses such as wood, straw, corn fodder (corncobs, stalks and leaves), bagasse from sugarcane or pulp, and algae are in further development and for the introduction of products for marketplaces. These days making of biomass from algal-based feedstock and many other valuable extracts from algal derivatives is gaining more attention due to their diversity, variable chemical composition and high photosynthetic rates. High value bioproducts made using algal biomass and their commercial utilization are gaining popularity.

Algal-based food additives and biopharmaceutical products will be the next-generation starting materials of omega-3 oils. In general, oils extracted through fishes are rich in omega-3 oils. In 2019, the Food and Drug Administration (FDA) declared that ω-3 fatty acids are qualified for fitness claims. Countries such as the USA and Canada are using omega-3 fatty acids, commercially named Algal-900 DHA composed of docosahexaenoic acid (DHA) and eicosapentaenoic acid (EPA) (Kris-Etherton et al., 2009). It is extracted from some types of algae oils. It is a well-known fact that ω-3 oils are essential structural components of the man's eye, brain, cerebral cortex and skin. When the human body cannot produce adequate amounts of ω-3 fatty oils, it has become a necessity to take them from an external source. In another study, it was reported that a supplement of DHA and arachidonic acid had reduced the occurrence of bronchitis (a respiratory illness) in newborn babies (Pastor et al., 2006). DHA was also used as a supplement in pregnancy to reduce the deficiency of ω-3 fatty oils in infant babies before birth (Mulder et al., 2014).

A large group of microalgae consists of *Chlorophyta* and *Streptophyta* with primarily *Chlorokybophyceae*, *Mesostigmatophyceae* and *Spirotaenia*. They are single-celled photosynthetic creatures typically found in fresh- and seawaters. Nevertheless, the algae are multicellular that can be further classified into microscopic (microalgae) and macroscopic (macroalgae) weeds. Out of the total population, the majority are microalgae. Algae such as *Chlorella* are a good research material that can be used in the investigation of photosynthesis. Algae are generally identified with green color, but they also happened to be in a different range of colors owing to the presence of different colors such as green (*chlorophylls*), red (*phycobilins*), yellow, brown and orange (*carotenoids*). Figure 10.1 shows the morphological, taxonomical and ecological diversity of different green algae.

All these are playing a significant role in several research fields such that the green algae can be used to help repossess saline and alkaline soils. Carrageenan, a polysaccharide found as the main component in the cell wall structures of *Chondrus* sp., is used in the textile, leather, cosmetic and brewing industries (Janarthanan and Senthil, 2017). These are widely used as a primary source of food and feed additives to enhance the growth of several aquatic creatures such as bony fishes, shellfish or crustaceans (Falaise et al., 2016).

The production of biomass by algae is much higher than that by other land plant species, but they are subjected to limited nutrients. The starch obtained from microalgae contain media components for cell culture with the help of enzymes, alkali, acids and cells of yeast, which are some of the raw materials used in biopharmaceutical manufacturing. The concept of algal-based biorefinery from cultivation, biomass processing to finished consumer products and chemicals is shown in Figure 10.2.

Microalgae are rich in carbohydrates, proteins and high-density lipoproteins (HDLs) (Muhammad et al., 2018). Across the globe, they will also play a very important role in man's food, nutraceuticals and other products. Besides, microalgal biomass obtained from wastewater streams gives a good potential for a variety of possible bioproducts such as biochar and biofertilizers (Santos et al., 2018), pigments (D'Alessandro et al., 2016), fatty acids and cattle feed (Wollmann et al., 2019). *Gracilaria* seaweeds are used in food for making salads and in chemical, textile and agro-based industries and also used as food for hens to get the yield of more eggs (Norziah and Ching, 2000). *Spirulina* can be used for feeding fish, poultry and cattle to increase their productivity.

Figure 10.1 Morphological, taxonomical and ecological diversity of different green algae. A: *Pterosperma* (Pyramimonadales); B: *Nephroselmis* (Nephroselmidophyceae); C: *Palmophyllum* (Palmophyllales); D: *Tetraselmis* (Chlorodendrophyceae); E: *Chlorella* (Trebouxiophyceae, Chlorellales); F: *Oocystis* (Trebouxiophyceae, Oocystaceae); G: *Haematococcus* (Chlorophyceae, Chlamydomonadales); H: *Pediastrum* (Chlorophyceae, Sphaeropleales); I: *Bulbochaete* (Chlorophyceae, Oedogoniales) J: *Chaetophora* (Chlorophyceae, Chaetophorales); K: *Ulothrix* (Ulvophyceae, Ulotrichales); L: *Ulva* (Ulvophyceae, Ulvales); M: *Cladophora* (Ulvophyceae, Cladophorales); N: *Boergesenia* (Ulvophyceae, Cladophorales); O: *Acetabularia* (Ulvophyceae, Dasycladales); P: *Caulerpa* (Ulvophyceae, Bryopsidophyceae); Q: *Klebsormidium* (Klebsormidiophyceae); R: *Spirotaenia*; S: *Nitella* (Charophyceae); T: *Micrasteria*s (Zygnematophyceae, Desmidiales); and U: *Coleochaete* (Coleochaetophyceae). Reproduced with permission from Taylor & Francis (Leliaert et al., 2012).

Diatoms and single-celled brown algae are used in oils, cleaning solvents and blast furnaces, and as an insulating material in boilers and to filter water. It is also used as an absorbent in nitroglycerin for the manufacture of dynamite. *Tolypothrix* is a genus of blue-green algae *Cyanobacteria* that can increase soil fertility and improve crop productivity in rice fields when

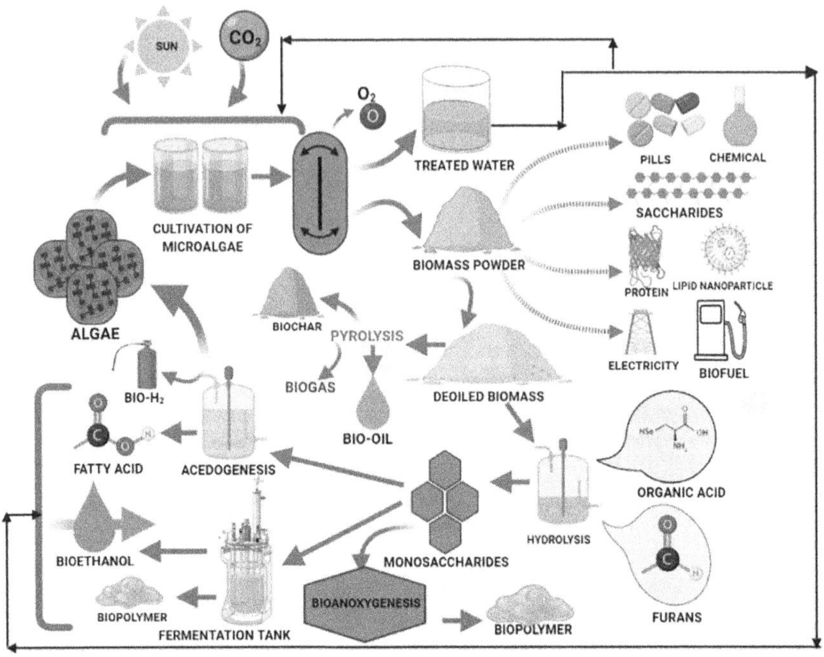

Figure 10.2 Algal-based biorefinery from cultivation, biomass processing to finished consumer products and chemicals. Redrawn from Mohan et al. (2020).

used as biofertilizers. Also, the use of these biofertilizers is inexpensive and environmentally friendly. Recent studies show that biofertilizers prepared from *Cyanobacteria* can also be used for growing crops such as radish, tomato, chili, lettuce, barley, oats, maize/corn, cotton and sugarcane (Thajuddin and Subramanian, 2005). The algae also help to build nitrogen (N) and phosphorus (P) fixation (Wollmann et al., 2019) to enhance soil fertility.

Since the mid-19th century, research has been underway in wastewater treatment processes using algae (Abdel-Raouf et al., 2012). The main important function of algae in the purification of effluents is one of the primary disciplines of the scientific community to provide a sustainable path and to maintain a safer environment. As reported in the literature, two significant goals for the utilization of algae in treating wastewater are as follows: (i) the direct intake or transformation of water pollutants and (ii) better purification ability of algae, bacteria and their aggregates by supplying extra oxygen from photosynthesis. Eventually, it will reduce the total energy price of direct or indirect oxygen supply to the system (Wollmann et al., 2019). Algae can help in water treatment for biochemical oxygen demand (BOD) reduction, inhibition of *Coliforms*, separation of heavy metals (Abdel-Raouf et al., 2012) and bacterial decomposition of sewage waste. For example, *Euglena* and

Scenedesmus are used for BOD reduction, bacterial decomposition of sewage waste and removal of effluents. The mineral-rich water is used for irrigation of crops and fish pond fertilization. Algal biomass is also used in the production of methane, fine chemicals and liquid fuels. On the other hand, it can also be used as composting/manure and cattle feed (Abdel-Raouf et al., 2012).

After 2002, half of the total carbon dioxide (CO_2) fixation is persuaded by algae as a primary product. CO_2 released by the cement-making plants is depicted as possible gas for algae culture. Subsequently, their biofixation efficiency is superior to terrestrial plants as reported in the literature, and in the same way, microalgae own a better capacity to fix CO_2 in comparison with C_4 plants (Cuellar-Bermudez et al., 2015).

A report recently reviewed by Alves et al. (2018) described that several algal-derived chemical compounds were demonstrated to have curative properties. These properties connected with several health-benefiting effects such as anti-obesity (Kim et al., 2018), anti-osteoporotic (Deng et al., 2018), antidiabetic, antioxidant (Pinteus et al., 2017), antihyperlipidemic (Sathivel et al., 2008), antihypertensive (Sivagnanam et al., 2015) and anticoagulant (Magalhaes et al., 2011). Similarly, medicines intended to reduce inflammation (De Souza et al., 2009), drugs used for the modulation of the immune system (Pérez-Recalde et al., 2014), estrogen blockers (Skibola, 2004), thyrotrophic hormones (Teas et al., 2007), drugs that help for the protection of neurons, anti-osteoarthritis medicines (Moon et al., 2018), drugs that treat viral infections (Aguilar-Briseño et al., 2015), germicidal drugs (Pinteus et al., 2015) and antineoplastic/cancer-treating agents were also reviewed substantially in the literature (Alves et al., 2018). Among the well-known biologically active agents are polysaccharides, brominated phenols, carotenoids and particularly many varieties of unsaturated hydrocarbon obtained from plant species (Gribble, 2015). Several researchers are working to find out marine-derived chemicals, and on their extensive characterization and the identification of bioactive compounds, yet more work on this domain is to be carried (Alves et al., 2018).

The recent innovations and developments for the extraction of drugs from marine species are a growing interest in producing biopharmaceuticals such as antibacterial, antiviral and antifungal medicines, curative proteins, neuroprotective compounds and drugs. Thousands of compounds from a variety of marine species, namely algae, bacteria, fungi, sponges and other sea creatures, were manifested and tested. Besides, some of these medicinal drugs derived from biological components are accessible in marketplaces for a cure (Falaise et al., 2016).

In this chapter, the focus is on the usage and application of several algal-derived pharmaceuticals and other value-added products and their commercial importance using algae and algal biomasses. This chapter also summarizes the most recent significant works that are possible via algae as a primary source of marine drugs such as antimicrobial, antiviral, antifungal and neuroprotective products and other therapeutics. Table 10.1 represents a different habitat of microalgae and their incorporation in several products with applications.

Table 10.1 Antiviral activities of selected algal polysaccharide extracts obtained from marine and/or freshwater sources

Antiviral polysaccharide	*Target microorganisms*	*Application/virus*	*References*
Blue-green algae			
Sulfated polysaccharide	Microalgae	Anti-HSV, IAV	Kanekiyo et al. (2007)
Ulvan	Green algae	Anti-IAV	Ivanova et al. (1994)
Red algae			
Alginate polysaccharide	*Phaeophyceae*	Anti-HIV, HBV, IAV	Xin et al. (2000)
ι-Carrageenan	*Chondrus crispus*	Anti-DENV, HRV, IAV, HPV, HAV	Talarico et al. (2007)
κ-Carrageenan	*Chondrus crispus*	Anti-enterovirus, HSV, HIV, IAV	Chiu et al. (2012)
λ-Carrageenan	*Chondrus crispus*	Anti-DENV, HSV, HPV, HIV, HAV	Talarico et al. (2007)
Brown algae			
Fucan	*Adenocystis utricularis, Undaria pinnatifida, Stoechospermum marginatum, Cystoseira indica, Cladosiphon okamuranus, Fucus vesiculosus*	HSV-1, HSV-2, HCMV, VSV, Sindbis virus, HIV-1	Queiroz et al. (2008)
Laminaran	*Fucus vesiculosus, Saccharina longicruris, Ascophyllum nodosum*	HIV	Muto et al. (1988)
Diatoms			
Dinoflagellates			
Marine and freshwater algae and marine plankton	*Vulcanodinium rugosum*	HIV-1/VSV-G	Izumida et al. (2019)

Note: Adapted and modified from Ahmadi et al. (2015) and Wang et al. (2012a,b).

Algal-derived pharmaceuticals: antimicrobial, antiviral and antifungal agents

The worldwide markets of pharmaceutical companies are major sources of medical innovation with an immense demand. These emerging markets signify great opportunities for pharmaceutical companies (Tannoury and Attieh, 2017). The main aim of algal-derived pharmaceuticals/biopharmaceuticals is to produce and develop bioactive metabolites and medicinal drugs. Ranges of enzymes, microorganisms, plant or animal cells, yeast and microbes are used to create medicines via a series of chemical reactions. Particularly, biopharmaceuticals obtained from algal chloroplast are related to macromolecules, large proteins and hormones. These are important biologically derived molecules and bioproducts that keep humans well and have the ability to cure several diseases. Examples are hormones, enzymes, cytokines, monoclonal antibodies (mAbs), vaccines, cell therapeutics, clotting factors and peptide therapeutics (Taunt et al., 2018). Recent technologies in the area of biopharmaceuticals are in the production of mAbs in protein-free media, better vaccine production methods, genomic technologies, making chemically modified cell media, novel cancer-treating agents and production of non-ribosomal peptides (NRPs) (Sekhon, 2010).

Different habitats of microalgae and their incorporation in several products and their applications were thoroughly reviewed by Caporgno and Mathys (2018) and Muhammad et al. (2018). In recent times, biopharmaceutical compounds have changed the course of medication methods for several illnesses. Along with the medicinal flora, naturally available bioactive compounds are rich in marine algae through possible active ingredients in curing type II diabetes mellitus (Unnikrishnan and Jayasri, 2018). Phlorotannin, a type of tannic acid found in brown algae (*Sargassacean*), is used to cure oxidative stress and diabetes. This oxidative stress is very intimately connected through the process of pathophysiological issues of diabetes mellitus (Wang et al., 2012a,b). However, technological information on the application of marine species in conventional medicines is increasing. The marine algae cooked in milk or water are applied for treating kidney problems and burns (Dias et al., 2012). Two different algal species (red), namely *Chondrus crispus* and *Mastocarpus stellatus,* are reported to be used for curing tuberculosis (TB), chest infections, severe colds and inflammatory diseases such as throat pain, and also as a beverage (Vickery, 1995). The marine red alga *Porphyra umbilicalis* (Linnaeus) Kützing was reported to be useful to treat breast cancer (Borlase, 1970).

Another biologically derived compound, oxytocin (which enhances the growth of green algae), is used to induce labor pain in childbirth. Antibiotics such as penicillin and other therapeutic medicines are saving countless lives across the globe. In the same way, the improved technologies of biotechnology developments are blood transfusions and organ transplant's lifesaving treatments. Biotechnology advancement made it easy that any drug can be modified genetically with the help of recombinant deoxyribonucleic acid (DNA) technologies to change the peculiarities of particular diseases with fewer side effects and powerful results on target cells (Khan et al., 2016).

Antimicrobial activity from microalgae

Antimicrobial substances are commonly used to kill, or inhibit the development of, pathogenic microorganisms such as bacteria, viruses, fungi and/or protozoan. The antimicrobial activity of these materials can prevent their pathogenic action. In another way, one can divide antimicrobial drugs into antibiotics and chemotherapeutics. These antimicrobial drugs are broadly divided as follows: i) synthetically prepared and ii) naturally occurring drugs. For example, penicillin is produced by a fungus, and polymyxin is obtained from a particular soil bacterium. Drugs such as sulfonamides, quinolones, cotrimoxazole and antiviral, antifungal, antileprotic and antiprotozoal drugs are examples of synthetic antimicrobial agents. The innovation of these antimicrobial drugs such as penicillin and tetracycline has lined the way for improved health for millions around the globe. These drugs can be extracted from living organisms that release products exogenously or from their lysate. Further, these bactericidal combinations are not limited to the production of new antibiotic drugs (Falaise et al., 2016), but they are also used for the preparation of food preservatives (parabens, sorbic acid, benzoic acid); in agriculture for growing plants and crops; as growth promoters in food animals; to prevent and treat infections; and to avoid the growth of spoilage and pathogenic microorganisms in food products (Lucera et al., 2012).

A variety of microalgae-derived substances and their extracts have ample antimicrobial behavior because of their antibacterial, antifungal, antiprotozoal and antialgal properties. Great attention was paid to the antibacterial chemical compounds from microalgae in the 19th century by Pratt, who led the way to show the antibacterial behavior of alga *Chlorella* (green) against various bacteria such as Gram-positive and Gram-negative (Pratt et al., 1944).

Antiviral activity from microalgae

Viral microorganisms are the main reason for human illnesses and increased fatality rate globally. The capability of microalgae as an antiviral agent has not yet been revealed fully for clinical use. Antiviral drugs derived from microalgae are primarily polysaccharide compounds. Their functions against pathogens are fully unknown. Polysaccharides have acquired attention in the biopharmaceutical and biomedical industries as they are abundantly available in nature. The majority of polysaccharides are non-poisonous, harmless, biologically compatible and environment-friendly (Ahmadi et al., 2015). Nonetheless, various polyoses with antiviral agents were not fully formulated for clinical use.

Till now, a very small number of reports have shown the antiviral potential of microalgae. Therefore, therapies against diseases caused by viruses have a great challenge to the human body's immune system. Some studies reported that the antiviral activity of marine algae was dependent on the sulfated polysaccharide content present in it (Falaise et al., 2016). Many infections by viral infectious agents such as human immunodeficiency virus (HIV), hepatitis C (HCV, a liver disease), dengue virus (DENV, which causes dengue fever), herpes simplex virus (HSV), respiratory syncytial virus (RSV), Ebola virus and flu (influenza) virus have affected the majority of the population in the world (Mayer et al., 2017). Some potential antiviral activity has been shown by microalgae such as blue-green algae *Cyanobacteria,* which have been shown to inhibit RSV, HSV-2 and HIV-1 diseases (Patterson et al., 1993). These are promising sources for the synthesis of antiviral agents because of their easy culture and growth, which make them suitable organisms for the manufacture of antiviral drugs at a large level (Falaise et al., 2016).

It was also stated that about 5% of *Cyanobacteria* algae have a virucidal property against HSV-2. Similarly, less than 5% of extracts can work against RSV, a widespread and very contagious virus that infects the lungs of most children within their two years of age and even adult humans (Rinehart et al., 1981). It was reported that both methanol and water extracts of *Spirulina* also can be beneficial sources of antiviral compounds (Abdo et al., 2012). Numerous studies have been reported in the literature to examine microalgae-derived drugs to fight against infectious human viruses as presented in Table 10.1.

Antifungal activity from microalgae

The investigation of antifungal agents from microalgae progressed much later than antibacterial drugs. Since the 1970s, fungi have been considered as infectious microorganisms *in vitro.* The mortality due to fungal contagions and the occurrence of nosocomial transmissions rose in patients admitted to hospitals (Falaise et al., 2016). The antifungal activity of algal species *Chlorella minutissima, Ulva Lactuca, Chlorella Vulgaris* and *Chlorella protothecoides* was experimented against fungi *Alternaria alternata, Aspergillus niger* and *Penicillium expansum* (Vehapi et al., 2020). In another work, three blue-green algae *Cyanobacteria* (*Nostoc, Microcystis* and *Oscillatoria geminata*) and *Chlorella vulgaris* and *Scenedesmus* (green algae) were examined with the agar well diffusion technique.

Cyanobacteria strains are a potential source of biologically effective compounds, such as ambiguine I isonitrile, nostocyclyne, fischambiguine B, NRPs, cyclic peptides and lipoproteins, which have shown antifungal and antibacterial activities (Mc Gee et al., 2020). The concentrates obtained from green algae, diatoms and dinoflagellates are rich in antifungal compounds. In the production of antimicrobial drugs, parameters such as temperature, incubation period, substrate pH, light intensity and the components present in the culture medium are the key factors that control the process (Muhammad et al., 2018). Several dinoflagellate metabolites are rich in antifungal activity. For instance, goniodomin A from *Goniodoma* sp. has shown powerful antifungal properties (Murakami et al., 1988). Another strong antimycotic agent (gambieric acid) was found in the culture media of *Gambierdiscus toxicus* (Yasumoto and Maurata, 1993). The potentiality of antibacterial and antifungal activities can change extensively depending on microalgae, the nature of the solvent utilized to separate the aggregates, and the bacterium examined. Table 10.2 is adopted and modified from Falaise et al. (2016) and indicates the antifungal activity from microalgae.

Table 10.2 Antifungal activity from microalgae

Microalgae type	*Antifungal compounds or fraction*	*Target microorganisms*	*References*
Green microalgae			
Chlorella vulgaris	Microalgal supernatant	Yeast: *Candida kefyr* Mold: *Aspergillus fumigatus, Aspergillus niger*	Ghasemi et al. (2007)
Chlorococcum humicola	Organic solvent extracts and pigments: beta-carotene, chlorophyll a and chlorophyll b	Yeast: *Candida albicans* Mold: *Aspergillus flavus, Aspergillus niger*	Bhagavathy et al. (2011)
Haematococcus pluvialis	Short-chain fatty acids	Yeast: *Candida albicans*	Santoyo et al. (2009)
Heterochlorella luteoviridis	Microalgal supernatant	Yeast: *Candida albicans*	Mudimu et al. (2014)
Scenedesmus quadricauda	Organic solvent extracts	Yeast: *Candida albicans, Saccharomyces cerevisiae* Mold: *Aspergillus flavus, Aspergillus niger, Penicillium herquei* Others: *Alternaria brassicae, Fusarium moniliforme, Helminthosporium* sp.	Abedin and Taha (2008)
Red microalgae			
Porphyridium aerugineum phycobiliproteins Yeast: *C. albicans*	Phycobiliproteins	Yeast: *Candida albicans*	Najdenski et al. (2013)
Porphyridium purpureum microalgal supernatant Yeast: C. albicans	Microalgal supernatant	Yeast: *Candida albicans*	Mudimu et al. (2014)
Rhodella reticulata exopolysaccharides Yeast: *C. albicans*	Exopolysaccharides	Yeast: *Candida albicans*	Najdenski et al. (2013)

Diatoms			
Chaetoceros lauderi	Polysaccharides	Mold: *Aspergillus fumigatus, Blastomyces dermatitidis, Emmonsia parva, Madurella mycetomatis, Sporothrix schenckii* Dermatophyte: *Epidermophyton floccosum, Microsporum audouinii, Microsporum canis, Microsporum ferrugineum, Microsporum gypseum, Microsporum nanum, Microsporum persicolor, Trichophyton* spp.	Gueho et al. (1977) and Viso et al. (1987)
Chaetoceros muelleri	Lipidic fractions: triglycerides and docosapentaenoic acid (DPA)	Yeast: *Candida albicans*	Mendiola et al. (2007)
Haslea karadagensis	Purified pigment (intra- and extracellular forms)	*Corollospora maritima, Lulworthia* sp., *Dendryphiella salina*	Gastineau et al. (2012)
Thalassiothrix frauenfeldii	Culture filtrates and organic solvent extracts	Yeast: *Candida albicans, Candida glabrata, Candida krusei, Candida tropicalis, Cryptococcus neoformans* Mold: *Aspergillus niger*	Walter and Mahesh (2000)
Dinoflagellates			
Amphidinium sp.	Polyols: karatungiols A(1)	Mold: *Aspergillus niger*	Washida et al. (2006)
Gambierdiscus toxicus	Gambieric acids A and B	Mold: *Aspergillus fumigatus, Aspergillus niger, Aspergillus oryzae, Penicillium citrinum, Penicillium chrysogenum, Paecilomyces variotii* Dermatophyte: *Epidermophyton floccosum, Trichophyton interdigitale*	Nagai et al. (1993)
Goniodoma pseudogonyaulax	Goniodomin A (polyether macrolide)	Yeast: *Candida albicans, Cryptococcus neoformans, Saccharomyces cerevisiae* Mold: *Penicillium* sp. Dermatophyte: *Trichophyton mentagrophytes*	Sharma et al. (1968) and Murakami et al. (1988)
Prorocentrum lima	Polyethers	Yeast: *Candida rugosa* Mold: *Aspergillus niger, Penicillium funiculosum*	Nagai et al. (1990)

Note: Adapted from Falaise et al. (2016).

Neuroprotective drugs

Neurodegenerative disorders, namely Parkinson's disease (PD) and Alzheimer's disease (AD), which impair the central nervous system (CNS), pose an extreme threat to the world population with the anticipation of turning into the second common cause of fatality after cancer among the elders by the end of the 2040s (Hirsch, 2009). Numerous strategies are using synthetic neuroprotective agents for the treatment (Carrera and Cacabelos, 2019). However, these strategies appear not to obviate the adverse side effects of these synthetic drugs such as sleepiness, drowsiness, anxiety or nervousness (Carrera and Cacabelos, 2019). Studies have disclosed that marine algae are significantly rich in bioactive compounds which are antioxidant, anti-inflammatory, anti-neurotoxin, neuroprotective and cholinesterase inhibitors in nature.

Anti-neuroinflammatory activity

Inflammation is the embedding cause of all the progressive neurodegenerative diseases such as AD and PD. A published report showed *Ecklonia cava* extract repressed the inflammation elicited by the major pro-inflammatory mediators (nitric oxide (NO), cyclooxygenase-2 (COX-2), interleukin-6/1β, tumor necrosis factor-α (TNF-α) and prostaglandin E2 (PGE2)) in lipopolysaccharide (LPS)-stimulated murine microglial BV2 cells. The mode of action was by stimulating mitogen-activated protein kinases (MAPKs) and blocking the nuclear factor-κB (NF-κB) (Jung et al., 2009a,b). Methanolic extracts of *Ulva conglobata* also displayed an anti-neuroinflammatory effect in murine microglial BV2 cells on interferon-gamma (IFN-γ) by totally suppressing pro-inflammatory enzymes, inducible nitric oxide synthase (iNOS) and COX-2 (Jin et al., 2006).

Antioxidant activity

Reactive nitrogen species (RNS) and reactive oxygen species (ROS) result from nitrosative stress and oxidative stress, respectively. They play a massive role in the onset of neurodegenerative diseases. Nevertheless, as rescuers, antioxidants play a monumental role in scavenging these reactive species and safeguarding the neurons from any damage (Di Meo et al., 2016). Antioxidants present in the extracts from seaweeds *Saccorhiza polyschides, Sargassum muticum, Codium tomentosum, Padina pavonica* and *Ulva compressa* significantly protect human neuroblastoma cells (SH-SY5Y) from neurotoxicity induced by 6-hydroxydopamine (6-OHDA) (Silva et al., 2018). Antioxidant and neuroprotective properties of aqueous extracts from *Halimeda incrassata* and *Bryothamniom triquetrum* were also reported. These extracts are capable of scavenging ROS in mouse hypothalamic GT1-7 cell line (Fallarero et al.,

2003). Similarly, *Chlorella sorokiniana* and *Chlorella minutissima* extracts having high phenol and flavonoid contents scavenged 2,2-diphenyl-1-picrylhydrazyl, hydroxyl radicals and 2,2-azinobis (3-ethylbenzothiazoline-6-sulfonate) (Olasehinde et al., 2019).

Cholinesterase inhibitory activity

The deficiency of the neurotransmitter acetylcholine (ACh) in the brain is the major cause of the irreversible neurodegenerative disease AD. Inhibiting the enzyme acetylcholinesterase (AChE) that catalyzes the breakdown of ACh is an efficient approach in the symptomatic treatment of AD (Tabet, 2006). Inhibitory activity of 27 seaweed extracts was screened for inhibition of AChE, and among those 27, *Ecklonia stolonifera* phlorofucofuroeckol A and eckstolonol inhibited AChE and butyrylcholinesterase (BChE) effectively (Yoon et al., 2009). One of the efficient treatments of AD is targeting BChE as it protects AChE from breakdown by scavenging it (Lockridge et al., 2016). Olasehinde et al. (2019) have shown that *Chlorella sorokiniana* ethanolic extract and *Chlorella minutissima* dichloromethane extract evinced the highest inhibitory effect on AChE and BChE activity, respectively.

Antineurotoxic activity

There are over 1,000 neurotoxins affecting humans by damaging the nervous system leading to neurodegeneration (Kim and Kim, 2012). Marine alga *Ulva conglobata* has significant neuroprotective effects on murine hippocampal HT22 cell line exposed to neurotoxicity caused by glutamate (Jin et al., 2006). Aggregation of the peptide β-amyloid (Aβ) is a major pathological marker of AD. *Chlorella vulgaris* biochanin A exhibits neuroprotection against neurotoxicity induced by Aβ in PC-12 cells (Tan and Kim, 2016). *Fucus vesiculosus* fucoidan (sulfated polysaccharide) blocks the Aβ neurotoxicity effect in rat cholinergic basal forebrain neuron cells by inhibiting caspase-dependent apoptosis (Jhamandas et al., 2005). *Laminaria japonica* fucoidan was shown to protect dopaminergic MN9D cells and C57/BL mouse model from neurotoxicity of 1-methyl-4-phenyl-1,2,3,6-tetrahydropyridine (MPTP) (Luo et al., 2009). *Bryothamnion triquetrum* aqueous extract can prevent mouse hypothalamic immortalized GT1-7 cells from apoptosis when exposed to 180 min of severe chemical hypoxia/aglycemia (Fallarero et al., 2006).

Therapeutic proteins

Expression systems for therapeutic proteins such as bacteria, yeast and mammalian cells have limitations such as the absence of posttranslational modification, hyperglycosylated, contamination and oncogenic sequences, respectively (Demain and Vaishnav, 2009). Although plants serve to be a better platform for producing the soluble human therapeutic proteins, they also face the risk of gene flow through transgenic pollens and difficulty in secreting recombinant proteins (Quist and Chapela, 2001). Algae have many advantages such as low growth cost, posttranscriptional modification, post-translational modification, safe, high yield and rapid growth under photoautotrophic condition (Potvin and Zhang, 2010). Hence, algae can be the most suitable candidate for recombinant protein expression.

The protein and amino acid contents in macroalgae are quite low, and many are difficult to extract due to the presence of polysaccharides (Øverland et al., 2019). However, microalgae have outwitted all these limitations as closed bioreactors. In recent days, transgenic microalgae have shown to be the platform to produce therapeutic human recombinant proteins using their nuclear and chloroplast genomes. Types of selection markers (Table 10.3) and promoters in the plasmid vectors play a vital role in the efficiency of transformation of the transgene into microalgae for protein expression (Doron et al., 2016). Several studies have revealed that efficient gene expression is also linked with the type of promoter and transformation location. Cauliflower mosaic virus (CaMV35S) promoter (Kathiresan et al., 2009), tobacco mosaic virus (TMV) promoter (Chen et al., 2001), *Chlamydomonas reinhardtii* rubisco small subunit (RbcS2) promoter (Kim et al., 2002) and maize ubiquitin 1 (Ubi1) gene promoter (Bai et al., 2013) are a few of the highly chosen promoters for nuclear transformation. Promoters for the genes encoding D1 and D2 proteins of the photosystem II reaction center (*psbA* and *psbD),* Y subunit of ATP synthase (*atpA)* and ribulose bisphosphate carboxylase/oxygenase large subunit (*rbcL)* are the choice for chloroplast transformation (Doron et al., 2016). Table 10.3 demonstrates the types of selective markers used for transforming genes of interest into microalgae. Commonly, four methods are employed for efficiently transforming genes of interest into microalgae: (i) particle bombardment, (ii) electroporation, (iii) vortexing the microalgae with glass beads and (iv) vortexing the microalgae with silicon (Figure 10.3) (Barrera and Mayfield, 2013).

To date, popular microalgae used for the transformation are *Chlamydomonas reinhardtii, Dunaliella, Chlorella* and *Scenedesmus* spp. Therapeutic proteins such as antibodies (Abs), vaccines, enzymes, antimicrobial agents and immunotoxins are successfully expressed in the microalgae (Yan et al., 2016). Table 10.4 details the high-value preclinical therapeutic vaccines expressed in microalgae. Similarly, Table 10.5 summarizes the production of Abs, enzymes, immunotoxins and antimicrobial agents using microalgae as an expression system.

Drugs

Amid the tremendous environmental stress and threats, algae have evolved their defense mechanism against all the menaces by producing bioactive compounds. It is reported that algae have over 3,000 natural products

Table 10.3 Types of selective markers used for transforming genes of interest into microalgae

Selection marker types	*Location*	*Marker gene example*	*Role*	*References*
Antibiotic resistance marker	Chloroplast	Bacterial-derived aminoglycoside 3'-adenyltransferase (aadA)	Resistance against spectinomycin	Gutiérrez et al. (2012)
Antibiotic resistance marker	Nucleus	*Streptomyces rimosus* aminoglycoside phosphotransferase gene (*aphVIII)*	Resistance against paromomycin	Sizova et al. (2001)
Auxotrophic growth marker	Nucleus	*Chlamydomonas reinhardtii* nitrate reductase (NIT1) gene	Promotion of auxotrophic growth	Kindle et al. (1989)
Herbicide resistance marker	Chloroplast	*Chlamydomonas reinhardtii* psbA mutant	Resistance against metribuzin	Przibilla et al. (1991)
Herbicide resistance marker	Nucleus	*Haematococcus pluvialis* phytoene desaturase (PDS) gene	Resistance against bleaching herbicide norflurazon	Steinbrenner and Sandmann (2006)
Metabolic enzyme marker	Chloroplast	*Arabidopsis thaliana* acetylornithine aminotransferase (ARG9)	The ability of growth in arginine-deficient media	Remacle et al. (2009)
Photoautotrophic growth marker	Chloroplast	*Chlamydomonas reinhardtii* psbA	Regaining photoautotrophic growth ability	Michelet et al. (2011)

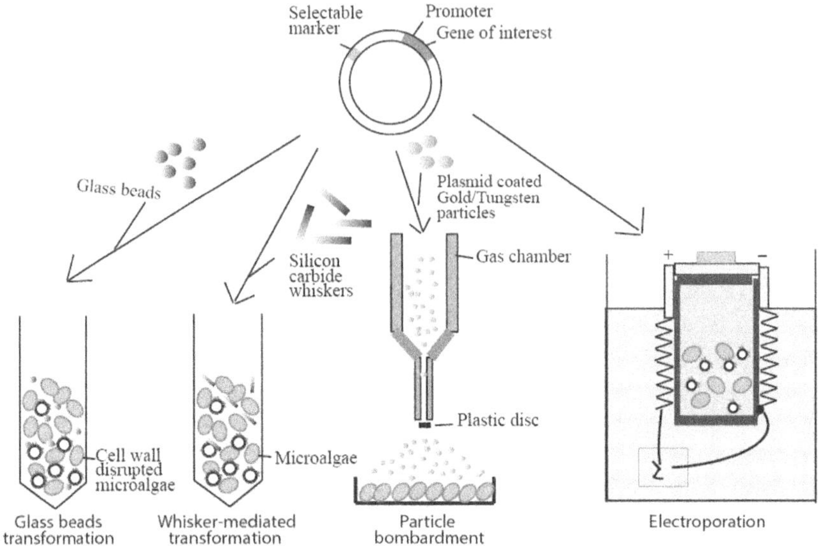

Figure 10.3 Transformation methods for transfer recombinant DNA into microalgae for recombinant protein expression.

(Leal et al., 2013). These bioactive compounds reported to have drug effects broadly come under classes of brominated phenols, polysaccharides, terpenoids and carotenoids (Alves et al., 2018). Many polysaccharides from algae have medicinal properties; however, few polysaccharides are used in formulating other drugs due to their excipient property. For example, agar (a jellylike polysaccharide) from red algae *Gelidium* and *Gracilaria* as well as alginate from brown algae *Laminaria digitata* have contributed to preparing drugs in tablet and capsule forms. Alginate aids in the controlled release of the drug as an excipient. Alginate is also reported to hasten wound healing due to its main components such as α-ketoglutarate, α-keto-isovalerate and α-keto-oxaloacetate (Zare-Gachi et al., 2020). Alginate is also used in combination with an antacid for digestive system healing Other polysaccharides such as sulfated polysaccharides have established an immense place in anticancer therapies, for example, DAEB from *Enteromorpha intestinalis* (Jiao et al., 2009), H3-1 from brown alga *Hydroclathrus clathratus* (Wang et al., 2010) and CF-PLS from red alga *C. feldmannii* (Lins et al., 2009).

Anti-obesity and antidiabetic drugs

Fucoxanthin, a potent antioxidant carotenoid from brown seaweed and diatoms (*Laminaria digitata*) and microalgae (*Isochrysis galbana*), is reported to have an anti-obesity effect (Gammone and D'Orazio, 2015). Fucoxanthinol, a metabolite of fucoxanthin, also exerts an anti-obesity effect in adipocyte cells by suppressing monocyte chemotactic protein-1 (MCP-1) and TNF-α,

Table 10.4 High-value preclinical therapeutic vaccines expressed in microalgae

Vaccine antigens	*Antigen source*	*Vaccine type*	*Location*	*Microalgae*	*Protein*	*References*
Angiotensin II–hepatitis B nucleocapsid antigen (HBcAg)	Hepatitis B virus	Hypertension	Chloroplast	*Chlamydomonas reinhardtii*	0.05%	Soria-Guerra et al. (2014)
Antigen P24	Human immunodeficiency virus	Acquired immunodeficiency syndrome	Nucleus	*Chlamydomonas reinhardtii*	0.25%	Barahimipour et al. (2016)
D2 fibronectin-binding domain–cholera toxin B subunit (CTB)	*Staphylococcus aureus*	Skin and respiratory diseases	Chloroplast	*Chlamydomonas reinhardtii*	0.7% TSP	Dreesen et al. (2010)
E2 protein	Classical swine fever virus	Swine fever	Chloroplast	*Chlamydomonas reinhardtii*	1.5–2%	He et al. (2007)
E7 oncoprotein	Human papillomavirus-16	Cancer	Chloroplast	*Chlamydomonas reinhardtii*	0.12%	Demurtas et al. (2013)
Granule-bound starch synthase (GBSS)–*Plasmodium berghei* apical major antigen (PbAMA1) and GBSS–*Plasmodium berghei* major surface protein (PbMSP1)	*Plasmodium berghei*	Malaria	Chloroplast	*Chlamydomonas reinhardtii*		Dauvillée et al. (2010)
Glutamic acid decarboxylase-65 (GAD65)	Human	Type 1 diabetes	Chloroplast	*Chlamydomonas reinhardtii*	0.25–0.3% TSP	Wang et al. (2008)
GBSS–*Plasmodium falciparum* apical major antigen (PfAMA1) and GBSS–*Plasmodium falciparum* major surface protein (PfMSP1)	*Plasmodium falciparum*	Malaria	Chloroplast	*Chlamydomonas reinhardtii*		Dauvillée et al. (2010)
Hepatitis B virus surface antigen (HBsAg)	Hepatitis B virus	Liver disease	Nucleus	*Dunaliella salina*	1.6–3.1 ng/mg	Geng et al. (2003)
Viral protein 28	White spot syndrome virus	Marine organism disease	Chloroplast	*Chlamydomonas reinhardtii*	3 ng/mg	Surzycki et al. (2009)
Viral protein 28 (VP28)	White spot syndrome virus	Marine organism disease	Nucleus	*Dunaliella salina*	21%	Feng et al. (2014)
VP1 protein–CTB	Foot-and-mouth disease virus	Livestock disease	Chloroplast	*Chlamydomonas reinhardtii*	3%	Sun et al. (2003)

Table 10.5 Production of antibodies, enzymes, immunotoxins and antimicrobial agents using microalgae as an expression system (*Phaeodactylum tricornutum*, *Pseudomonas aeruginosa*, *Chlorella ellipsoidea* and *Nannochloropsis oculata*)

Therapeutic recombinant proteins	*Type of protein*	*Microalgae*	*Location*	*References*
Monoclonal antibodies (mAbs) and their antigen	Antibodies	*Phaeodactylum tricornutum*	Nucleus	Hempel et al. (2011)
Large complex mammalian single-chain Abs	Antibody	*Chlamydomonas reinhardtii*	Chloroplast	Mayfield et al. (2003)
Single-chain Abs	Antibody	*Chlamydomonas reinhardtii*	Chloroplast	Franklin and Mayfield (2005)
Full-length mAbs	Antibody	*Chlamydomonas reinhardtii*	Chloroplast	Tran et al. (2009)
B-cell surface epitope CD22–*Pseudomonas aeruginosa* exotoxin A enzymatic domain	Immunotoxin	*Chlamydomonas reinhardtii*	Chloroplast	Tran et al. (2013)
Gelonin–anti-CD22 single-chain Abs	Immunotoxin	*Chlamydomonas reinhardtii*	Chloroplast	Tran et al. (2013)
Mature rabbit neutrophil peptide-1 (NP-1)	Antimicrobial agent	*Chlorella ellipsoidea*	Chloroplast	Bai et al. (2013)
Bovine lactoferricin (LFB)	Antibacterial agent	*Nannochloropsis oculata*	Nucleus	Li and Tsai (2009)
Xylanase and α-galactosidases	Enzymes	*Dunaliella tertiolecta*	Chloroplast	Georgianna et al. (2013)
Phytase (AppA)	Enzyme	*Chlamydomonas reinhardtii*	Chloroplast	Yoon et al. (2011)

resulting in a reduction of the inflammation induced by obesity (Maeda et al., 2015). Fucoidan, a highly sulfated polysaccharide, is a potent anticancer agent but also has an anti-obesity effect. For instance, *Undaria pinnatifida* fucoidans have been revealed to stimulate lipolysis and inhibit lipid accumulation by down-regulating adipogenic marker genes at mRNA level in 3T3-L1 adipocytes (Park et al., 2011). A patent has been *licensed* for the treatment method for type 2 diabetes and its complications for the combination of seaweed extracts mainly consisting of polysaccharides, namely rhamnose and xylose. This includes brown algae such as *Ascophyllum nodosum* and *Fucus vesiculosus* and green algae such as *Cladophora* sp., *Ulva compressa, Codium* sp. and *Monostroma* sp. Phloroglucinol (a polyphenol) and phlorotannins are generally extracted from brown seaweeds such as *Sargassum thunbergii, Eisenia bicyclis, Laminaria japonica, Ecklonia stolonifera, Ecklonia cava, Ishigeo kamurae* and *Undaria pinnatifida.*

Phlorotannins have been reported to alleviate obesity through pancreatic lipase inhibition and inhibition of adipocyte differentiation (Jung et al., 2014). It mainly acts on adipose tissues by suppressing the adipogenesis through the down-regulation of CCAAT/enhancer-binding protein alpha (C/EBPα) and peroxisome proliferator-activated receptor gamma (PPARγ) pathway (Ko et al., 2013). Significant antidiabetic properties have been detected in *Ascophyllum nodosum* phlorotannin extract (Hernández-Corona et al., 2014). Enzymatically digested alginate was fed simultaneously to the high-fat diet-fed mice also resulted in reducing adipose tissue mass, which in turn resulted in a bodyweight reduction of the mice (Nakazono et al., 2016).

Antioxidant drugs

The extract of the polyphenol phlorotannin from *Ecklonia cava* is a potential antioxidant, which works as a protective agent against photocarcinogenesis by protecting the skin against the adverse effect of ultraviolet-B (UV-B) radiation by inhibiting cell proliferation and suppressing COX-2 expression. Sulfated polysaccharides containing heterofucans, especially sulfate, glucose, fucose, uronic acid, galactose and xylose, have shown significant iron-chelating effects and inhibition of proliferation of cervical cancer HeLa cells (Magalhaes et al., 2011). Among the large-scale screening of 27 seaweeds from Portugal Peniche coast, the highest cytoprotective potential of *Ulva compressa* was observed in MCF-7 cells due to its antioxidant property (Pinteus et al., 2017).

Antihyperlipidemic drugs

Ethanol extracts of green alga *Cladophora koeiei* efficiently decreased the triglyceride and low-density lipoprotein (LDL) levels in albino rats. The blood lipid profiles also showed an increase in the level of HDLs (Abdel-Raouf et al., 2018). Oral administration of Seapolynol™ (polyphenol extract) from marine brown alga *Ecklonia cava* and its main component dieckol to the high-fat diet-fed mice resulted in a significant reduction in cholesterol, LDLs and triglycerides (Yeo et al., 2012). Ethanolic extracts of *Ecklonia stolonifera* and its phlorotannin constituents (eckol and dieckol) have exhibited antihyperlipidemic effects in rats (Yoon et al., 2009).

Immunomodulatory drugs

Macroalgae are abundant resources of immunomodulating compounds such as sulfated polysaccharides and anti-inflammatory and immunomodulating molecules (Raposo et al., 2013), e.g., *Monostroma nitidum* (Cao et al., 2019) and red algae *Gelidium corneum* (Abdala Díaz et al., 2019). Enzymatic protein hydrolysates from various algae, such as *Porphyra columbia*, *Porphyra tenera*, *Spirulina maxima* and *Chlorella vulgaris*, have been shown to have immunomodulation property by stimulating the immune system against the pathogenic predator (Kang et al., 2019). Sulfolipids mainly concentrated in algal chloroplast have been used in generating vaccine adjuvant synthetic sulfoglycolipid (β-SQDG18) to trigger an immune response by activating dendritic cells in B16 mouse melanoma cells (Manzo et al., 2017). Microalgal polyunsaturated fatty acids (PUFAs) (Khozin-Goldberg et al., 2016) and carotenoid pigment astaxanthin (ASX) (Galasso et al., 2017) were also reported to exhibit immunomodulatory effects.

Anticancer drugs

Algae consist of various components that have been shown to have anticancer activity. Their mechanism is found to be diverse including cell cycle arrest, induction of apoptosis via caspase system activation and disruption of mitochondrial membrane potential, leakage of the cell membrane, and inhibition of the angiogenesis process. For example, few apoptosis-inducing bioactive compounds from algae, such as lophocladine B from *Lophocladia* sp. (Gross et al., 2006), caulerpenyne from *Caulerpa taxifolia* (Fischel et al., 1995), hexadecyl-1-O-α-1-arabinopyranoside from *Laurencia majuscula* (Du et al., 2010) and laminarin from *Laminaria digitata* (Park et al., 2013), induce apoptosis via cell cycle regulation. Many studies have reported that several identified compounds from algae have antiangiogenic properties. For instance, *Ascophyllum nodosum* ascophyllan showed its antiangiogenic activity on B16 melanoma cells (Abu et al., 2015).

Other identified compounds showed antiangiogenic effects, for example, marine brown alga *Ecklonia cava* 6-6'-bieckol on human fibrosarcoma cell line (HT1080) (Zhang et al., 2010), *Sargassum siliquastrum* 9'-cis-(6'R) fucoxanthin (FcA) on human fibrosarcoma cells (Nguyen et al., 2014) and brown marine alga *Sargassum stenophyllum* SargA in chick embryo and mice (Dias et al., 2005). Due to abundant evidence of the potential anticancer effect of algae, several studies revealed that algal bioactive compounds also induce apoptosis by altering the intracellular signaling pathways such as down-regulation of Bcl-2 or Bcl-xL, cleavage of PARP, caspase-3/9 activation-mediated apoptosis and up-regulation of Bax. Few examples of drugs inducing apoptosis via the abovementioned mechanisms are as follows:

i Sulfated carrageenan from *Laurencia papillosa* prompted apoptosis (programmed cell death) in breast cancer MDA-MB-231 cell line (Murad et al., 2014).
ii Mertensene, a halogenated monoterpene from red alga *Pterocladiella capillacea*, led to caspase-dependent apoptosis and cell cycle (G2/M) arrest in human colon adenocarcinoma HT29 cell line (Tarhouni-Jabberi et al., 2017).
iii The carotenoid siphonaxanthin from marine green algae induced apoptosis via caspase-3 activation and down-regulation of Bcl-2 in human leukemia HL-60 cells (Ganesan et al., 2011)
iv Marine alga *Codium fragile* induced apoptotic cell death in human melanoma A2058 cells (Kim et al., 2013)
v *Enteromorpha intestinalis* sulfated polysaccharide induced apoptosis in human hepatoma HepG2 cells through caspase-mediated mitochondrial signaling pathway releasing mitochondrial cytochrome C into the cytosol (Wang et al., 2014).

Dioxinodehydroeckol from brown alga *Ishige okamurae* also induced apoptosis in HL60 cells through mitochondrial dysfunction (Kang et al., 2012). Additionally, *Eucheuma serra* agglutinin (ESA), a lectin, has been reported to induce apoptosis in mouse colon adenocarcinoma colon-26 cells by translocating the cell membrane phosphatidylserine and activating caspase-3. ESA has also been found to be more effective as a result of *in vivo* antitumor activity on colon-26 tumors in BALB/c mice (Fukuda et al., 2006). Interestingly, marine algal-derived compounds have also emerged as a promising coadjuvant in reducing undesirable side effects of the known anticancer drugs. Red alga *Chondrus ocellatus* λ-carrageenan and *Porphyra haitanensis* porphyran in combination with 5-FU enhanced the antitumor activity of prodrug 5-fluorouracil (5-FU) on transplanted H-22 tumor mice and transplanted S180 tumor mice, respectively, whereas the combination of both also helped in improving the 5-FU-induced immunocompetence damage (Wang and Zhang, 2014). Cyclic depsipeptides, kahalalide F and iso-kahalalide F derived from green macroalgae *Bryopsis pennata* were tested in anticancer clinical trials due to their potent cytotoxic effect. Nevertheless, they were dropped during the phase II clinical trial due to poor solubility, lack of efficacy and poor patient response. However, these peptides are now under the modification process for their improved effectiveness and better solubility (Murphy et al., 2014).

Other promising drugs

Microalgae extracts are having several other promising drugs such as anti-osteoporotic and anti-osteoarthritic (Deng et al., 2018; Moon et al., 2018). For instance, dictyol, a key component from *Padina pavonica* extract, in combination with alginate is used in dietary supplements in osteoporosis conditions (Chaugule et al., 2017). Although there are no marine algal-derived products that have reached clinical assessment, there are several marine algae extracts that have lead compounds with potential antiprotozoal activity against *Leishmania, Trypanosoma cruzi* and *Trypanosoma brucei* (Torres et al., 2014). The consumption of edible kelp species such as *Fucus vesiculosus* reduces the levels of cholesterol, which in turn reduces estrogen as cholesterol is the precursor of estrogen biosynthesis. This anti-estrogenic effect of kelp is beneficial to women who are highly vulnerable to diseases related to high estrogen levels (Skibola, 2004). Apart from this, heterofucans extracted from *Dictyopteris delicatula* have shown anticoagulant activity similar to a heparin (Clexane®) at the same concentration (Magalhaes et al., 2011). The extracts of *Saccharina japonica* and *Sargassum horneri* have shown potential antihypertensive activity (Sivagnanam et al., 2015).

Conclusions

This chapter is primarily focused on the review of biopharmaceuticals such as antimicrobial, antiviral, antifungal and neuroprotective products, therapeutic proteins and drugs that are derived from algae. It also accounts for the general applications of algae in everyday life, for example, in food products and salads, soups, high-grade protein, B-complex vitamins and minerals (*Porphyra, Chlorella*), manure for the cultivation of crops (*Spirulina*), animal/stock feed and poultry feed, fertilizers (*Gracilaria, Laminaria* and *Spirulina*), atmospheric N fixation (*Nostoc, Anabaena*), paints and plastics, thickeners in cosmetics, leather and textile industries. Furthermore, it is also demonstrated that microalgae can be used in applications such as CO_2 fixation and wastewater treatment.

The chemical compounds that are derived from marine polysaccharides are summarized in figures and tables for potential explorations in the future in the therapeutic fields. The antifungal activity of algal species *Chlorella minutissima, Ulva lactuca, Chlorella vulgaris* and *Chlorella protothecoides* was demonstrated. The blue-green algae *Cyanobacteria* (*Nostoc, Microcystis* and *Oscillatoria geminata*) and *Chlorella vulgaris* and *Scenedesmus* were reviewed and reported with the agar well diffusion technique for their antimicrobial activities. They are the potential source of biologically effective compounds such as ambiguine I isonitrile, nostocyclyne, fischambiguine B, NRPs, cyclic peptides and lipoproteins which have shown antifungal and antibacterial activities. In the literature, strains of blue-green algae *Cyanobacteria* were reported to inhibit RSV, HSV-2 and HIV-1. These are competent sources for making antiviral drugs on a large scale. This work also summarizes the applications of the extract of *Chlorella vulgaris* biomass cultured in the post-fermentation process which can productively work as an excellent source of natural antioxidants, feed, food constituents and antifungal and antibacterial agents. Marine and freshwater eukaryotic algae such as *Bacillariophyceae* and *Chrysophyceae* are rich in antimicrobial activity.

The algal-derived neuroprotective agents, therapeutic proteins and drugs are briefly discussed. Marine algae *Ulva conglobata* extracts showing anti-neuroinflammatory effects and neuroprotective effects in murine microglial BV2 and murine hippocampal HT22 cell lines, respectively, are also reviewed. Algae have also become an important focus for deriving drugs for various diseases – cancer, diabetes and obesity, to name a few. From this review, it is also revealed that transgenic microalgae are the best choice for the production of therapeutic human recombinant proteins using their nuclear and chloroplast genomes. In short, the transition of these preclinical assessments of the therapeutic effects of these algal-derived compounds, such as antimicrobial activity and activity against various diseases, to clinical trials will serve major advantages in further research and in the medicinal field.

Acknowledgments

Santhi Raju Pilli would like to thank Jazan University for providing Saudi Digital Library facilities for accessing journals for writing this chapter. Santhi Raju Pilli also greatly acknowledges Dr. Sudhakar Beedam, Department of English, Faculty of Arts and Humanities, Jazan University, for his valuable time and suggestions and for reviewing and correcting the manuscript's English language.

References

Abdala Díaz, R.T., Casas Arrojo, V., Arrojo Agudo, M.A., Cárdenas, C., Dobretsov, S. and Figueroa, F.L. (2019). Immunomodulatory and antioxidant activities of sulfated polysaccharides from *Laminaria ochroleuca, Porphyra umbilicalis* and *Gelidium corneum*. *Marine Biotechnology* 21:577–587.

Abdel-Raouf, N., Al-Enazi, N.M., Ibraheem, I.B.M., Alharbi, R.M. and Alkhulaifi, M.M. (2018). Antibacterial and anti-hyperlipidemic activities of the green alga *Cladophora koeiei*. *Beni-Suef University Journal of Basic and Applied Sciences* 7:158–164.

Abdel-Raouf, N., Al-Homaidan, A.A. and Ibraheem, I.B.M. (2012). Microalgae and wastewater treatment. *Saudi Journal of Biological Sciences* 19:257–275.

Abdo, S.M., Hetta, M., El-Senousy, W.M., Salah El Din, R.A. and Ali, G. (2012). Antiviral activity of freshwater algae. *Journal of Applied Pharmaceutical Science* 2:21–25.

Abedin, R.M.A. and Taha, H.M. (2008). Antibacterial and antifungal activity of cyanobacteria and green microalgae. Evaluation of medium components by Plackett–Burman design for antimicrobial activity of *Spirulina platensis*. *Global Journal of Biotechnology & Biochemistry* 3:22–31.

Abu, R., Jiang, Z., Ueno, M., Isaka, S., Nakazono, S., Okimura, T., Cho, K., Yamaguchi, K., Kim, D. and Oda, T. (2015). Anti-metastatic effects of the sulfated polysaccharide ascophyllan isolated from *Ascophyllum nodosum* on B16 melanoma. *Biochemical and Biophysical Research Communications* 458:727–732.

Aguilar-Briseño, J., Cruz-Suarez, L., Sassi, J.F., Ricque-Marie, D., Zapata-Benavides, P., Mendoza-Gamboa, E., Rodríguez-Padilla, C. and Trejo-Avila, L.M. (2015). Sulphated polysaccharides from *Ulva clathrata* and *Cladosiphon okamuranus* seaweeds both inhibit viral attachment/entry and cell–cell fusion, in NDV infection. *Marine Drugs* 13:697.

Ahmadi, A., Zorofchian Moghadamtousi, S., Abubakar, S. and Zandi, K. (2015). Antiviral potential of algae polysaccharides isolated from marine sources: A review. *BioMed Research International* 2015:1–10.

Alves, C., Silva, J., Pinteus, S., Gaspar, H., Alpoim, M.C., Botana, L.M. and Pedrosa, R. (2018). From marine origin to therapeutics: The antitumor potential of marine algae-derived compounds. *Frontiers in Pharmacology* 9:777.

Bai, L.L., Yin, W.B., Chen, Y.H., Niu, L.L., Sun, Y.R., Zhao, S.M., Yang, F., Wang, R.R.C., Wu, Q., Zhang, X. and Hu, Z.M. (2013). A new strategy to produce a defensin: Stable production of mutated NP-1 in nitrate reductase-deficient *Chlorella ellipsoidea. PLoS One* 8:e54966.

Barahimipour, R., Neupert, J. and Bock, R. (2016). Efficient expression of nuclear transgenes in the green alga *Chlamydomonas*: Synthesis of an HIV antigen and development of a new selectable marker. *Plant Molecular Biology* 90:403–418.

Barrera, D.J. and Mayfield, S.P. (2013). High value recombinant protein production in microalgae. In: Richmond, A. and Hu, Q. (eds.) *Handbook of Microalgal Culture: Applied Phycology and Biotechnology.* Oxford, UK: John Wiley and Sons, Ltd, pp. 532–544.

Bhagavathy, S., Sumathi, P. and Jancy Sherene, B.I. (2011). Green algae *Chlorococcum humicola* – A new source of bioactive compounds with antimicrobial activity. *Asian Pacific Journal of Tropical Biomedicine* 1:S1–S7.

Borlase, W. (1970). *The Natural History of Cornwall.* Oxford, UK: E. & W. Books.

Cao, S., He, X., Qin, L., He, M., Yang, Y., Liu, Z. and Mao, W. (2019). Anticoagulant and antithrombotic properties in vitro and in vivo of a novel sulfated polysaccharide from marine green alga *Monostroma nitidum. Marine Drugs* 17:247.

Caporgno, M.P. and Mathys, A. (2018). Trends in microalgae incorporation into innovative food products with potential health benefits. *Frontiers in Nutrition* 5:1–10.

Carrera, I. and Cacabelos, R. (2019). Current drugs and potential future neuroprotective compounds for Parkinson's disease. *Current Neuropharmacology* 17:295–306.

Chaugule, S.R., Indap, M.M. and Chiplunkar, S.V. (2017). Marine natural products: New avenue in treatment of osteoporosis. *Frontiers in Marine Science* 4:384.

Chen, Y., Wang, Y., Sun, Y., Zhang, L. and Li, W. (2001). Highly efficient expression of rabbit neutrophil peptide-1 gene in Chlorella ellipsoidea cells. *Current Genetics* 39:365–370.

Cherubini, F. (2010). The biorefinery concept: Using biomass instead of oil for producing energy and chemicals. *Energy Conversion and Management* 51:1412–1421.

Chiu, Y.H., Chan, Y.L., Tsai, L.W., Li, T.L. and Wu, C.J. (2012). Prevention of human enterovirus 71 infection by kappa carrageenan. *Antiviral Research* 95:128–134.

Cuellar-Bermudez, S.P., Garcia-Perez, J.S., Rittmann, B. and Parra-Saldivar, R. (2015). Photosynthetic bioenergy utilizing CO_2: An approach on flue gases utilization for third generation biofuels. *Journal of Cleaner Production* 98:53–65.

D'Alessandro, E.B. and Antoniosi, F.N.R. (2016). Concepts and studies on lipid and pigments of microalgae: A review. *Renewable and Sustainable Energy Reviews* 58:832–841.

Dauvillée, D., Delhaye, S., Gruyer, S., Slomianny, C., Moretz, S.E., d'Hulst, C., Long, C.A., Ball, S.G. and Tomavo, S. (2010). Engineering the chloroplast targeted malarial vaccine antigens in *Chlamydomonas* starch granules. *PLoS One* 5:e15424.

Demain, A.L. and Vaishnav, P. (2009). Production of recombinant proteins by microbes and higher organisms. *Biotechnology Advances* 27:297–306.

Demurtas, O.C., Massa, S., Ferrante, P., Venuti, A., Franconi, R. and Giuliano, G. (2013). A *Chlamydomonas*-derived Human Papillomavirus 16 E7 vaccine induces specific tumor protection. *PLoS One* 8:e61473.

Deng, R., Zhou, B., Guan, B., Chan, G.C.F. and Shen, J. (2018). Marine algae extract attenuated osteoporosis in OVX mice, enhanced osteogenesis on human mesenchymal stem cells and promoted OPG expression. *Journal of Functional Foods* 40:229–237.

De Souza, É.T., Pereira de Lira, D., Cavalcanti de Queiroz, A., Costa da Silva, D.J., Bezerra de Aquino, A. and Campessato Mella, E.A., Prates Lorenzo, V., De Miranda, G.E.C., Araújo-Júnior, D., Xavier, J. and De Oliveira Chaves, M.C. (2009). The antinociceptive and anti-inflammatory activities of caulerpin, a bisindole alkaloid isolated from seaweeds of the genus. *Caulerpa*. *Marine Drugs* 7:689–704.

Dias, D.A., Urban, S. and Roessner, U. (2012). A historical overview of natural products in drug discovery. *Metabolites* 2:303–336.

Dias, P.F., Siqueira, J.M., Jr., Vendruscolo, L.F., de Jesus Neiva, T., Gagliardi, A.R., Maraschin, M. and Ribeiro-do-Valle, R.M. (2005). Antiangiogenic and antitumoral properties of a polysaccharide isolated from the seaweed *Sargassum stenophyllum*. *Cancer Chemotherapy and Pharmacology* 56:436–446.

Di Meo, S., Reed, T.T., Venditti, P. and Victor, V.M. (2016). Role of ROS and RNS sources in physiological and pathological conditions. *Oxidative Medicine and Cellular Longevity* 2016:44.

Doron, L., Segal, N. and Shapira, M. (2016). Transgene expression in microalgae—From tools to applications. *Frontiers in Plant Science* 7:505.

Dreesen, I.A., Charpin-El Hamri, G. and Fussenegger, M. (2010). Heat-stable oral alga-based vaccine protects mice from *Staphylococcus aureus* infection. *Journal of Biotechnology* 145:273–280.

Du, B., Zhong, X., Liao, X., Xu, W., Zhou, X. and Xu, S. (2010). A new antitumor arabinopyranoside from *Laurencia majuscula* induces G2/M cell cycle arrest. *Phytotherapy Research* 24:1447–1450.

Falaise, C., François, C., Travers, M.A., Morga, B., Haure, J., Tremblay, R., Turcotte, F., Pasetto, P., Gastineau, R., Hardivillier, Y., Leignel, V. and Mouget, J.L. (2016). Antimicrobial compounds from eukaryotic microalgae against human pathogens and diseases in aquaculture. *Marine Drugs* 14:159.

Fallarero, A., Loikkanen, J.J., Männistö, P.T., Castañeda, O. and Vidal, A. (2003). Effects of aqueous extracts of *Halimeda incrassata* (Ellis) Lamouroux and *Bryothamnion triquetrum* (S.G. Gmelim) Howe on hydrogen peroxide and methyl mercury-induced oxidative stress in GT1-7 mouse hypothalamic immortalized cells. *Phytomedicine* 10:39–47.

Fallarero, A., Peltoketo, A., Loikkanen, J., Tammela, P., Vidal, A. and Vuorela, P. (2006). Effects of the aqueous extract of *Bryothamnion triquetrum* on chemical hypoxia and aglycemia-induced damage in GT1-7 mouse hypothalamic immortalized cells. *Phytomedicine* 13:240–245.

Feng, S., Feng, W., Zhao, L., Gu, H., Li, Q., Shi, K., Guo, S. and Zhang, N. (2014). Preparation of transgenic *Dunaliella salina* for immunization against white spot syndrome virus in crayfish. *Archives of Virology* 159:519–525.

Fischel, J.L., Lemee, R., Formento, P., Caldani, C., Moll, J.L., Pesando, D., Meinesz, A., Grelier, P., Pietra, P. and Guerriero, A. (1995). Cell growth inhibitory effects of caulerpenyne, a sesquiterpenoid from the marine algae *Caulerpa taxifolia*. *Anticancer Research* 15:2155–2160.

Franklin, S.E. and Mayfield, S.P. (2005). Recent developments in the production of human therapeutic proteins in eukaryotic algae. *Expert Opinion on Biological Therapy* 5:225–235.

Fukuda, Y., Sugahara, T., Ueno, M., Fukuta, Y., Ochi, Y., Akiyama, K., Miyazaki, T., Masuda, S., Kawakubo, A. and Kato, K. (2006). The anti-tumor effect of *Euchema serra* agglutinin on colon cancer cells in vitro and in vivo. *Anticancer Drugs* 17:943–947.

Galasso, C., Corinaldesi, C. and Sansone, C. (2017). Carotenoids from marine organisms: Biological functions and industrial applications. *Antioxidants* 6:96.

Gammone, M.A. and D'Orazio, N. (2015). Anti-obesity activity of the marine carotenoid fucoxanthin. *Marine Drugs* 13:2196–2214.

Ganesan, P., Noda, K., Manabe, Y., Ohkubo, T., Tanaka, Y., Maoka, T., Sugawara, T. and Hirata, T. (2011). Siphonaxanthin, a marine carotenoid from green algae, effectively induces apoptosis in human leukemia (HL-60) cells. *Biochimica et Biophysica Acta* 1810:497–503.

Gastineau, R., Hardivillier, Y., Leignel, V., Tekaya, N., Morançais, M., Fleurence, J., Davidovich, N., Jacquette, B., Gaudin, P., Hellio, C. and Bourgougnon, N. (2012). Greening effect on oysters and biological activities of the blue pigments produced by the diatom *Haslea karadagensis* (Naviculaceae). *Aquaculture* 368:61–67.

Geng, D., Wang, Y., Wang, P., Li, W. and Sun, Y. (2003). Stable expression of hepatitis B surface antigen gene in *Dunaliella salina* (Chlorophyta). *Journal of Applied Phycology* 15:451–456.

Georgianna, D.R., Hannon, M.J., Marcuschi, M., Wu, S., Botsch, K., Lewis, A.J., Hyun, J., Mendez, M. and Mayfield, S.P. (2013). Production of recombinant enzymes in the marine alga *Dunaliella tertiolecta*. *Algal Research* 2:2–9.

Ghasemi, Y., Moradian, A., Mohagheghzadeh, A., Shokravi, S. and Morowvat, M.H. (2007). Antifungal and antibacterial activity of the microalgae collected from paddy fields of Iran: Characterization of antimicrobial activity of *Chroococcus dispersus*. *Journal of Biological Sciences* 7:904–910.

Gribble, W.G. (2015). Biological activity of recently discovered halogenated marine natural products. *Marine Drugs* 13:4044.

Gross, H., Goeger, D.E., Hills, P., Mooberry, S.L., Ballantine, D.L., Murray, T.F., Valeriote, F.A. and Gerwick, W.H. (2006). Lophocladines, bioactive alkaloids from the red alga *Lophocladia* sp. *Journal of Natural Products* 69:640–644.

Gueho, E., Pesando, D. and Barelli, M. (1977). Propriétés antifongiques d'une diatomeé *Chaetoceros lauderi* ralfs C C. *Mycopathologia* 60:105–107.

Gutiérrez, C.L., Gimpel, J., Escobar, C., Marshall, S.H. and Henríquez, V. (2012). Chloroplast genetic tool for the green microalgae *Haematococcus Pluvialis* (Chlorophyceae, Volvocales)1. *Journal of Phycology* 48:976–983.

He, D.M., Qian, K.X., Shen, G.F., Zhang, Z.F., Li, Y.N., Su, Z.L. and Shao, H.B. (2007). Recombination and expression of classical swine fever virus (CSFV) structural protein E2 gene in *Chlamydomonas reinhardtii* chroloplasts. *Colloids and Surfaces. B, Biointerfaces* 55:26–30.

Hempel, F., Lau, J., Klingl, A. and Maier, U.G. (2011). Algae as protein factories: Expression of a human antibody and the respective antigen in the diatom *Phaeodactylum tricornutum*. *PLoS One* 6:e28424.

Hernández-Corona, D.M., Martínez-Abundis, E. and González-Ortiz, M. (2014). Effect of fucoidan administration on insulin secretion and insulin resistance in overweight or obese adults. *Journal of Medicinal Food* 17:830–832.

Hirsch, M.A. (2009). Muscle strength in Parkinson's disease. Commentary on Pang and Mak. *Journal of Rehabilitation Medicine* 41:291–292. Available online at: https://www.fda.gov/Food/LabelingNutrition/ucm072756.htm [Accessed on 11 August 2020].

Ivanova, V., Rouseva, R., Kolarova, M., Serkedjieva, J., Rachev, R. and Manolova, N. (1994). Isolation of a polysaccharide with antiviral effect from *Ulva lactuca*. *Preparative Biochemistry & Biotechnology* 24:83–97.

Izumida, M., Suga, K., Ishibashi, F. and Kubo, Y. (2019). The spirocyclic imine from a marine *Benthic Dinoflagellate*, Portimine, is a potent anti-human immunodeficiency virus Type 1 therapeutic lead compound. *Marine Drugs* 17:495.

Janarthanan, M. and Kumar, M.S. (2017). The properties of bioactive substances obtained from seaweeds and their applications in textile industries. *Journal of Industrial Textiles* 48:361–401.

Jhamandas, J.H., Wie, M.B., Harris, K., MacTavish, D. and Kar, S. (2005). Fucoidan inhibits cellular and neurotoxic effects of beta-amyloid (A beta) in rat cholinergic basal forebrain neurons. *European Journal of Neuroscience* 21:2649–2659.

Jiao, L., Li, X., Li, T., Jiang, P., Zhang, L., Wu, M. and Zhang, L. (2009). Characterization and anti-tumor activity of alkali-extracted polysaccharide from *Enteromorpha intestinalis*. *International Journal of Immunopharmacology* 9:324–329.

Jin, D.Q., Lim, C.S., Sung, J.Y., Choi, H.G., Ha, I. and Han, J.S. (2006). *Ulva conglobata*, a marine algae, has neuroprotective and anti-inflammatory effects in murine hippocampal and microglial cells. *Neuroscience Letters* 402:154–158.

Jung, H.A., Jung, H.J., Jeong, H.Y., Kwon, H.J., Ali, M.Y. and Choi, J.S. (2014). Phlorotannins isolated from the edible brown alga *Ecklonia stolonifera* exert anti-adipogenic activity on 3T3-L1 adipocytes by downregulating C/EBPα and PPARγ. *Fitoterapia* 92:260–269.

Jung, W.K., Ahn, Y.W., Lee, S.H., Choi, Y.H., Kim, S.K., Yea, S.S., Choi, I., Park, S., Seo, S., Lee, S. and Choi, I.W. (2009a). *Ecklonia cava* ethanolic extracts inhibit lipopolysaccharide-induced cyclooxygenase-2 and inducible nitric oxide synthase expression in BV2 microglia via the MAP kinase and NF-kappaB pathways. *Food and Chemical Toxicology* 47:410–417.

Jung, W.K., Heo, S.J., Jeon, Y.J., Lee, C.M., Park, Y.M., Byun, H.G., Choi, Y.H., Park, S. and Choi, I.W. (2009b). Inhibitory effects and molecular mechanism of dieckol isolated from marine brown alga on COX-2 and iNOS in microglial cells. *Journal of Agricultural and Food Chemistry* 57:4439–4446.

Kanekiyo, K., Hayashi, K., Takenaka, H., Lee, J.B. and Hayashi, T. (2007). Anti-herpes simplex virus target of an acidic polysaccharide, nostoflan, from the edible blue-green alga *Nostoc flagelliforme*. *Biological and Pharmaceutical Bulletin* 30:1573–1575.

Kang, H.K., Lee, H.H., Seo, C.H. and Park, Y. (2019). Antimicrobial and immunomodulatory properties and applications of marine-derived proteins and peptides. *Marine Drugs* 17:350.

Kang, S.M., Kim, A.D., Heo, S.J., Kim, K.N., Lee, S.H., Ko, S.C. and Jeon, Y.J. (2012). Induction of apoptosis by diphlorethohydroxycarmalol isolated from brown alga, *Ishige okamurae*. *Journal of Functional Foods* 4:433–439.

Kathiresan, S., Chandrashekar, A., Ravishankar, G.A. and Sarada, R. (2009). Agrobacterium-mediated transformation in the green alga *Haematococcus Pluvialis* (Chlorophyceae, Volvocales)(1). *Journal of Phycology* 45:642–649.

Khan, S., Ullah, M.W., Siddique, R., Nabi, G., Manan, S., Yousaf, M. and Hou, H. (2016). Role of recombinant DNA technology to improve life. *International Journal of Genomics* 2016:2405954.

Khozin-Goldberg, I., Leu, S. and Boussiba, S. (2016). Microalgae as a source for VLC-PUFA production. *Sub-cellular Biochemistry* 86:471–510.

Kim, A.D., Lee, Y., Kang, S.H., Kim, G.Y., Kim, H.S. and Hyun, J.W. (2013). Cytotoxic effect of clerosterol isolated from *Codium fragile* on A2058 human melanoma cells. *Marine Drugs* 11:418–430.

Kim, D.H., Kim, Y.T., Cho, J.J., Bae, J.H., Hur, S.B., Hwang, I. and Choi, T.J. (2002). Stable integration and functional expression of flounder growth hormone gene in transformed microalga, *Chlorella ellipsoidea. Marine Biotechnology (NY)* 4:63–73.

Kim, I.H., Choi, J.W., Lee, M.K., Kwon, C.J. and Nam, T.J. (2018). Anti-obesity effects of pectinase and cellulase enzyme-treated *Ecklonia cava* extract in high-fat diet-fed C57BL/6N mice. *International Journal of Molecular Medicine* 41:924–934.

Kim, Y. and Kim, J.W. (2012). Toxic encephalopathy. *Safety and Health at Work* 3:243–256.

Kindle, K.L., Schnell, R.A., Fernández, E. and Lefebvre, P.A. (1989). Stable nuclear transformation of *Chlamydomonas* using the *Chlamydomonas* gene for nitrate reductase. *Journal of Cell Biology* 109:2589–2601.

Ko, S.C., Lee, M., Lee, J.H., Lee, S.H., Lim, Y. and Jeon, Y.J. (2013). Dieckol, a phlorotannin isolated from a brown seaweed, *Ecklonia cava,* inhibits adipogenesis through AMP-activated protein kinase (AMPK) activation in 3T3-L1 preadipocytes. *Environmental Toxicology and Pharmacology* 36:1253–1260.

Kris-Etherton, P.M., Grieger, J.A. and Etherton, T.D. (2009). Dietary reference intakes for DHA and EPA. 1070. *Prostaglandins, Leukotrienes and Essential Fatty Acids* 81:99–104.

Leal, M.C., Munro, M.H.G., Blunt, J.W., Puga, J., Jesus, B., Calado, R., Rosa, R. and Madeira, C. (2013). Biogeography and biodiscovery hotspots of macroalgal marine natural products. *Natural Product Reports* 30:1380–1390.

Leliaert, F., Smith, D.R., Moreau, H., Herron, M., Verbruggen, H., Delwiche, C.F. and De Clerck, O. (2012). Phylogeny and molecular evolution of the green algae. *Critical Reviews in Plant Sciences* 31:1–46.

Li, S.S. and Tsai, H.J. (2009). Transgenic microalgae as a non-antibiotic bactericide producer to defend against bacterial pathogen infection in the fish digestive tract. *Fish and Shellfish Immunology* 26:316–325.

Lins, K.O.A.L., Bezerra, D.P., Alves, A.P.N.N., Alencar, N.M.N., Lima, M.W., Torres, V.M., Farias, W.R.L., Pessoa, C., Moraes, M.O.de. and Costa-Lotufo, L.V. (2009). Antitumor properties of a sulfated polysaccharide from the red seaweed *Champia feldmannii* (Diaz-Pifferer). *Journal of Applied Toxicology* 29:20–26.

Lockridge, O., Norgren, R.B., Jr., Johnson, R.C. and Blake, T.A. (2016). Naturally occurring genetic variants of human acetylcholinesterase and butyrylcholinesterase and their potential impact on the risk of toxicity from cholinesterase inhibitors. *Chemical Research in Toxicology* 29:1381–1392.

Lucera, A., Costa, C., Conte, A. and Del Nobile, M.A. (2012). Food applications of natural antimicrobial compounds. *Frontiers in Microbiology* 3:287.

Luo, D., Zhang, Q., Wang, H., Cui, Y., Sun, Z., Yang, J., Zheng, Y., Jia, J., Yu, F., Wang, X. and Wang, X. (2009). Fucoidan protects against dopaminergic neuron death in vivo and in vitro. *European Journal of Pharmacology* 617:33–40.

Maeda, H., Kanno, S., Kodate, M., Hosokawa, M. and Miyashita, K. (2015). Fucoxanthinol, metabolite of fucoxanthin, improves obesity-induced inflammation in adipocyte cells. *Marine Drugs* 13:4799–4813.

Magalhaes, K.D., Costa, L.S., Fidelis, G.P., Oliveira, R.M., Nobre, L.T.D.B., Dantas-Santos, N., Camera, R.B.G., Albuquerque, I.R.L., Cordeiro, S.L., Sabry, D.A., Costa, M.S.S.P., Alves, L.G. and Rocha, H.A.O. (2011). Anticoagulant, antioxidant and antitumor activities of heterofucans from the seaweed *Dictyopteris delicatula*. *International Journal of Molecular Sciences* 12:3352–3365.

Manzo, E., Cutignano, A., Pagano, D., Gallo, C., Barra, G., Nuzzo, G., Sansone, C., Ianora, A., Urbanek, K., Fenoglio, D., Ferrera, F., Bernardi, C., Parodi, A., Pasquale, G., Leonardi, A., Filaci, G., De Palma, R. and Fontana, A. (2017). A new marine-derived sulfoglycolipid triggers dendritic cell activation and immune adjuvant response. *Scientific Reports* 7:6286.

Mayer, A., Rodríguez, A., Taglialatela-Scafati, O. and Fusetani, N. (2017). Marine pharmacology in 2012–2013: Marine compounds with antibacterial, antidiabetic, antifungal, anti-inflammatory, antiprotozoal, antituberculosis, and antiviral activities affecting the immune and nervous systems, and other miscellaneous mechanisms of action. *Marine Drugs* 15:273.

Mayfield, S.P., Franklin, S.E. and Lerner, R.A. (2003). Expression and assembly of a fully active antibody in algae. *Proceedings of the National Academy Sciences* 100:438–442.

Mc Gee, D., Archer, L., Smyth, T., Fleming, G. and Touzet, N. (2020). Bioprospecting and LED-based spectral enhancement of antimicrobial activity of microalgae isolated from the west of Ireland. *Algal Research* 45:101704.

Mendiola, J.A., Torres, C.F., Toré, A., Martín-Álvarez, P.J., Santoyo, S., Arredondo, B.O., Señoráns, F.J., Cifuentes, A. and Ibáñez, E. (2007). Use of supercritical CO_2 to obtain extracts with antimicrobial activity from *Chaetoceros muelleri* microalga. A correlation with their lipidic content. *European Food Research and Technology* 224:505–510.

Michelet, L., Lefebvre-Legendre, L., Burr, S.E., Rochaix, J.D. and Goldschmidt-Clermont, M. (2011). Enhanced chloroplast transgene expression in a nuclear mutant of *Chlamydomonas*. *Plant Biotechnology Journal* 9:565–574.

Mohan, S.V., Hemalathaa, M., Chakraborty, D., Chatterjeea, S., Ranadheera, P. and Rajesh, K. (2020). Algal biorefinery models with self-sustainable closed loop approach: Trends and prospective for blue-bioeconomy. *Bioresource Technology* 295:122–128.

Moon, S.M., Lee, S.A., Han, S.H., Park B.R., Choi M.S., Kim, J.S. Kim, S.G., Kim, H.J., Chun, H.S., Kim, D.K. and Kim, C.S. (2018). Aqueous extract of *Codium fragile* alleviates osteoarthritis through the MAPK/NF-κB pathways in IL-1β-induced rat primary chondrocytes and a rat osteoarthritis model. *Biomedicine & Pharmacotherapy* 97:264–270.

Mudimu, O., Rybalka, N., Bauersachs, T., Born, J., Friedl, T. and Schulz, R. (2014). Biotechnological screening of microalgal and cyanobacterial strains for biogas production and antibacterial and antifungal effects. *Metabolites* 4:373–393.

Muhammad, R., Ghulam, M., Sheraz A.M., Kisay, L. and Naim, R. (2018). Exploring the potential of microalgae for new biotechnology applications and beyond: A review. *Renewable and Sustainable Energy Reviews* 92:394–404.

Mulder, K.A., King, D.J. and Innis, S.M. (2014). Omega-3 fatty acid deficiency in infants before birth identified using a randomized trial of maternal DHA supplementation in pregnancy. *PLoS One* 9:e83764.

Murad, H., Ghannam, A., Al-Ktaifani, M., Abbas, A. and Hawat, M. (2014). Algal sulfated carrageenan inhibits proliferation of MDA-MB-231 cells via apoptosis regulatory genes. *Molecular Medicine Reports* 11:2153–2158.

Murakami, M., Makabe, K., Yamaguchi, K., Konosu, S. and Walehli, M.R. (1988). Goniodomin A, a novel polyether macrolide from the dinoflagellate *Goniodoma pseudogonyaulax*. *Tetrahedron Letters* 29:1149–1152.

Murphy, C., Hotchkiss, S., Worthington, J. and McKeown, S.R. (2014). The potential of seaweed as a source of drugs for use in cancer chemotherapy. *Journal of Applied Phycology* 26:2211–2264.

Muto, S., Niimura, K., Oohara, M., Oguchi, Y., Matsunaga, K., Hirose, K., Kakuchi, J., Sugita, N. and Furusho, T. (1988). Polysaccharides from marine algae and antiviral drugs containing the same as active ingredients. *European Patent* EP295956.

Nagai, H., Mikami, Y., Yazawa, K., Gonoi, T. and Yasumoto, T. (1993). Biological activities of novel polyether antifungals, gambieric acids A and B from a marine dinoflagellate *Gambierdiscus toxicus*. *Journal of Antibiotics* 46:520–522.

Nagai, H., Satake, M. and Yasumoto, T. (1990). Antimicrobial activities of polyether compounds of dinoflagellate origins. *Journal of Applied Phycology* 2:305–308.

Najdenski, H.M., Gigova, L.G., Iliev, I.I., Pilarski, P.S., Lukavský, J., Tsvetkova, I.V., Ninova, M.S. and Kussovski, V.K. (2013). Antibacterial and antifungal activities of selected microalgae and cyanobacteria. *International Journal of Food Science Technology* 48:1533–1540.

Nakazono, S., Cho, K., Isaka, S., Abu, R., Yokose, T., Murata, M., Ueno, M., Tachibana, K., Hirasaka, K., Kim, D. and Oda, T. (2016). Anti-obesity effects of enzymatically-digested alginate oligomer in mice model fed a high-fat-diet. *Bioactive Carbohydrates and Dietary Fibre* 7:1–8.

Nguyen, V., Qian, Z., Lee, B., Heo, S., Kim, K., Jeon, Y., Park, W.S., Choi, I., Jang, C.H., Ko, S., Park, S., Kim, Y., Kim, G., Lee, D., Yim, M., Je, J. and Jung, W. (2014). Fucoxanthin derivatives from *Sargassum siliquastrum* inhibit matrix metalloproteinases by suppressing NF-kB and MAPKs in human fibrosarcoma cells. *Algae* 29:355–366.

Norziah, M.H. and Ching, C.Y. (2000). Nutritional composition of edible seaweed Gracilaria changgi. *Food Chemistry* 68:69–76.

Olasehinde, T.A., Odjadjare, E.C., Mabinya, L.V., Olaniran, A.O. and Okoh, A.I. (2019). *Chlorella sorokiniana* and *Chlorella minutissima* exhibit antioxidant potentials, inhibit cholinesterases and modulate desegregation of β-amyloid fibrils. *Electronic Journal of Biotechnology* 40:1–9.

Øverland, M., Mydland, L.T. and Skrede, A. (2019). Marine macroalgae as sources of protein and bioactive compounds in feed for monogastric animals. *Journal of the Science of Food and Agriculture* 99:13–24.

Park, H.K., Kim, I.H., Kim, J. and Nam, T.J. (2013). Induction of apoptosis and the regulation of ErbB signaling by laminarin in HT-29 human colon cancer cells. *International Journal of Molecular Medicine* 32:291–295.

Park, M.K., Jung, U. and Roh, C. (2011). Fucoidan from marine brown algae inhibits lipid accumulation. *Marine Drugs* 9:1359–1367.

Pastor, N., Soler, B., Mitmesser, S.H., Ferguson, P. and Lifschitz, C. (2006). Infants fed docosahexaenoic acid- and arachidonic acid-supplemented formula have decreased incidence of bronchiolitis/bronchitis the first year of life. *Clinical Pediatrics (Phila)* 45:850–855.

Patterson, G.M., Baker, K.K., Baldwin, C.L., Bolis, C.M., Caplan, F.R., Larsen, L.K. Levine, I.A., Moore, R.E., Moore, E., Nelson, C.S. and Tschappat, K.D. (1993). Antiviral activity of cultured blue-green algae (Cyanophyta). *Journal of Phycology* 29:125–130.

Pérez-Recalde, M., Matulewicz, M.C., Pujol, C.A. and Carlucci, M.J. (2014). *In vitro* and *in vivo* immunomodulatory activity of sulfated polysaccharides from red seaweed *Nemalion helminthoides*. *International Journal of Biological Macromolecules* 63:38–42.

Pinteus, S., Alves, C., Monteiro, H., Araújo, E., Horta, A. and Pedrosa, R. (2015). *Asparagopsis armata* and *Sphaerococcus coronopifolius* as a natural source of antimicrobial compounds. *World Journal of Microbiology and Biotechnology* 31:445–451.

Pinteus, S., Silva, J., Alves, C., Horta, A., Fino, N., Rodrigues, A.I., Mendes, S. and Pedrosa, R. (2017). Cytoprotective effect of seaweeds with high antioxidant activity from the Peniche coast (Portugal). *Food Chemistry* 21:591–599.

Potvin, G. and Zhang, Z. (2010). Strategies for high-level recombinant protein expression in transgenic microalgae: A review. *Biotechnology Advances* 28:910–918.

Pratt, R., Daniels, T.C., Eiler, J.J., Gunnison, J.B., Kumler, W.D., Oneto, J.F., Strait, L.A., Spoehr, H.A., Hardin, G.J., Milner, H.W., Smith, J.H. and Strain, H.H. (1944). Chlorellin, an antibacterial substance from *Chlorella*. *Science* 99:351–352.

Przibilla, E., Heiss, S., Johanningmeier, U. and Trebst, A. (1991). Site-specific mutagenesis of the D1 subunit of photosystem II in wild-type Chlamydomonas. *Plant Cell* 3:169–174.

Queiroz, K.C.S., Medeiros, V.P., Queiroz, L.S., Abreu, L.R.D., Rocha, H.A.O., Ferreira, C.V., Juca, M.B., Aoyama, H. and Leite, E.L. (2008). Inhibition of reverse transcriptase activity of HIV by polysaccharides of brown algae. *Biomedicine & Pharmacotherapy* 62:303–307.

Quist, D. and Chapela, I.H. (2001). Transgenic DNA introgressed into traditional maize landraces in Oaxaca, Mexico. *Nature* 414:541–543.

Raposo, M.F., de Morais, R.M. and Bernardo de Morais, A.M. (2013). Bioactivity and applications of sulphated polysaccharides from marine microalgae. *Marine Drugs* 11:233–252.

Remacle, C., Cline, S., Boutaffala, L., Gabilly, S., Larosa, V., Barbieri, M.R., Coosemans, N. and Hamel, P.P. (2009). The ARG9 gene encodes the plastid-resident N-acetyl ornithine aminotransferase in the green alga *Chlamydomonas reinhardtii*. *Eukaryotic Cell* 8:1460–1463.

Rinehart, K.L., Jr., Shaw, P.D., Shield, L.S., Gloer, J.B., Harbour, G.C., Koker, M.E.S., Samain, D., Schwartz, R.E., Tymiak, A.A., Weller, D.L., Carter, G.T., Munro, M.H.G., Hughes, R.G., Jr., Renis, H.E., Swynenberg, E.B., Stringfellow, D.A., Vavra, J.J., Coats, J.H., Zurenko, G.E., Kuentzel, S.L., Li, L.H., Bakus, G.J., Brusca, R.C., Craft, L.L., Young, D.N. and Connor, J.L. (1981). Marine natural products as sources of antiviral, antimicrobial, and antineoplastic agents. *Pure and Applied Chemistry* 53:795–817.

Ruiz, L., Hector, A., Thomsen, M. and Trajano, H. (eds.) (2017) *Hydrothermal Processing in Biorefineries*. Springer Publishing. ISBN 978-3-319-56457-9.

Santos, F.M. and Pires, J.C.M. (2018). Nutrient recovery from wastewaters by microalgae and its potential application as bio-char. *Bioresource Technology* 267:725–731.

Santoyo, S., Rodríguez-Meizoso, I., Cifuentes, A., Jaime, L., García-Blairsy Reina, G., Señorans, F.J. and Ibáñez, E. (2009). Green processes based on the extraction with pressurized fluids to obtain potent antimicrobials from *Haematococcus pluvialis* microalgae. *LWT Food Science and Technology* 42:1213–1218.

Sathivel, A., Raghavendran, H.R.B., Srinivasan, P. and Devaki, T. (2008). Anti-0peroxidative and anti-hyperlipidemic nature of *Ulva lactuca* crude polysaccharide on d-galactosamine induced hepatitis in rats. *Food and Chemical Toxicology* 46:3262–3267.

Sekhon, B. (2010). Biopharmaceuticals: An overview. *Thai Journal of Pharmaceutical Sciences* 34:1.

Sharma, G.M., Michaels, L. and Burkholder, P.R. (1968). Goniodomin, a new antibiotic from a dinoflagellate. *Journal of Antibiotics* 21:659–664.

Silva, J., Alves, C., Pinteus, S., Mendes, S. and Pedrosa, R. (2018). Neuroprotective effects of seaweeds against 6-hydroxidopamine-induced cell death on an in vitro human neuroblastoma model. *BMC Complementary and Alternative Medicine* 18:58.

Sivagnanam, S.P., Yin, S., Choi, J.H., Park, Y.B., Woo, H.C. and Chun, B.S. (2015). Biological properties of fucoxanthin in oil recovered from two brown seaweeds using supercritical CO_2 extraction. *Marine Drugs* 13:3422–3442.

Sizova, I., Fuhrmann, M. and Hegemann, P. (2001). A *Streptomyces rimosus* aphVIII gene coding for a new type phosphotransferase provides stable antibiotic resistance to *Chlamydomonas reinhardtii. Gene* 277:221–229.

Skibola, C.F. (2004). The effect of *Fucus vesiculosus*, an edible brown seaweed, upon menstrual cycle length and hormonal status in three pre-menopausal women: A case report. *BMC Complementary and Alternative Medicine* 4:10.

Soria-Guerra, R.E., Ramírez-Alonso, J.I., Ibáñez-Salazar, A., Govea-Alonso, D.O., Paz-Maldonado, L.M.T., Bañuelos-Hernández, B., Korban, S.S. and Rosales-Mendoza, S. (2014). Expression of an HBcAg-based antigen carrying angiotensin II in *Chlamydomonas reinhardtii* as a candidate hypertension vaccine. *Plant Cell, Tissue and Organ Culture (PCTOC)* 116:133–139.

Steinbrenner, J. and Sandmann, G. (2006). Transformation of the green alga *Haematococcus Pluvialis* with a phytoene desaturase for accelerated astaxanthin biosynthesis. *Applied and Environmental Microbiology* 72:7477–7484.

Sun, M., Qian, K., Su, N., Chang, H., Liu, J. and Shen, G. (2003). Foot-and-mouth disease virus VP1 protein fused with cholera toxin B subunit expressed in *Chlamydomonas reinhardtii* chloroplast. *Biotechnology Letters* 25:1087–1092.

Surzycki, R., Greenham, K., Kitayama, K., Dibal, F., Wagner, R., Rochaix, J.D., Ajam, T. and Surzycki, S. (2009). Factors effecting expression of vaccines in microalgae. *Biologicals* 37:133–138.

Tabet, N. (2006). Acetylcholinesterase inhibitors for Alzheimer's disease: Anti-inflammatories in acetylcholine clothing! *Age Ageing* 35:336–338.

Talarico, L.B. and Damonte, E.B. (2007). Interference in dengue virus adsorption and uncoating by carrageenans. *Virology* 363:473–485.

Tan, J.W. and Kim, M.K. (2016). Neuroprotective effects of Biochanin A against β-amyloid-induced neurotoxicity in PC12 cells via a mitochondrial-dependent apoptosis pathway. *Molecules* 21:548.

Tannoury, M. and Attieh, Z. (2017). The influence of emerging markets on the pharmaceutical industry. *Current Therapeutic Research* 86:19–22.

Tarhouni-Jabberi, S., Zakraoui, O., Ioannou, E., Riahi-Chebbi, I., Haoues, M., Roussis, V., Kharrat, R. and Essafi-Benkhadir, K. (2017). Mertensene, a halogenated monoterpene, induces G2/M cell cycle arrest and caspase dependent apoptosis of human colon adenocarcinoma HT29 cell line through the modulation of ERK-1/-2, AKT and NF-κB signaling. *Marine Drugs* 15:221.

Taunt, H.N., Stoffels, L. and Purton, S. (2018). Green biologics: The algal chloroplast as a platform for making biopharmaceuticals. *Bioengineered* 9:48–54.

Teas, J., Braverman, L.E., Kurzer, M.S., Pino, S., Hurley, T.G. and Hebert, J.R. (2007). Seaweed and soy: Companion foods in Asian cuisine and their effects on thyroid function in American women. *Journal of Medicinal Food* 10:90–100.

Thajuddin, N. and Subramanian, G. (2005). Cyanobacterial biodiversity and potential applications in biotechnology. *Current Science* 89:47–57.

Torres, F.A.E., Passalacqua, T.G., Velásquez, A.M.A., de Souza, R.A., Colepicolo, P. and Graminha, M.A.S. (2014). New drugs with antiprotozoal activity from marine algae: A review. *Revista Brasileira de Farmacognosia* 24:265–276.

Tran, M., Henry, R.E., Siefker, D., Van, C., Newkirk, G., Kim, J., Bui, J. and Mayfield, S.P. (2013). Production of anti-cancer immunotoxins in algae: Ribosome inactivating proteins as fusion partners. *Biotechnology and Bioengineering* 110:2826–2835.

Tran, M., Van, C., Barrera, D.J., Pettersson, P.L., Peinado, C.D., Bui, J. and Mayfield, S.P. (2013). Production of unique immunotoxin cancer therapeutics in algal chloroplasts. *Proceedings of the National Academy of Sciences* 110:E15–E22.

Tran, M., Zhou, B., Pettersson, P.L., Gonzalez, M.J. and Mayfield, S.P. (2009). Synthesis and assembly of a full-length human monoclonal antibody in algal chloroplasts. *Biotechnology and Bioengineering* 104:663–673.

Unnikrishnan, P.S. and Jayasri, M.A. (2018). Marine algae as a prospective source for antidiabetic compounds – A brief review. *Current Diabetes Reviews* 14:237–245.

Vehapi, M., Koçer, A.T., Yılmaz, A. and Özçimen, D. (2020). Investigation of the antifungal effects of algal extracts on apple-infecting fungi. *Archives of Microbiology* 202:455–471.

Vickery, R. (1995). *A Dictionary of Plant-Lore*. Oxford, UK: Oxford University Press.

Viso, A.C., Pesando, D. and Baby, C. (1987). Antibacterial and antifungal properties of some marine diatoms in culture. *Botanica Marina* 30:41–46.

Walter, C.S. and Mahesh, R. (2000). Antibacterial and antifungal activities of some marine diatoms in culture. *Indian Journal of Geo-Marine Sciences* 29:238–242.

Wang, H., Chiu, L.C., Ooi, V.E. and Ang, P.O., Jr. (2010). A potent antitumor polysaccharide from the edible brown seaweed *Hydroclathrus clathratus*. *Botanica Marina* 53:265–274.

Wang, T., Jónsdóttir, R., Liu, H., Liwei, G., Hordur, G.K., Sivakumar, R. and Gudrún, O. (2012a). Antioxidant capacities of phlorotannins extracted from the brown algae *Fucus vesiculosus*. *Journal of Agricultural and Food Chemistry* 60:5874–5883.

Wang, W., Wang, S.X. and Guan, H.S. (2012b). The antiviral activities and mechanisms of marine polysaccharides: An overview. *Marine Drugs* 10:2795–2816.

Wang, X., Brandsma, M., Tremblay, R., Maxwell, D., Jevnikar, A.M., Huner, N. and Ma, S. (2008). A novel expression platform for the production of diabetes-associated autoantigen human glutamic acid decarboxylase (hGAD65). *BMC Biotechnology* 8:87.

Wang, X., Chen, Y., Wang, J., Liu, Z. and Zhao, S. (2014). Antitumor activity of a sulfated polysaccharide from *Enteromorpha intestinalis* targeted against hepatoma through mitochondrial pathway. *Tumour Biology* 35:1641–1647.

Wang, X. and Zhang, Z. (2014). The antitumor activity of a red alga polysaccharide complexes carrying 5-fluorouracil. *International Journal of Biological Macromolecules* 69:542–545.

Washida, K., Koyama, T., Yamada, K., Kita, M. and Uemura, D. (2006). Karatungiols A and B, two novel antimicrobial polyol compounds, from the symbiotic marine dinoflagellate *Amphidinium* sp. *Tetrahedron Letters* 47:2521–2525.

Wollmann, F., Dietze, S., Ackermann, J.U., Bley, T., Walther, T., Steingroewer, J. and Krujatz, F. (2019). Microalgae wastewater treatment: Biological and technological approaches. *Engineering Life Science* 19:860–871.

Xin, X.L., Geng, M.Y., Guan, H.S. and Li, Z.L. (2000). Study on the mechanism of inhibitory action of 911 on replication of HIV-1 *in vitro*. *Chinese Journal of Marine Drugs* 19:15–18.

Yan, N., Fan, C., Chen, Y. and Hu, Z. (2016). The potential for microalgae as bioreactors to produce pharmaceuticals. *International Journal of Molecular Sciences* 17:962.

Yasumoto, T. and Maurata, M. (1993). Marine toxins. *Chemical Reviews* 93:1897–1909.

Yeo, A.R., Lee, J., Tae, I.H., Park, S.R., Cho, Y.H., Lee, B.H., Shin, H.C., Kim, S.H. and Yoo, Y.C. (2012). Anti-hyperlipidemias effect of polyphenol extract (Seapolynol (™)) and dieckol isolated from *Ecklonia cava* in in vivo and in vitro models. *Preventive Nutrition and Food Science* 17:1–7.

Yoon, N.Y., Lee, S.H., Yong, L. and Kim, S.K. (2009). Phlorotannins from Ishige okamurae and their acetyl- and butyrylcholinesterase inhibitory effects. *Journal of Functional Foods* 1:331–335.

Yoon, S.M., Kim, S.Y., Li, K.F., Yoon, B.H., Choe, S. and Kuo, M.M. (2011). Transgenic microalgae expressing Escherichia coli AppA phytase as feed additive to reduce phytate excretion in the manure of young broiler chicks. *Applied Microbiology Biotechnology* 91:553–563.

Zare-Gachi, M., Daemi, H., Mohammadi, J., Baei, P., Bazgir, F., Hosseini-Salekdeh, S. and Baharvand, H. (2020). Improving anti-hemolytic, antibacterial and wound healing properties of alginate fibrous wound dressings by exchanging counter-cation for infected full-thickness skin wounds. *Materials Science and Engineering: C* 107:110321.

Zhang, C., Li, Y., Shi, X. and Kim, S.K. (2010). Inhibition of the expression on MMP-2, 9 and morphological changes via human fibrosarcoma cell line by 6,6'-bieckol from marine alga *Ecklonia cava*. *BMB Reports* 43:62–68.

11 Techno-economic analysis of algal biorefineries

Shima Masoumi, Venu Babu Borugadda, and Ajay K. Dalai

Introduction

The world relies on energy supply, which can be divided into non-renewable and renewable sources. Due to the incremental non-renewable fossil fuel consumption and global population, many studies have been focused on finding renewable and alternative energy sources. Also, the consumption of fossil fuels leads to a drastic increase in greenhouse gas emissions which have endangered the world (Gollakota et al., 2018). To overcome the elevated energy demand as well as greenhouse gas emissions, researchers have focused on developing sustainable biofuel production. Biomass is considered one of the most promising alternatives to produce biofuels which can be broken down into three main generations. First-generation feedstocks can also be consumed as human foods such as wheat and soybeans. Second-generation feedstocks include woody crops and agricultural wastes which are not considered human food, but oil extraction needs advanced conversion technologies. The third generation of feedstocks is marine biomasses or algae, whose conversion to biofuels is considered a promising method due to the advantages of algae such as fast growth rate, having to overlap with food resources and ability to sequester CO_2 (Masoumi et al., 2020).

Algae contain elements such as nitrogen, oxygen, hydrogen and carbon in the form of three main components of lipid, protein and carbohydrates. Further, algae are classified into two categories of micro- and macroalgae. For transportation liquid biofuel production, the algae with high lipid content are important. However, due to the low amount of lipid content and high ash and carbohydrate contents of most macroalgal species, it is relatively unexplored as a source of bio-oil production (Alfonsín et al., 2019).

Besides the production of algal-based fuels, algal derivatives found favorable applications as value-added chemicals, fertilizers, pharmaceuticals, nutraceuticals, pigments and fish and animal feed, and in water treatment and protein industries. To promote the sustainability of algal-derived products, innovative approaches need to be practiced for the production and purification techniques. Also, a techno-economic assessment and a lifecycle assessment (LCA) need to be carried out to evaluate the economic feasibility of each process.

Techno-economic analysis (TEA) is considered a significant analytical tool to understand the economic feasibility of algal biomass-derived products. Several types of research have been focused on economic analysis to report the algal biofuels' selling cost that ranges between $1.64 and 30/gal. The variable results are associated with various systems used for algal cultivation systems, different reaction pathways, product distributions and coproduct utilization (Quinn and Davis, 2015).

The following sections are focused on the production of biofuels, chemicals, pharmaceuticals, nutraceuticals, bioplastics, fatty acids, vitamins, elastomers and fish feed from micro- and macroalgae as renewable sources. The minimum fuel selling price (MFSP) of algal-derived products was estimated through their production routes, growth conditions, lipid extraction techniques, and product development technologies. Techno-economic feasibility of each product and process is described relative to the present market price and production technologies. In addition, current challenges involved in commercialization technologies and prospects are discussed critically for the algal biorefinery.

Biofuel production from microalgae

Microalgae are unicellular species, which exist in different groups or chains such as *Chlorella algae species*, *Spirulina* and *Nannochloropsis*. The productivity of microalgae-derived biofuels could be approximately twice the yield of oil obtained from terrestrial crops. Due to the different growing conditions, the lipid and protein contents in each microalgal species are different (Demirbas and Demirbas, 2011). Furthermore, the energy required for the dewatering or drying step, which is needed for algae, constituted up to 84.9% of the total required energy (Kroumov et al., 2017). All around the world, a lot of researchers have set targets to improve the energy sector demands and climate change situation. The European Union launched the strategy in 2010 aiming to achieve 20% of the climate and energy package from renewables. In the transportation sector, biofuels obtained from renewables were set to 10%. Also, since the first oil crisis in the mid-1970s, a lot of researchers have been interested in biofuel production. In 2005, Canada initiated a strategy to cut greenhouse gas emissions by 30% in 25 years (Haugstad, 2017).

Algal biofuel production and utilization of coproducts require downstream processing steps (Figure 11.1). Extensive studies are focused on all the major steps of algal biofuel production such as cultivation methods (open pond or photobioreactor, PBR), harvesting technology and conversion processes. Further, the energy required for the dewatering or drying step, which is needed for some of the conversion processes, constituted up to 84.9% of the total required energy (Kroumov et al., 2017).

The development of algal biofuel technology through different conversion processes has been a top research topic in the field of biorefinery in recent years. One of the important steps to have a cost-effective biofuel production process

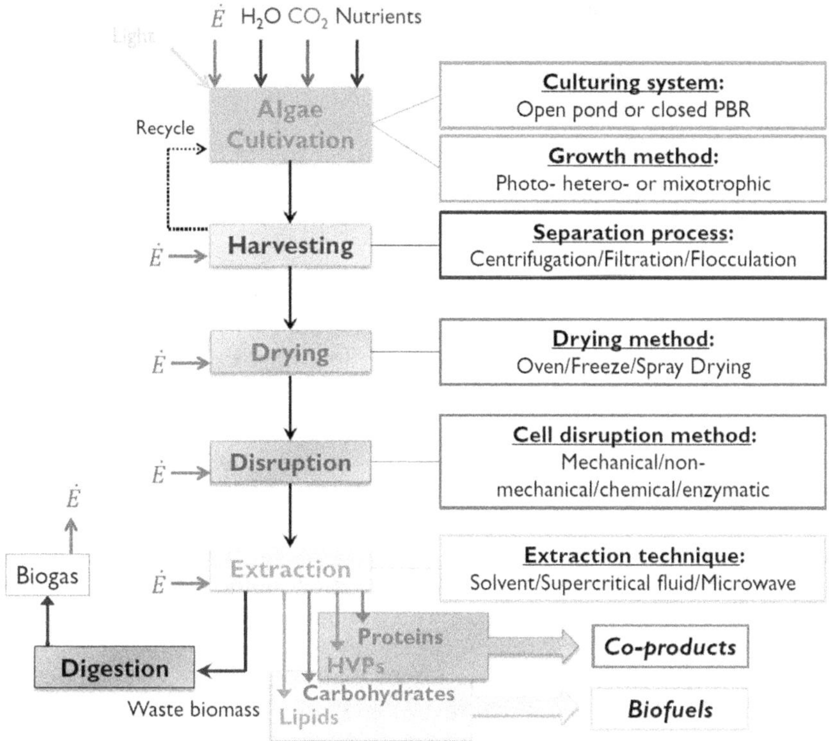

Figure 11.1 Schematic structure of algal biofuel production and coproduct utilization (Ė, energy flux). Reproduced with permission from Kroumov et al. (2017).

is to find a highly efficient and economical conversion process. There is a lot of literature related to different conversion technology including supercritical processes, pyrolysis and hydrothermal liquefaction (HTL) followed by catalytic upgrading to produce algal biofuels and bioproducts (Ansari et al., 2020).

Supercritical processes require severe operating conditions resulting in higher capital costs. Supercritical water gasification (SCWG) is a technology to convert the microalgae with lower heating value to high-value gaseous products under high-temperature and high-pressure conditions (Chakinala et al., 2010). Thus, sustainable and low-cost feedstocks such as microalgae are necessary for biofuel production. In severe process conditions, the physical properties of water change significantly, resulting in the decrease of dielectric constant and thermal conductivity. By using different process conditions and catalysts, the selectivity of the gas produced from the process can change toward H_2, syngas or methane.

Brandenberger et al. (2013) investigated TEA for production of the synthetic natural gas (SNG) from microalgae through SCWG with a capacity of 86,500 tons/year of feed in Switzerland. Their objective was to study the impact of water and nutrients flow for the production of microalgae and SNG production. They studied different methods for microalgae production, such as raceway ponds (RPs) and flat-panel airlift photobioreactors (FPA-PBRs). The production cost and energy return on energy investment (EROEI) are defined in Eq. (11.1):

$$\text{EROEI} = \text{Energy return after the process} / \text{Energy required for the process} \quad (11.1)$$

According to the sensitivity analysis results by using an annual algal productivity of 38.5 tons/ha in RPs, an EROEI of 1.84 and an SNG production cost of €194/GJ were obtained. Regarding FPA-PBRs, for microalgae productivity of 79 tons/ha, an EROEI of 1.01 and an SNG production cost of €266/GJ were calculated. However, their assumptions to calculate the investment for algal biomass production step, as well as novel reactor designs, are not easily achievable. Compared to the present price for natural gas, due to high algal biomass production cost, the calculated production cost of SNG from algae is still too high.

To achieve complete renewable transportation biofuels, some researchers have focused on a new technology, which is the integration of concentrated solar thermal (CST) heat with other processes such as SCWG, or the integration of algal cultivation system with SCWG. The use of the integrated system of CST with other processes that require high temperature and energy such as Fischer–Tropsch (FT) synthesis (known as a technology to produce a wide range of transportation fuels) and SCWG has raised a great interest to produce renewable transportation fuels (Campanario and Ortiz, 2017).

The techno-economic assessment and steady-state model for a combined system of the solar–SCWG–reforming of microalgae and FT process in Australia were investigated by Rahbari et al. (2019). 15.2 wt.% concentration of algae slurries using an optimized heat exchanger to achieve the maximum syngas production results in maximum liquid fuels (gasoline and diesel) for the plant lifetime of 30 years. The total cost of the plant includes the total capital investment of approximately AUD 160 million for fuel production of 7,600 tons/year and the cost related to CO_2 emissions. According to the sensitivity analysis results, the algal feedstock is considered the most important variable, followed by the discount rate and the capital cost of the FT unit. The whole system proposed a levelized cost of fuel (LCOF) as low as US$2.44/L of gasoline equivalent. Compared to the petroleum-based products, the estimated LCOF economically seems high; however, with a predicted decrease in the cost of algae production on a large scale, it could be economically viable.

Mian et al. (2015) investigated the modeling and TEA of SNG production via supercritical hydrothermal gasification of microalgae in Brazil. Their complete modeling includes all the steps of microalgae cultivation, settling ponds and catalytic hydrothermal gasification (CHG) as well as SNG purification system. Through this system, microalgae (carbon source) can be converted into methane/syngas, and the nutrients, water and CO_2 can be recovered. They concluded that integration of cultivation system with hydrothermal gasification as well as energy recovery could lead to high energy efficiency and increased sustainability. For the economic evaluation, it was assumed that the main factors affecting the operating costs of the plant were the operation hours and price of the electricity; however, they did not use the sensitivity analysis to justify their assumptions. TEA results showed that the total cost of biomass varied between US$10/GJ and 16/GJ/year for a lifetime of 25 years with an annual operation of 8,000 h including a 6% interest rate.

HTL is considered a promising process for the production of algal biofuels as it is suitable for the high-moisture-content feedstocks such as microalgae (Hu et al., 2017). The HTL process involves the reaction temperature and pressure in the range of 200–370°C and 4–20 MPa, respectively, with or without another solvent with lower critical temperature and pressure such as alcohols to moderate reaction conditions (Feng et al., 2018). It should be highlighted that the HTL process requires energy to provide heat for the reactor (Pearce et al., 2016). Also, since HTL bio-crude oil has low heating value because of its high oxygen contents, it should be further processed to remove heterogeneous compounds and increase the heating value to be able to use it as a transportation fuel (Haider et al., 2018).

Hognon et al. (2015) investigated the comparative TEA through HTL and pyrolysis of microalgae (*Chlamydomonas reinhardtii*). Biofuel production was simulated from the cultivation of microalgae to the final step of the upgrading of bio-crude oil to produce high-value liquid fuels. HTL of microalgae at 14% dry basis of biomass was carried out at a temperature and pressure of 310°C and 10.4 MPa, respectively, resulting in a bio-crude yield of 57.8 wt.%. Pyrolysis of microalgae was performed at a temperature and pressure of 500°C and 0.102 MPa, respectively, to obtain the bio-crude oil mass yield of 60.2%. The biofuel production costs from either HTL or pyrolysis (average of €70.4/GJ) are significantly higher than the production costs of petroleum-based fuels. To decrease the biofuel production costs and energy consumption, alternative dewatering technologies were evaluated. It was shown that the most expensive step for algal biodiesel production was the costs of microalgae cultivation related to raceway capital cost and also nutrient cost.

DeRose et al. (2019) studied the TEA of two different pathways: (1) biochemical processing, which includes fermentation of proteins and carbohydrates to produce fuel alcohol products; and (2) thermal–chemical processing through HTL of whole algae. Techno-economic results showed MFSPs of $12.85/gasoline gallon equivalent (GGE) and $10.41/GGE for the biochemical and thermal–chemical pathways, respectively. Sensitivity analysis on the techno-economic model showed that an MFSP of $3.85/GGE could be achieved by lower biomass cost and higher yield and ash removal.

Gu et al. (2020) studied the comparative TEA of algal biofuel production through the HTL process in two different pathways – two-stage sequential hydrothermal liquefaction (SEQHTL) and conventional one-stage direct hydrothermal liquefaction (DHTL) – in Washington State, USA. They considered that SEQHTL could be an alternative to facilitate the production of coproducts as well as carry out the process at milder process conditions. Aspen Plus simulation with comprehensive TEA was conducted for conversion processes of the same algal stain (*Chlorella sorokiniana*). Compared to DHTL with an MFSP of \$2.1/L, SEQHTL showed a lower amount of MFSP of \$1.61/L due to its milder operating conditions, which resulted in lower capital and operating costs. Also, their study showed that the total installed costs of the facilities for SEQHTL and DHTL were \$89 million and \$112 million, respectively, for processing 1,215 metric tons/d of algal feedstock. The sensitivity analysis results confirmed that the yield of the biofuels and the feed concentration are the most influential factors on production cost. Higher operational EROEI was \$6.73/L and \$5.31/L for SEQHTL and DHTL, respectively.

Zhu et al. (2019) investigated TEA of three different treatment methods of aqueous phase for HTL of microalgae and bio-crude upgrading system: (1) direct recycle of aqueous phase obtained from HTL to the algae farm, (2) anaerobic digestion (AD) and (3) CHG. Techno-economic assessments of these methods showed that the cost of direct recycling was lower compared to other cases, resulting in a lower MFSP. No methane was produced through direct combustion of the aqueous phase; thus unlike CHG and AD, during this method, all the natural gas was consumed. In this study, compared to the benefits regarding methane generation with different treatment methods, the cost benefits of the removal process of the aqueous phase were higher. The MFSP of direct recycling, CHG and AD was \$12.5/GGE, \$13.8/GGE and \$12.9/GGE, respectively. Compared to AD and CHG, the direct combustion method required pumps and buffer tankage to recycle the aqueous phase, which resulted in the lowest cost. According to the sensitivity analysis, feedstock cost, the plant scale, bio-crude upgrading and the treatment method of the aqueous phase are the important factors affecting the total cost of algal biofuel production.

Pearce et al. (2016) investigated the techno-economic assessment of integrated biofuel technology using HTL and concentrated solar power (CSP) in the UK. Integration of HTL and CSP resulted in a reduction in biofuel production cost due to the absence of additional equipment cost. The plant size capacity was based on the HTL feedstock production of 1,000 L/d with 20% of microalgae. TEA results showed that the MFSP of \$1.23/kg due to the nutrient recycling and system integration (HTL/CSP) could be economically viable for algal biofuel production. Sensitivity analysis showed that an increase of 20% in biofuel yield resulted in a lower MFSP (16.7% reduction).

Van Doren et al. (2017) studied two process conversion technologies: integrated HTL with AD and integrated HTL with CHG. Each of the pathways was followed by heat–power combination, and combustion in a boiler for gas utilization. To improve overall system efficiency, pinch analysis was applied. Van Doren et al. (2017) also highlighted that conversion of waste to usable energy should focus not only on bio-crude oil production but also on the by-product utilization to recover energy. As a result, the overall process could improve economically and environmentally through the recovery of internal heat and power, which leads to a reduction in external energy demands. TEA results showed that minimum bio-crude oil selling prices based on the CHG pathway were twice as high compared to that of AD. AD pathways under two different scenarios were profitable but still were higher than the price of conventional fossil fuels. Considering the market prices of electricity or natural gas, the cogeneration engine or boiler could be considered as an economical route.

Ou et al. (2015) studied TEA of transportation fuel production via HTL of defatted microalgae and subsequently hydrotreating of produced bio-crude oil to evaluate the system's economic feasibility. The lower price of defatted algal feedstock makes it a good candidate for HTL feedstock. Figure 11.2 shows the schematic diagram of the process including HTL, hydrotreating, hydrogen generation and product refining.

The HTL process was carried out at a temperature of 350°C and pressure of 18 MPa. After a residence time of 30 min, the products were separated into gas, solid and liquid. Bio-crude having high oxygen and nitrogen contents with a yield of 36% was obtained. The hydrotreatment process was carried out at 200°C and 11.7 MPa for stage 1 to stabilize the bio-crude, and at 400°C and 11.7 MPa for stage 2 of hydroprocessing. The authors determined

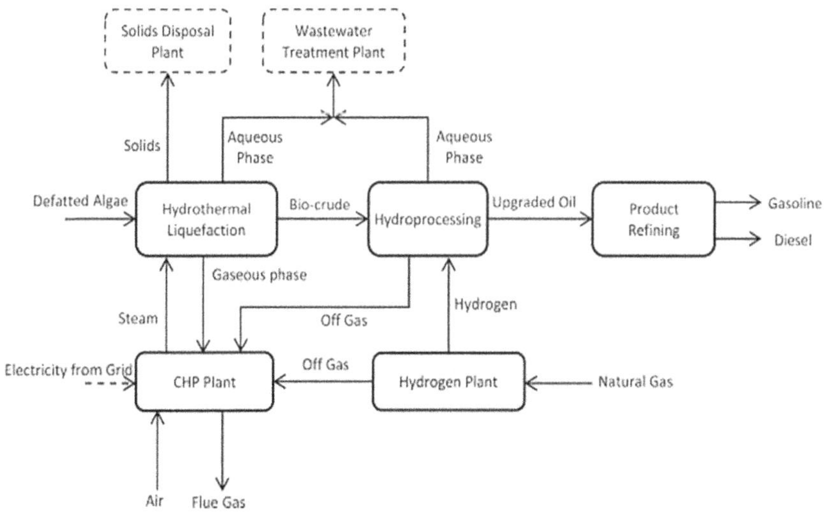

Figure 11.2 Schematic diagram of the HTL and hydrotreating processes to produce diesel and gasoline. Reproduced with permission from Ou et al. (2015).

that the MFSP was in the range of \$2.30–\$3.15/gal with optimum feedstock cost. The study showed that the MFSP is economically competitive with petroleum-based liquid fuels. Sensitivity analysis showed that the fuel yield, fixed capital investment, internal rate of return (IRR) and cost of feedstock significantly influenced MFSP.

The results of the sensitivity analysis done by Zhu et al. (2013) also showed that the feedstock cost, the final product yield and the upgrading equipment cost were the main factors affecting MFSP. Zhu et al. (2013) investigated TEA for conversion of lipid-extracted microalgae (LEA) to liquid hydrocarbon fuels through HTL and the hydrotreating process based on the optimum experimental data. The LEA feedstock with a moisture content of 83 wt.% contained 19.9 ± 3 wt.% residual lipids. The HTL tests were carried out with a temperature and pressure of 348°C and 21 MPa, respectively. The upgrading test was run using presulfided CoMo/γ-Al_2O_3 for 12 h for 0.045 g/g dry bio-oil catalyst loading. The authors assumed a 608 dry metric tons/d for feedstock feed rate in which algal biomass was transformed to liquid fuels with an annual yield of 26.9 million GGE. The authors determined that the MFSP would be in the range of \$2.07–\$7.11/GGE with optimum feedstock cost.

Jones et al. (2014) evaluated TEA of HTL followed by catalytic upgrading of bio-crude oil obtained from HTL of whole *Nannochloropsis* and *Chlorella algae*. The product yields using HTL of bio-crude were modeled at a temperature of 349°C and pressure of 21 MPa and 51 wt.% using 20 wt.% solid concentration (ash-free basis). Hydrotreating was carried out at a temperature of 403°C and pressure of 10.5 MPa using CoMo/alumina as a catalyst. The yield of hydrotreated oil was 77 wt.% containing 86 wt.% of carbon compared to bio-crude oil with 77 wt.% of carbon content. For this scenario, TEA results showed the MFSP of \$4.77/gal for diesel using the feedstock cost of \$430/ton.

Dutta et al. (2016) evaluated a comparative study on TEA of two conversion pathways of microalgae to biofuels; one of them was studied at the University of Aveiro (UA), Portugal, and the other one was studied at the National Renewable Energy Laboratory (NREL), Golden, Colorado, USA. Microalgae cultivation for both the cases was based on open-pond technology, and the feedstock costs for the UA and NREL studies were calculated to be \$1,279/ton and \$430/ton of biomass, respectively. Their study showed that the utilization of coproducts plays an important role to reduce the final fuel selling price. The technologies evaluated in UA and NREL were solvent extraction, transesterification and product purification for biodiesel with an MFSP of \$10.55/GGE and fermentation, distillation and hydrodeoxygenation (HDO) for renewable diesel with an MFSP of \$4.35/GGE.

Based on the techno-economic assessment and sensitivity analysis for these two in different scenarios, the decrease in the cost of algal feedstock was considered as one of the main factors, which impacted the total cost of large-scale algal biofuel production to make it economically viable (Xin et al., 2016). Barlow et al. (2016) studied techno-economic assessment of the production of renewable diesel through HTL of biomass feedstock obtained from a rotating algal biofilm reactor (RABR). They developed the model including RABR integrated with wastewater treatment plant in Utah, USA, and conversion of algae to HTL bio-crude. HTL was carried out in a 500-mL reactor operating at a temperature of 325°C and pressure of 14.5–16.2 MPa for 60 min. This resulted in a bio-crude yield of 32 wt.%. The results showed that algal productivity and the capital cost of the RABR significantly affected the MFSP. The base scenario considered the annual output of 270,000 gallons of biodiesel with an optimized fuel selling price of $11.80/gallon.

Albrecht et al. (2016) studied the impact of two different methods of algal feedstock preparation: (1) a biofilm-based approach to grow an algal strain phototrophically and (2) heterotrophically stressing of an algal strain to have high lipid concentration. The HTL process was carried out at 350°C and 20.7 MPa to convert algal feedstock to bio-crude. The bio-crudes were upgraded to liquid hydrocarbon fuel through catalytic hydrotreating at 400°C and 10.34 MPa. The high lipid concentration of algae obtained from heterotrophic stressing of algae resulted in high bio-crude and diesel fuel yield; however, it is costly. The results of TEA showed that the significant increase in bio-crude and fuel yield could significantly affect the economics of the system more than feedstock production costs. The price of feedstock cultivation varied from $0 to $1,200/dry ash-free metric ton. The MFSP for standard lipid varied from $3.5 to $22/gallon. For the same feedstock cost, the MFSP for high lipid varied from $1.2 to $7.7/gallon.

The studies on TEA confirmed that to make the biofuel production from microalgae to be economically feasible, the algal biomass production cost should be reduced significantly. To address the challenges related to the cultivation of microalgae in RP systems with nutrients and CO_2, the algal biofuel production from wastewater has received much attention in recent years (Mehrabadi et al., 2015).

Ranganathan and Savithri (2019) investigated TEA of algal biofuel production through HTL and hydrotreating of wastewater-based microalgae. An MFSP of the hydrocarbon liquid fuels was calculated at $4.3/GGE using a discounted cash flow analysis. This study considered the input capacity of 800 Ml/d of wastewater with an algal concentration of 20 wt.% fed into the HTL reactor. The temperature and pressure in the HTL process were set at 350°C and 20 MPa, respectively. The product yields for gas, solid, aqueous and organic phases were 4, 11.5, 33.5 and 51 wt.%, respectively.

Xin et al. (2016) studied the comprehensive TEA of the wastewater-based algal biofuel production system with the combination of algal cultivation and wastewater treatment and considered it to be economically viable. The system boundary used for their study included an algal cultivation system and pyrolysis process to produce bio-oil. The TEA results showed that the break-even selling price of produced biofuel was $2.23/gallon.

Juneja and Murthy (2017) evaluated the potential of biodiesel production from microalgae grown on wastewater. In their study, 20% solid concentrated microalgae (*Chlorella vulgaris*) were used for biodiesel production. They used HTL to convert microalgae into bio-oil at a temperature and pressure of 340°C and 20 MPa, respectively. The bio-oil with a high heating value was further upgraded to renewable diesel. They studied the economic analysis on a commercial scale with a processing capacity of 60 million gallon wastewater per day. The unit price of renewable diesel production was $1.75/L. According to sensitivity analysis, high lipid amounts present in microalgae and increased plant size would reduce the production cost.

The algal strain (*Chlorella* sp.) was grown from a municipal wastewater plant. Microwave-assisted pyrolysis (MAP) was employed for the conversion of microalgal biomass. Compared to conventional pyrolysis, MAP conversion showed several advantages, including uniform internal heating and lack of agitation or fluidization (Du et al., 2011). Xin et al. (2018) investigated an innovative system of sludge treatment and biofuel production. They employed the MAP technology as a conversion system to produce algal biofuels. They conducted a comprehensive TEA to evaluate the integrated system considering the biofuel production and wastewater treatment improvement. Through the system used for their study, three main products were produced: bio-oil, syngas and biochar. These products could be sold to a refinery or partly used to provide heat for the drying process. According to their study, an estimated breakeven selling price was approximately $1.85/gallon, which was close to the 5-year average of crude oil price.

Microalgae have also gained attention to produce gaseous fuels through different processes such as gasification and digestion. SCWG and thermal gasification can convert wet biomass such as microalgae and wastewater sludge into gaseous fuels such as hydrogen. Hydrogen as an energy source is mainly produced by reforming a natural gas, which leads to significant greenhouse gas emissions. So, there is an interest and great need to find a renewable and sustainable alternative source for biohydrogen production.

Kumar et al. (2019) studied the comprehensive techno-economic assessment of large-scale hydrogen production from microalgae. They developed models for TEA of the hydrogen production through the SCWG and thermal gasification processes. The results showed that 2,000 dry tons/d plant required the total capital cost of $277.8 and $215 M for the SCWG and thermal gasification processes, respectively. Also, due to a slightly higher yield of hydrogen obtained from SCWG compared to thermal gasification, hydrogen product values were relatively higher ($4.59 ± 0.10/kg) as compared to those ($5.66 ± 0.10/kg) for thermal gasification. The costs are approximately four times higher than those of hydrogen produced from natural gas. The sensitivity analysis showed that the microalgal feedstock cost and hydrogen yield were the key factors that affected the economics of the process.

Wu et al. (2019) studied the TEA of biogas production from microalgae (*Cyanothece* BG0011) through AD. AD could be economically viable if the whole microalgae were converted into biofuels. Three different scenarios were investigated for biogas production: (1) mesophilic AD; (2) low-temperature

AD, which requires heating; and (3) covered anaerobic lagoon, which does not require additional energy due to the lack of heating and mixing. Also, in this study, a chemical scrubbing method using monoethanolamine (MEA) was used for the removal of acidic gas. The production cost for biomethane was $14.8/MMBtu using the covered anaerobic lagoon method for AD and with scrubbing for purification of biogas using high-pressure water. The production cost for electricity was $0.13/kWh. The biomethane production and electricity costs could be reduced to $12.16/MMBtu and $0.11/kWh, respectively, by sparging air with 1% CO_2, which led to an increase in the productivity by 33%. Table 11.1 shows the comparison of some of the TEA results of algal biofuel production from microalgae through different production techniques.

Microalgae as the third generation of feedstocks to produce biofuels are considered a promising renewable feedstock, due to their fast growth rate, having no overlap with food crops and the ability to sequester CO_2. Based on the TEA results reviewed in the current chapter, the cultivation of microalgae is costly. The use of wastewater-based microalgae to produce algal biofuels could reduce the production costs significantly. There are different technologies such as HTL and pyrolysis for the conversion of wastewater-based microalgae to produce bio-oil or biodiesel, which seem economically viable. However, utilization of coproducts such as biochar and syngas and combination of different processes to minimize energy consumption could greatly affect process cost and efficiency.

Table 11.1 Techno-economic results of algal biofuel production via different technologies

Feedstock	*Production technique*	*Products*	*MFSP*	*Country*	*References*
Microalgae (€4.95/kg)	HTL and CSP	Biofuels	$1.23/kg	UK	Pearce et al. (2016)
Microalgae ($430/ton)	HTL and hydrotreating	Biodiesel	$4.77/gallon	USA	Jones et al. (2014)
Microalgae ($430/ton)	HDO	Biodiesel	$4.35/GGE	USA	Dutta et al. (2016)
Microalgae ($1,279/ton)	Transesterification	Biodiesel	$10.55/GGE	Portugal	Dutta et al. (2016)
Microalgae ($1,200/ton)	HTL and hydrotreating	Biodiesel	$3.5–$22/gallon	USA	Albrecht et al. (2016)
Sludge	Microwave-assisted pyrolysis	Bio-oil, syngas, biochar	$1.85/gallon	USA	Xin et al. (2018)
Wastewater	HTL	Biodiesel	$11.80/gallon	USA	Barlow et al. (2016)
Wastewater	HTL and hydrotreating	Biofuels	$4.3/GGE	India	Ranganathan and Savithri (2019)
Wastewater	Pyrolysis	Bio-oil	$2.23/gallon	China and the USA	Xin et al. (2016)
Wastewater	HTL	Biodiesel	$1.75/L	USA	Juneja and Murthy (2017)

Chemical production from microalgae

With limited conventional fossil fuel reservoirs, there is a great shift toward renewable and sustainable resources. According to the literature, microalgae with a high amount of lipids could be considered as a sustainable and economical feedstock for biofuel production through different conversion strategies. This section is focused on using microalgae as a promising renewable feedstock to produce value-added chemicals. Microalgae that contain a high amount of carbohydrates could be used extensively as feedstock to produce bioethanol and biobutanol as biofuels. Microalgae with a higher amount of proteins could be an economically viable source to produce high-quality bioplastics with high mechanical properties compared with petroleum-based plastics; however, there are some drawbacks and complications since the presence of lipids in microalgae causes unpleasant odors through bioplastic production (Wang et al., 2016). By 2050, the production of petroleum-based plastic will consume 20% of global annual oil reservoirs, and microalgae-based bioplastics are considered promising alternatives to petroleum-based plastics (Beckstrom et al., 2020).

Beckstrom et al. (2020) investigated TEA of bioplastic feedstock production from microalgae through nine scenarios, which were through combinations of three different methods for microalgae growth and three different conversion pathways. The three different microalgae growth methods were open raceway ponds (ORPs), cyclic-flow photobioreactors (CF-PBRs) and a combined CF-PBR/ORP system. The three different conversion technologies were drying, lipid extraction and fractionation. The simplest conversion method was drying of whole microalgae without production of fuel coproducts. In the lipid extraction method, after extracting the lipid component for utilizing fuels, the remaining microalgae were used as a bioplastic feedstock. For the fractionation pathway, both carbohydrate and lipid constituents of microalgae were extracted to utilize fuel. TEA results showed that the combination of ORPs with drying scenario was an economically viable pathway for bioplastic feedstock production with a minimum selling price of \$970/ton. While this scenario was considered as the most profitable feedstock price, other scenarios with ORP with a minimum selling price of \$1,370/ton and \$1,460/ton for lipid extraction and fractionation were considered as favorable feedstock prices. Thus, in their study, fuel production did not result in a decrease in the minimum selling price of bioplastic feedstock.

Canada as the third-largest oil reserve in the world is anticipated to have a capacity of 3.8 million barrels/day by 2022 (Dyer et al., 2013). But, to recover the highly viscous bitumen to desirable products, diluents need to be used to reduce the viscosity and let it flow through pipelines. Kumar et al. (2020) studied a comparative TEA of using microalgae to produce chemicals (diluting agent) for bitumen transport in Alberta, through two different thermochemical technologies: HTL and fast pyrolysis. The capacity of microalgal feedstock assumed for the plant was 2,000 tons (dry basis) per day.

The process conditions through the HTL process were a near-critical state, a temperature of 374°C and a pressure of 22.1 MPa. The block diagram of the algal HTL pathway to produce diluent is presented in Figure 11.3. This reaction condition leads to a reduction in the water dielectric constant and an improvement in the solubility of organics. The fast pyrolysis operates at atmospheric pressure and at a temperature of 300–700°C, which leads to thermal disintegration of main compounds of microalgae, in the order of carbohydrates followed by proteins and lipids. The values of products from pyrolysis and HTL were \$1.69 ± 0.11/L and 1.6 ± 0.09/L, respectively. The sensitivity analysis showed that product yield and biomass cost impacted the end product's value. However, the industrialization of microalgae-based chemical production was inhibited by the current high cost of microalgae cultivation. In comparison with inexpensive petroleum-based chemicals, there is still a need to develop suitable technology to provide low-cost chemicals with the same or higher quality through the conversion of low-cost microalgae.

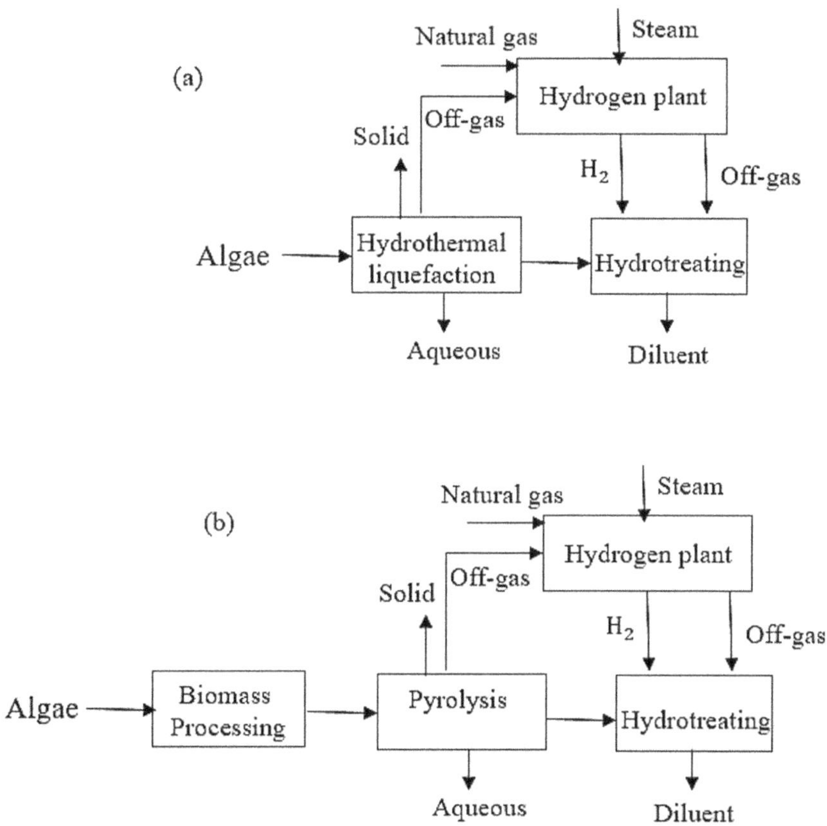

Figure 11.3 Block diagram for the algal (a) HTL and (b) pyrolysis pathway. Redrawn from Kumar et al. (2020).

Utilizing the non-protein component of microalgae to effectively produce valuable coproducts could make the process economically viable.

Bioethanol production from macroalgae

Currently, the most feasible biofuel production is bioethanol from corn as a feedstock due to its ease of well-known production techniques. However, utilizing large-scale production of biofuels (e.g., bioethanol) from food crops cannot satisfy the incremental demand for renewable biofuel production. Thus, there is a high demand to find alternative renewable feedstocks (Wei et al., 2013).

Macroalgae (seaweed) are multicellular organisms with different species that possess more plant-like characteristics. They can grow rapidly in various sizes, even in non-arable land without requiring freshwater for cultivation. As there is no competition with food crops, they can be considered as renewable feedstocks for biofuel production. Macroalgae are classified into three main groups of red, brown and green based on their photosynthetic pigmentation process. Macroalgae are mainly used in food, mineral, pharmaceutical and biomedical applications (Chen et al., 2015).

Compared to microalgae as discussed earlier, macroalgae are rich in carbohydrates and generally do not contain triglycerides, which could not be economical for biodiesel or bio-oil production. It should be highlighted that macroalgae contain high ash content and carbohydrate with no lignin, which is suitable for biogas production and fermentable for biobutanol and bioethanol production. However, to make macroalgae-based biorefineries commercially feasible, the economic aspects of the whole processes including cultivation and conversion of microalgae to biofuels or chemicals at a large scale should be investigated. Macroalgae are mainly produced in Asian countries, such as China, the Philippines, the Republic of Korea, Japan and Malaysia (Borines et al., 2011; Suutari et al., 2015).

Soleymani and Rosentrater (2017) studied TEA of biofuel production from macroalgae (seaweed). Different scenarios including different cultivation methods and production of ethanol from fermentation or electricity from AD are shown in Figure 11.4. This system includes seaweed production and two conversion pathways: first, AD integrated with combined heat and power (CHP), in which biogas was produced through AD and burned in the CHP system for electricity production; and second, ethanol production using the fermentation process, in which ethanol was produced as the main product and by-products were used as animal feed or for electricity production. According to the TEA, the best results were obtained from the fermentation process for ethanol production, and subsequently, residuals went through AD. Considering the best management of cost components, the breakeven selling prices of dry seaweed were \$0.93/L for ethanol and \$0.07/kWh for electricity.

Konda et al. (2015) investigated the detailed economic feasibility of macroalgae for biorefineries. Their results showed that the minimum ethanol

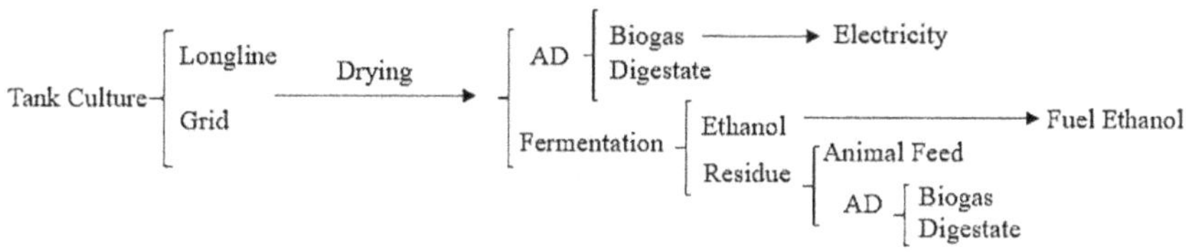

Figure 11.4 Overview of biofuel production methods from macroalgae. Redrawn from Soleymani and Rosentrater (2017).

selling price (MESP) was up to \$8.5/gal, which was influenced by the price of macroalgal feedstock, the overall yield of the process, enzyme loading and the solid loading. Considering a feedstock price of \$100/MT, the MFSP was in the range of \$3.6–8.5/gal. By reducing the price of feedstock to \$50/MT, the MFSP was reduced to \$2.9–7.5/gal. By improving the yield from 50% to 80%, the MESP was reduced by \$1.8–3.0/gal, depending on the price of feedstock. Also, by increasing the solid loading from 5% to 20%, the MESP was reduced by \$2.7/gal. The MESP would be reduced by \$0.8/gal by increasing the enzyme loading from 20 to 10 mg/g of hydrolyzable polysaccharide.

Chong et al. (2020) investigated the techno-economic assessment of bioethanol production through simultaneous saccharification and fermentation (SSF) conversion technology of macroalgae cellulosic residue (MCR). MCR contained a high cellulose content of ~99 wt.% dry basis; thus, it could be used as an alternative feedstock for bioethanol production. The total solid and cellulose enzyme loadings were 20 wt.% and 20 mg/g cellulose, which resulted in a bioethanol yield of 85.5%. Based on an optimized design using Aspen Plus, a feed rate of 15,833 kg/h was required to produce 7,626 kg/h of bioethanol. Considering 20 years of plant life and an IRR of 23.2%, the minimum selling price of the product (anhydrous bioethanol) was calculated to be \$0.54/kg, which could be competitive with the market price of bioethanol obtained from other biomasses. It was also identified that Kota Kinabalu Industrial Park at Sabah (Malaysia) should be designated as the best location for anhydrous bioethanol production. It was also found that using process optimization and heat exchanger network synthesis resulted in a reduction of up to 32% of energy utilization with a net energy ratio of 0.53.

In summary, macroalgal biomass has gained interest as an alternative renewable feedstock for bioethanol production due to its high carbohydrate availability which is considered a great fermentable feedstock. However, according to the literature on the economic viability of biofuel production from macroalgae, the minimum selling price of bioethanol highly depends on the macroalgae purchasing price, process parameters such as solid and enzyme loading and produced biofuel yield, and utilization of by-products through the fermentation of macroalgae such as glycerol.

Techno-economic feasibility of micro- and macroalgal-derived high-value pharmaceuticals, nutraceuticals and other value-added products

Owing to the uneconomical feasibility of the algal biofuels to enhance the economics of micro- and macroalgal biorefinery, several high-value products such as pigments, antioxidants, proteins, carbohydrates, fatty acids, vitamins, fish and animal feed can be produced. This section unlocks the techno-economic feasibility for the production of high-value-added algal-derived products and the challenges in integration with the biorefinery. New sources of fatty acids eicosapentaenoic acid (EPA) and docosahexaenoic acid (DHA) were developed to meet the global demand using marine algal biomass. Based on the biological and technical parameters, TEA was conducted by varying the geographical locations and production technology to lower the production cost.

Chauton et al. (2015) assumed a commercial plant in 100 ha with seven operators for economical-scale production with 10 years of average lifetime with 4% maintenance cost and 10% depreciation rate. In the base case scenario, EPA and DHA production cost was found to be US$39/kg of EPA and DHA. This study concluded that a potential improvement in all the fields that are represented in Figure 11.5 would improve the algal productivity and reduce the cost of the production of fatty acids EPA and DHA to compete with fish oil. However, the optimization of the engineering parameters and biological productivity would reduce the cost to US$11.9/kg of EPA and DHA. A lower production cost was found using flat-panel PBRs with clear-sky conditions.

Economic assessment for the coproduction of fucoxanthin and EPA was made using *Phaeodactylum tricornutum* and commercial FPA-PBRs (Derwenskus et al., 2020). Fucoxanthin has several health-promoting characteristics such as anticancer, anti-oxidative, anti-obesity and antiproliferative. Base, industrial and hypothetical industrial-scale scenarios were evaluated to understand these scenarios via upstream (dewatering, cell disruption and drying) and downstream processes (biomass extraction) for the minimum selling price. Major equipment cost, capital and operating expenditures were calculated for all three scenarios to determine the product price. Product price per kg of dry biomass decreased from €882/kg to €392/kg when the annual production capacity was increased to 170 t for the industrial-scale scenario from 0.7 t in the base case scenario. On the other hand, similar results were obtained when the reactor volume was increased to 180 L from 30 L.

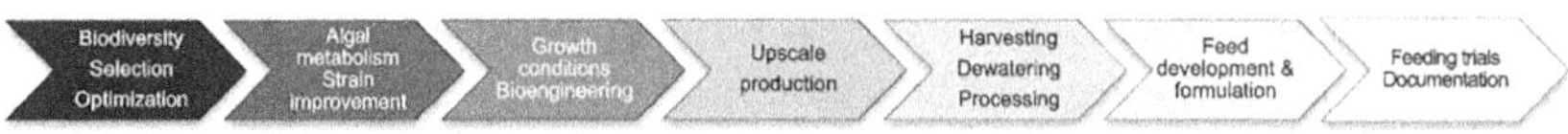

Figure 11.5 Strategies to improve the economic feasibility by adopting advanced technologies for microalgae as an aquafeed resource. Reproduced with permission from Chauton et al. (2015).

This resembled the production price of €7,343/kg for concentrated EPA and €32,042/kg for fucoxanthin with >90% purity for the hypothetical industrial scenario. From the entire production process, electricity and energy requirements were found to be the major bottlenecks to compete with the commercial production of EPA and fucoxanthin.

Stokes et al. (2020) used *Nannochloropsis salina* algae to produce fatty acid methyl esters (FAME) as biodiesel. However, their techno-economic study found that FAME production is not economically feasible. Therefore, a portion of the FAME was further processed into ω-3 fatty acids that were sold as valuable nutraceuticals and food supplements at a significantly higher price relative to biodiesel. Their study adopted a liquid–liquid extraction to separate EPA methyl esters (methyl-EPA) and DHA methyl esters (methyl-DHA) from biodiesel. Further, the extracted methyl-EPA and methyl-DHA were base-catalyzed by ester hydrolysis to produce long-chain ω-3 marketable fatty acids to sell them as carboxylic acids. Profitability analysis of the entire process was evaluated using Aspen Plus to obtain operating expenses, equipment sizes and costs. Profitability analysis was calculated based on the 20-year project life including one design and one construction year, 300 days of operating the plant annually. During the economic study, FAME were purchased at 507 MM kg/year at $1.102/kg ($3.75/gal) and the ω-3 marketable fatty acids were sold at $30/gal. Therefore, the total costs were reduced to $272.6 M with a 15% investor rate of return. Based on the profitability analysis from this study, the production of ω-3 marketable fatty acids as a by-product from the *Nannochloropsis salina* algal-derived biodiesel significantly reduced the biodiesel price in the market. Thus, the widespread use of ω-3 marketable fatty acids as a valuable by-product is promising as potential nutraceuticals and food supplements. Therefore, their production and commercialization significantly reduce the biofuel prices in the market.

Grima et al. (2003) studied the bottlenecks in producing and commercializing high-value EPA using *Phaeodactylum tricornutum*. Capital investments and operating expenses were calculated for the economic evaluation, and all the prices were calculated. The basis for the economic analysis was the production of 430 kg of EPA of >96% purity via a chromatographic technique. Major equipment, total and annual fixed capital costs were found to be US$509,706, $2,045,620 and $230,028, respectively. Among the major equipment costs, the chromatographic system cost was 51.1% of the equipment cost. Downstream recovery of EPA production accounted for 70–80% of the total cost of production, which suggested that this process is very expensive.

Bioactive compounds and biodiesel were produced from the *Scenedesmus obliquus* microalgal community (bloom). After extracting the lipids for biodiesel production, phenolics (5.3 mg/g), flavonoids (1.5 mg/g), proteins (600 mg/g) and carbohydrates (78.57 mg/g) were further extracted from the residual biomass. Abdo et al. (2020) evaluated the economic modeling of the process via the Aspen HYSYS (version 8.4) model. At the end of the economic evaluation, total manufacturing costs, total capital investment, the return on investment and breakeven point were calculated to estimate the economic feasibility with an operating time of the plant set to 8,000 h (333 days) annually. During this study, biodiesel was considered as the main product with a production cost of \$1,235/ton. The leftover biomass and glycerol were considered as by-products of the process. A critical examination of the economic assessment results revealed that the biodiesel production from this process was not feasible economically; therefore, relative to the carotenoids and lipids, carotenoids must be extracted first to use them in the cosmetic industry to enhance the economic feasibility of the process. Besides, it was concluded that the other value-added chemicals such as total phenolic, flavonoids, carbohydrates and proteins could be produced as they are potential candidates for different pharmaceutical applications to enhance antioxidant, antimicrobial, anticancer and antidiabetic activities and to reduce anti-oxidative stress, gastric ulcers, wounds and constipation.

Renewable diesel was produced via HTL of two macroalgal species (*Nereocystis luetkeana* and *Saccharina latissima*) with emerging technologies and hatchery configurations by Greene et al. (2020). During HTL, nitrogen (urea) and phosphorus (superphosphate) were extracted from the aqueous phase as a potential source to be sold as fertilizer for terrestrial crops. The proposed system in this study was designed down the west coast of the USA; owing to the uncontrollable atmospheric conditions such as nutrient availability, the water temperature was limited to roughly 100 operational days of cultivation. During HTL, 26% of the aqueous phase was produced relative to the input feedstock from the post-HTL aqueous waste phase, which contained 36% of nitrogen and 0.017 kg of phosphorus. During the development of a techno-economic model, a 30-year discounted cash flow rate of return was adopted to determine the minimum phosphorus and nitrogen selling price. The cost of the nitrogen and phosphorus extraction was estimated to be \$12.17/kg (with 846.3 g nitrogen per 1.8 kg bag) and \$8.94 kg (with 480.2 g phosphorus per 1.8 kg bag), respectively.

Ansari et al. (2019) investigated the dual application of *Scenedesmus obliquus* microalgae for wastewater phycoremediation, i.e., to remove nitrogen- and phosphorus-related chemical species, chemical oxygen demand (COD) and accumulation of biochemical components (lipids and carbohydrates) for fuel production. In the current study, *Scenedesmus obliquus* microalgae growth was found to be 0.42 ± 0.021/d (0.88 ± 0.04 g/L) in municipal wastewater with a pollutant removal efficiency of 71.2 ± 3.5 for COD, 81.9 ± 3.8 and ~100 for nitrogen compounds (NH_4^+ and NH_3^-, respectively) and 94.1 ± 4.7 for phosphorus compounds (PO_4^{3-}). The economic feasibility of the phycoremediation system was designed with a water discharge of 150 m^3/d. The total cost of the phycoremediation process was developed based on the amortization cost with remediation time of 16 h, daily working hours of 24 h. By converting the phycoremediation outputs (fuel production and removal of pollutants) into cost, a net profit of US$16,885 was estimated annually with 14.8 years of the payback period.

A wide variety of conventional plastics such as commodity thermoplastics and biodegradable and engineered resins could be a promising substitute for bioplastics derived from algae. Beckstrom et al. (2020) investigated several scenarios for the economic viability of algal biorefinery for bioplastic feedstock and fuel production. The performance of each scenario was evaluated by the mass and energy balances of the process that was involved in the production line. The analysis of discounted cash flow rate of return was utilized for the economic analysis with an algal production facility of 3,000 ha. The minimum bioplastic feedstock price was calculated by considering the capital and operational costs for 30 years of plant life, 10% IRR and discount rate with a debt-to-equity ratio of 60:40. The impact of each economic factor such as taxes, variable operating costs, capital cost, fixed operating cost and by-product credit was considered to calculate the minimum selling price. Among all the scenarios, utilization of ORPs coupled with drying scenario was found to be effective in reducing the minimum selling price and economically competitive for dedicated algae to bioplastic feedstock with a minimum selling price of $970/ton, which was considered as a profitable feedstock price. Further enhancement in the technologies or production of high-value-added products would lower the minimum selling prices to compete with the conventional feedstocks.

The carotenoid pigment in astaxanthin is a significant antioxidant in nature as an efficient free radical scavenger to enhance oxidative stability in the human body. Panis and Carreon (2016) evaluated the economic performance of the commercial-scale astaxanthin production from *Haematococcus pluvialis* for two cities in Europe with identical climatic zones with varying environments (Livadeia, Greece; and Amsterdam, the Netherlands) by applying there different scenarios: best, base and worst case. Capital, annual operating expenditures and the return on investment were calculated for different market prices of astaxanthin production for the project lifetime of 10 years. During the study, solar power was used as a primary source of energy to reduce the cost, and it was found to be €0.13/kWh and €0.089/kWh for Livadeia and Amsterdam, respectively. By employing the eco-friendly strategies and water recycling, production costs were reduced by 18% and 5% for Livadeia and Amsterdam, respectively. Finally, economic evaluation for the production of astaxanthin was estimated to be €1,536/kg_{ASTAX} and €6,403/kg_{ASTAX} for the base case scenario representing that the present technologies are unable to compete with the traditional (€880/kg_{ASTAX}) astaxanthin.

Charoensiddhi et al. (2018) carried out a case study for the production of value-added functional food products derived from the *Ecklonia radiata* brown seaweed algae. Derived functional food derivatives have potential applications in the food and nutraceuticals as antioxidant, neuroprotective, antibacterial, prebiotic, anti-inflammatory, antidiabetic, anticancer, immune modulation and anticoagulant agents. The economic feasibility of the batch process was calculated using SuperPro simulation software with 2,000 kg of seaweed with an annual operating time of 330 days. The present case studies followed three scenarios: enzyme-assisted extraction of seaweed extract (scenario 1); scenario 1 followed by crude fractionation based on the molecular weight (scenario 2); and ethanolic extraction followed by enzyme-assisted extraction and separation of crude extract into high and low molecular weight fractions (scenario 3) (Figure 11.6). Capital investment, total plant direct and indirect costs and annual operating costs were calculated with a project lifetime of 15 years for 30 months of construction period with 4 months of the start-up period. Among all the scenarios, scenario 2 was found to be economically profitable with a payback time of 1.6 years, and scenario 1 and 3 payback periods were found to be 1.5 and 3 times higher than scenario 1 payback period.

In a study by Bochenski et al. (2018), polyurethane-inspired elastomers were produced via co-polymerization of microalgal proteins combined with ethylene glycol methyl ether. Through economic feasibility, the technology readiness level was estimated for the said process through the SuperPro simulation designer. The elastomer production process was modeled to produce 40 kg/h of algal protein with 7,920 h of operating time annually, and to meet this demand, 316.8 tons of protein was required. Total capital investment, plant direct and indirect costs, direct fixed capital cost, operating cost, return on investment and payback times were evaluated by SuperPro. The manufacturing price for 1 ton of algal protein was \$6,490 with a profit margin of 7.8% (production price of \$7,000/ton), resulting in an 11.9% return on investment with 8.4 years of the payback period. Total capital investment was predicted to be \$16.2 million with \$4.8 million for plant indirect costs and \$8 million for plant direct costs. During the process, \$882,000 surplus was received between the operating costs and revenues. Sensitivity analysis of the process concluded that this process had a high potential to be economically feasible.

Techno-economic assessment for the production of plasticizer, lactic acid and animal feed was simulated using SuperPro with a capacity of 1 metric ton/d of food waste valorization via fungal hydrolysis and microalgae (*Chlorella pyrenoidosa*) cultivation (Kwan et al., 2015). Two scenarios were proposed (Figure 11.7) during the study: plasticizer and lactic acid production in scenario 1 and plasticizer and animal feed production in scenario 2. The plant was assumed to be based in Hong Kong including 1 year for construction and start-up phase with 20 years of plant life in a batch mode with 312 days/year of operating time (85% of the working capacity). Furthermore, capital cost, operation cost, profitability and sensitivity were determined for the comparative analysis of the two scenarios. From the detailed economic analysis, only scenario 1 was found to be economically viable with an annual net profit of \$422,700 with a payback period of 7.6 years, and the IRR was found to be 19%. Scenario 2 was not economical owing to the deficit of \$42,600 annually. Sensitivity analysis had shown that the economic value of the lactic acid determined the profitability factor.

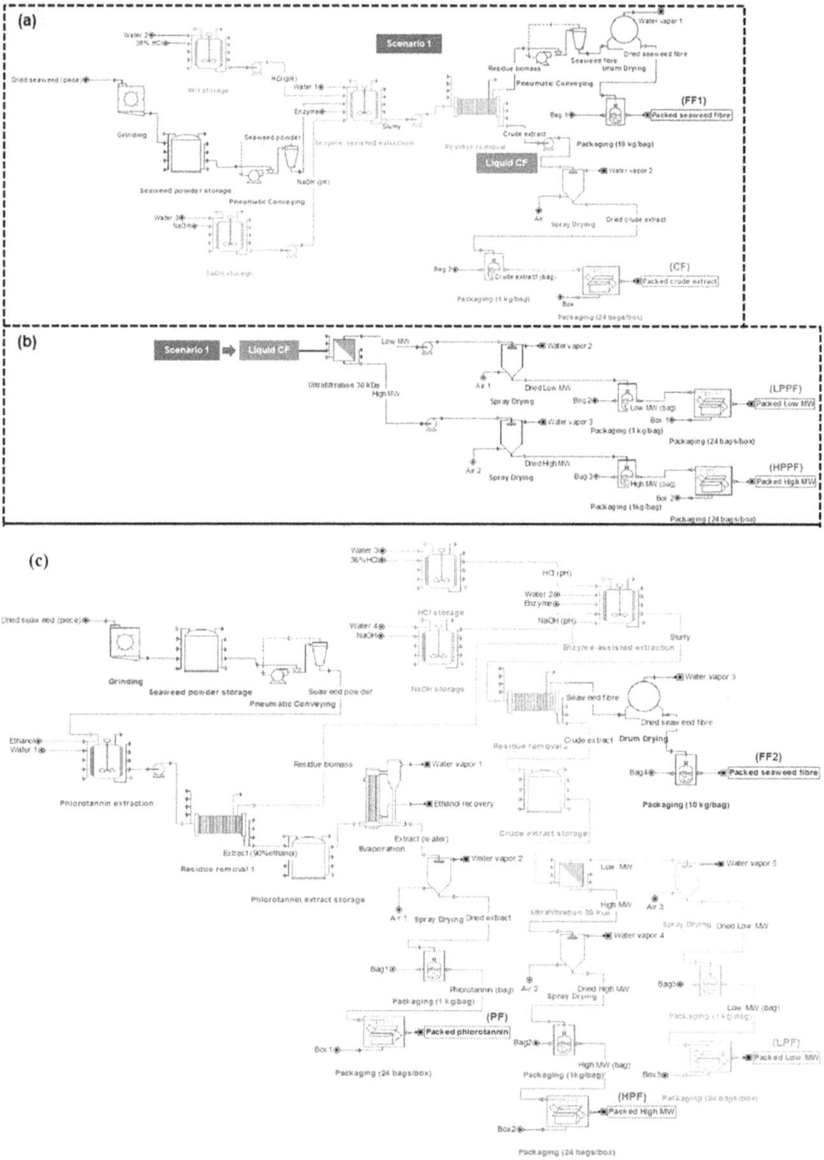

Figure 11.6 Projected process layout to extract the functional foods from *Ecklonia radiata* via (a) fiber-enriched and crude-extracted fractions 1 production – scenario 1; (b) high and low molecular weight polysaccharide- and phlorotannin-enriched fraction and fiber-enriched fraction 1 production – scenario 2; and (c) phlorotannin-enriched fraction, high and low MW polysaccharide-enriched fraction and fiber-enriched fraction 2 production – scenario 3. Reproduced with permission from Charoensiddhi et al. (2018).

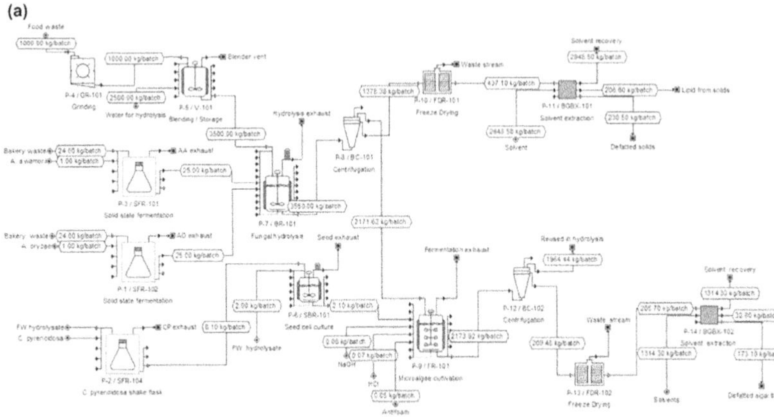

Figure 11.7 Process flow diagram for the food waste biorefinery via (a) microalgal fermentation and fungal hydrolysis, (b) fermentation of lactic acid and (c) production of plasticizers. Reproduced with permission from Kwan et al. (2015).

The application of *Scenedesmus obliquus* for feeding Nile tilapia (*Oreochromis niloticus*) fish (7.5%) was examined by Ansari et al. (2020). Nile tilapia fish was grown with lipids (scenario 1) and lipid-extracted algae (scenario 2) (Figure 11.8) from *Scenedesmus obliquus*. These two scenarios were economically evaluated to determine their commercial feasibility in an algae aquaculture system with 44 weeks of fish cultivation system. The initial investment, operating cost and profit were calculated for both the scenarios for the economic feasibility with 15 years of project life span. Considering all the expenses, payback periods of 7.5 and 6.8 years were evaluated for scenarios 1 and 2. Additionally, scenario 2 was found to be more profitable as the extracted lipids were used to produce biodiesel and the lipid-extracted biomass was fed to the Nile tilapia fish. However, both the scenarios were found to be profitable as they could generate the revenue within the half-life of the project (7.5 years).

Succinic acid production was studied by Marinho et al. (2016) using *Saccharina latissimi* macroalgae as a potential feedstock. Succinic acid was found in several applications as an ion chelator and additive in pharmaceuticals and foods and surfactants. Further, 1,4-butanediol, diethyl succinate, ethylene diamine disuccinate and adipic acid were sourced from succinic acid. After extraction of succinic acid via enzymatic hydrolysis, residual macroalgal biomass was found to be higher in phenolic compounds, macronutrients (Ca, Fe, K, Mg, N, Na and P) and micronutrients. Therefore, their presence made the production of antioxidants and fertilizers as a by-product alongside succinic acid. A simplified economic assessment was performed for the advanced

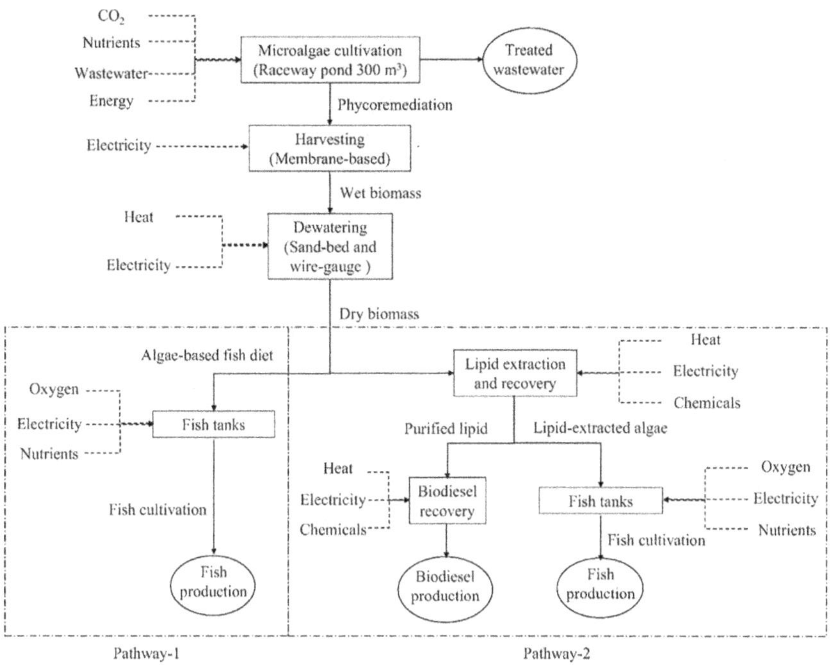

Figure 11.8 Techno-economic assessment for biodiesel and fish feed production from algal aquaculture (pathways 1 and 2). Reproduced with permission from Ansari et al. (2020).

macroalgal biorefinery concept. During the TEA studies, production for fixed costs and return on investment were not considered. The following equation (Eq. 11.2) was used to calculate the gross profit:

$$\text{Gross Profit} = \Sigma\ \text{Revenues} - \Sigma\ \text{Raw Material Costs} - \Sigma\ \text{Energy Costs} \tag{11.2}$$

where revenues are the combination of income gained from the sales of succinic acid as a main product and revenues from by-products (antioxidants and fertilizers). The current macroalgal feedstock production price was €1.757/kg, and the market price of succinic acid was €6.96/kg. At this price, the profit was negative. Further, to reduce the price, phenolics were extracted before extracting succinic acid and the residual biomass was used as a fertilizer, and thus, the succinic acid price was reduced to €4.77/kg from €6.96/kg, which was a profitable process. The investigation concluded that in the next 10 years, adopting advanced techniques could reduce the succinic acid production price to €3.12/kg.

Conclusions

Microalgae and microalgae have been considered promising renewable feedstocks for the production of different kinds of biofuels through different technologies, because there is no competition with food crops and also because of their higher photosynthetic efficiencies and ability to sequester CO_2. The use of two-step biofuel production through HTL or MAP of microalgae grown in wastewater to produce bio-crude oil and subsequently HDO over heterogeneous catalysts for biofuel production as well as utilization of the by-products could be considered economical. This needs moderate process conditions to convert the whole biomass into high-quality biofuel. The use of macroalgae to produce bioethanol, biobutanol and biogas through AD and fermentation technologies is considered a promising method. However, some technological challenges have not yet been solved to make these methods economically and commercially viable. Also, more research is needed focusing on cultivation methods to reduce the production cost of microalgae- and macroalgae-derived products with high value.

Acknowledgments

We thank the financial support by the Natural Sciences and Engineering Research Council of Canada (NSERC) for this study.

References

Abdo, S.M., Ei-Diwani, G.I., Ei-Khatib, K.M., Ei-Enin, S.A.A., Ei-Galad, M.I., Basily, H.S. and Ali, G.H. (2020). Primitive techno-economic study of biodiesel and bio-active compound production from microalgae. *Bulletin of the National Research Centre* 44:169.

Albrecht, K.O., Zhu, Y., Schmidt, A.J., Billing, J.M., Hart, T.R., Jones, S.B., Maupin, G., Hallen, R., Ahrens, T. and Anderson, D. (2016). Impact of heterotrophically stressed algae for biofuel production via hydrothermal liquefaction and catalytic hydrotreating in continuous-flow reactors. *Algal Research* 14:17–27.

Alfonsín, V., Maceiras, R. and Gutiérrez, C. (2019). Bioethanol production from industrial algae waste. *Waste Management* 87:791–797.

Ansari, F.A., Nasr, M., Guldhe, A., Gupta, S.K., Rawat, I. and Bux, F. (2020). Techno-economic feasibility of algal aquaculture via fish and biodiesel production pathways: A commercial-scale application. *Science of the Total Environment* 704:135259.

Ansari, F.H., Ravindran, B., Gupta, S.K., Nasr, M., Rawat, I. and Bux, F. (2019). Techno-economic estimation of wastewater phycoremediation and environmental benefits using *Scenedesmus obliquus* microalgae. *Journal of Environmental Management* 240:293–302.

Barlow, J., Sims, R.C. and Quinn, J.C. (2016). Techno-economic and life-cycle assessment of an attached growth algal biorefinery. *Bioresource Technology* 220:360–368.

Beckstrom, B.D., Wilson, M.H., Crocker, M. and Quinn, J.C. (2020). Bioplastic feedstock production from microalgae with fuel co-products: A techno-economic and life cycle impact assessment. *Algal Research* 46:101769.

Bochenski, T., Chan, W.Y., Olsen, B.D. and Schmidt, J.E. (2018). Techno-economic analysis for the production of novel, bio-derived elastomers with modified algal proteins as a reinforcing agent. *Algal Research* 33:337–344.

Borines, M.G., De Leon, R.L. and McHenry, M.P. (2011). Bioethanol production from farming non-food macroalgae in Pacific island nations: Chemical constituents, bioethanol yields, and prospective species in the Philippines. *Renewable and Sustainable Energy Reviews* 15(9):4432–4435.

Brandenberger, M., Matzenberger, J., Vogel, F. and Ludwig, C. (2013). Producing synthetic natural gas from microalgae via supercritical water gasification: A techno-economic sensitivity analysis. *Biomass and Bioenergy* 51:26–34.

Campanario, F.J. and Ortiz, F.G. (2017). Techno-economic assessment of bio-oil aqueous phase-to-liquids via Fischer-Tropsch synthesis and based on supercritical water reforming. *Energy Conversion and Management* 154:591–602.

Chakinala, A.G., Brilman, D.W., van Swaaij, W.P. and Kersten, S.R. (2010). Catalytic and non-catalytic supercritical water gasification of microalgae and glycerol. *Industrial & Engineering Chemistry Research* 49(3):1113–1122.

Charoensiddhi, S., Lorbeer, A.J., Franco, C.M.M., Su, P., Conlon, C.A. and Zhang, W. (2018). Process and economic feasibility for the production of functional food from the brown alga *Ecklonia radiata*. *Algal Research* 29:80–91.

Chauton, M.S., Reitan, K.I., Norsker, N.H., Tveteras, R. and Kleivdal, H.T. (2015). A techno-economic analysis of industrial production of marine microalgae as a source of EPA and DHA-rich raw material for aquafeed: Research challenges and possibilities. *Aquaculture* 436:95–103.

Chen, H., Zhou, D., Luo, G., Zhang, S. and Chen, J. (2015). Macroalgae for biofuels production: Progress and perspectives. *Renewable and Sustainable Energy Reviews* 47:427–437.

Chong, T.Y., Cheah, S.A., Ong, C.T., Wong, L.Y., Goh, C.R., Tan, I.S., Foo, H.C.Y., Lam, M.K. and Lim, S. (2020). Techno-economic evaluation of third-generation bioethanol production utilizing the macroalgae waste: A case study in Malaysia. *Energy* 210:118491.

Demirbas, A. and Demirbas, M.F. (2011). Importance of algae oil as a source of biodiesel. *Energy Conversion and Management* 52(1):163–170.

DeRose, K., DeMill, C., Davis, R.W. and Quinn, J.C. (2019). Integrated techno economic and life cycle assessment of the conversion of high productivity, low lipid algae to renewable fuels. *Algal Research* 38:101412.

Derwenskus, F., Weickert, S., Lewandowski, I. and Schmid-Staiger, U. (2020). Economic evaluation of up- and downstream scenarios for the co-production of fucoxanthin and eicosapentaenoic acid with *P. tricornutum* using flat-panel airlift photobioreactors with artificial light. *Algal Research* 51:102078.

Du, Z., Li, Y., Wang, X., Wan, Y., Chen, Q., Wang, C., Lin, X., Liu, Y., Chen, P. and Ruan, R. (2011). Microwave-assisted pyrolysis of microalgae for biofuel production. *Bioresource Technology* 102(7):4890–4896.

Dutta, S., Neto, F. and Coelho, M.C. (2016). Microalgae biofuels: A comparative study on techno-economic analysis & life-cycle assessment. *Algal Research* 20:44–52.

Dyer, S., Grant, J., and Angen, E. (2013). Forecasting the impacts of oilsands expansion. *Pembina Institute*. https://www.pembina.org/pub/forecasting-impacts-of-oilsands-expansion (accessed on January 21, 2021).

Feng, H., Zhang, B., He, Z., Wang, S., Salih, O. and Wang, Q. (2018). Study on co-liquefaction of *Spirulina* and *Spartina alterniflora* in ethanol-water co-solvent for bio-oil. *Energy* 155:1093–1101.

Gollakota, A.R.K., Kishore, N. and Gu, S. (2018). A review on hydrothermal liquefaction of biomass. *Renewable and Sustainable Energy Reviews* 81:1378–1392.

Greene, J.M., Gulden, J., Wood, G., Huesemann, M. and Quinn, C.J. (2020). Techno-economic analysis and global warming potential of a novel offshore macroalgae biorefinery. *Algal Research* 51:102032.

Grima, E.M., Belarbi, E.H., Fernandez, F.G.A., Medina, A.R. and Chisti, Y. (2003). Recovery of microalgal biomass and metabolites: Process options and economics. *Biotechnology Advances* 20:491–515.

Gu, X., Yu, L., Pang, N., Martinez-Fernandez, J.S., Fu, X. and Chen, S. (2020). Comparative techno-economic analysis of algal biofuel production via hydrothermal liquefaction: One stage versus two stages. *Applied Energy* 259:114115.

Haider, M.S., Castello, D., Michalski, K.M., Pedersen, T.H. and Rosendahl, L.A. (2018). Catalytic hydrotreatment of microalgae biocrude from continuous hydrothermal liquefaction: Heteroatom removal and their distribution in distillation cuts. *Energies* 11(12):3360.

Haugstad, B.H. (2017). Fuel Characterization and Process Analysis of Hydrothermal Liquefaction of Algae (Master's thesis, University of Agder).

Hognon, C., Delrue, F. and Boissonnet, G. (2015). Energetic and economic evaluation of *Chlamydomonas reinhardtii* hydrothermal liquefaction and pyrolysis through thermochemical models. *Energy* 93:31–40.

Hu, Y., Feng, S., Xu, C.C. and Bassi, A. (2017). Production of low-nitrogen bio-crude oils from microalgae pre-treated with pre-cooled NaOH/urea solution. *Fuel* 206:300–306.

Jones, S.B., Zhu, Y., Anderson, D.B., Hallen, R.T., Elliott, D.C., Schmidt, A.J., Albrecht, K.O., Hart, T.R., Butcher, M.G., Drennan, C. and Snowden-Swan, L.J. (2014). Process design and economics for the conversion of algal biomass to hydrocarbons: Whole algae hydrothermal liquefaction and upgrading (No. PNNL-23227). Pacific Northwest National Lab. (PNNL), Richland, WA (United States).

Juneja, A. and Murthy, G.S. (2017). Evaluating the potential of renewable diesel production from algae cultured on wastewater: Techno-economic analysis and life cycle assessment. *Aims Energy* 5(2):239–257.

Konda, N.M., Singh, S., Simmons, B.A. and Klein-Marcuschamer, D. (2015). An investigation on the economic feasibility of macroalgae as a potential feedstock for biorefineries. *BioEnergy Research* 8(3):1046–1056.

Kroumov, A.D., Scheufele, F.B., Trigueros, D.E.G., Modenes, A.N., Zaharieva, M. and Najdenski, H. (2017). Modeling and technoeconomic analysis of algae for bioenergy and coproducts. In: Rastogi, R.P., Madamwar, D., Pandey, A. (Eds.), *Algal Green Chemistry*, Elsevier, pp. 201–241.

Kumar, M., Oyedun, A.O. and Kumar, A. (2019). A comparative analysis of hydrogen production from the thermochemical conversion of algal biomass. *International Journal of Hydrogen Energy* 44(21):10384–10397.

Kumar, M., Oyedun, A.O. and Kumar, A. (2020). A comparative technoeconomic analysis of algal thermochemical conversion technologies for diluent production. *Energy Technology* 8:1900828.

Kwan, T.H., Pleissner, D., Lau, K.Y., Venus, J., Pommeret, A. and Lin, C.S.K. (2015). Techno-economic analysis of a food waste valorization process via microalgae cultivation and co-production of plasticizer, lactic acid and animal feed from algal biomass and food waste. *Bioresource Technology* 198:292–299.

Marinho, G.S., Alvarado-Morales, M. and Angelidaki, I. (2016). Valorization of macroalga *Saccharina latissima* as novel feedstock for fermentation-based succinic acid production in a biorefinery approach and economic aspects. *Algal Research* 16:102–109.

Masoumi, S., Borugadda, V.B. and Dalai, A.K. (2020). Biocrude oil production via hydrothermal liquefaction of algae and upgradation techniques to liquid transportation fuels. In: Nanda, S., Vo, D.V.N., Sarangi, P.K. (eds.), *Biorefinery of Alternative Resources: Targeting Green Fuels and Platform Chemicals.* Singapore: Springer, pp. 249–270.

Mehrabadi, A., Craggs, R. and Farid, M.M. (2015). Wastewater treatment high rate algal ponds (WWT HRAP) for low-cost biofuel production. *Bioresource Technology* 184:202–214.

Mian, A., Ensinas, A.V. and Marechal, F. (2015). Multi-objective optimization of SNG production from microalgae through hydrothermal gasification. *Computers & Chemical Engineering* 76:170–183.

Ou, L., Thilakaratne, R., Brown, R.C. and Wright, M.M. (2015). Techno-economic analysis of transportation fuels from defatted microalgae via hydrothermal liquefaction and hydroprocessing. *Biomass and Bioenergy* 72:45–54.

Panis, G. and Carreon, J.R. (2016). Commercial astaxanthin production derived by green alga *Haematococcus pluvialis*: A microalgae process model and a techno-economic assessment all through production line. *Algal Research* 18:175–190.

Pearce, M., Shemfe, M. and Sansom, C. (2016). Techno-economic analysis of solar integrated hydrothermal liquefaction of microalgae. *Applied Energy* 166:19–26.

Quinn, J.C. and Davis, R. (2015). The potentials and challenges of algae based biofuels: A review of the techno-economic, life cycle, and resource assessment modeling. *Bioresource Technology* 184:444–452.

Rahbari, A., Shirazi, A., Venkataraman, M.B. and Pye, J. (2019). A solar fuel plant via supercritical water gasification integrated with Fischer–Tropsch synthesis: Steady-state modelling and techno-economic assessment. *Energy Conversion and Management* 184:636–648.

Ranganathan, P. and Savithri, S. (2019). Techno-economic analysis of microalgae-based liquid fuels production from wastewater via hydrothermal liquefaction and hydroprocessing. *Bioresource Technology* 284:256–265.

Soleymani, M. and Rosentrater, K.A. (2017). Techno-economic analysis of biofuel production from macroalgae (seaweed). *Bioengineering* 4(4):92.

Stokes, J., Tu, R., Peters, M., Yadav, G., Fabiano, A.L. and Seider, D.W. (2020). Omega-3 fatty acids from algae produced biodiesel. *Algal Research* 51:102047.

Suutari, M., Leskinen, E., Fagerstedt, K., Kuparinen, J., Kuuppo, P. and Blomster, J. (2015). Macroalgae in biofuel production. *Phycology Research* 63:1–18.

Van Doren, L.G., Posmanik, R., Bicalho, F.A., Tester, J.W. and Sills, D.L. (2017). Prospects for energy recovery during hydrothermal and biological processing of waste biomass. *Bioresource Technology* 225:67–74.

Wang, K., Mandal, A., Ayton, E., Hunt, R., Zeller, M.A., Sharma, S. (2016). Modification of protein rich algal-biomass to form bioplastics and odor removal. In: Dhilon, G.S., (ed.), *Protein Byproducts: Transformation from Environmental Burden into Value-Added Products.* Academic Press, pp. 107–117.

Wei, N., Quarterman, J. and Jin, Y.S. (2013). Marine macroalgae: An untapped resource for producing fuels and chemicals. *Trends in Biotechnology* 31(2):70–77.

Wu, N., Moreira, C., Zhang, Y., Doan, N., Yang, S., Phlips, E., Svoronos, S. and Pullammanappallil, P. (2019). Techno-economic analysis of biogas production from microalgae through anaerobic digestion. In: Banu, J.R. (eds.), *Anaerobic Digestion*. IntechOpen. Doi: 10.5772/intechopen.86090.

Xin, C., Addy, M.M., Zhao, J., Cheng, Y., Cheng, S., Mu, D., Liu, Y., Ding, R., Chen, P. and Ruan, R. (2016). Comprehensive techno-economic analysis of wastewater-based algal biofuel production: A case study. *Bioresource Technology* 211:584–593.

Xin, C., Addy, M.M., Zhao, J., Cheng, Y., Ma, Y., Liu, S., Mu, D., Liu, Y., Chen, P. and Ruan, R. (2018). Waste-to-biofuel integrated system and its comprehensive techno-economic assessment in wastewater treatment plants. *Bioresource Technology* 250:523–531.

Zhu, Y., Albrecht, K.O., Elliott, D.C., Hallen, R.T. and Jones, S.B. (2013). Development of hydrothermal liquefaction and upgrading technologies for lipid-extracted algae conversion to liquid fuels. *Algal Research* 2:455–464.

Zhu, Y., Jones, S.B., Schmidt, A.J., Albrecht, K.O., Edmundson, S.J. and Anderson, D.B. (2019). Techno-economic analysis of alternative aqueous phase treatment methods for microalgae hydrothermal liquefaction and biocrude upgrading system. *Algal Research* 39:101467.

12 Lifecycle assessment of microalgal biorefinery

Dipesh Kumar and Vaibhav V. Goud

Introduction

The concept of bioeconomy has recently gained significant attention as it extends multiple benefits on strategic, economic and environmental fronts (D'Amato et al., 2017). It involves the reliance on renewable biomass resources for the production of energy, fertilizers, industrially relevant chemicals and a host of other high-value commodities. To realize a bioeconomy, the processing of biomass in a biorefinery is envisaged (Ubando et al., 2020). The idea of a biorefinery is akin to that of a petroleum refinery in which biomass is processed into a range of products, including fuels, platform/industrial chemicals, nutraceuticals and other value-added commodities (Sadhukhan et al., 2018). The overall appeal of a biorefinery is greatly dependent on the type of biomass being processed. Microalgae have emerged as one of the most promising feedstocks for a biorefinery owing primarily to: (i) high biomass and lipid productivity, (ii) no-gestation period, (iii) short production cycles, (iv) avoidance of food *vs.* fuel controversy, and (v) ability to uptake nutrients from waste streams and industrial discharges, among several others (Chew et al., 2017).

One of the key drivers for bioeconomy is the realization of a carbon-neutral economy, and to this end, lifecycle assessment (LCA)-based appraisal of alternative products and processes is expected to play a pivotal role (Seghetta et al., 2016). It particularly holds for alternative energy production programs. Despite the significant appeal, there is a great deal of uncertainty regarding the feasibility and environmental credentials of a microalgal biorefinery (Moreno-Garcia et al., 2017; Javed et al., 2019). These primarily arise from the dilemma of multiple (alternative) processing routes, uncertainties in biomass yield, composition (in the absence of an optimized process), market demand of bio-based products and knowledge gaps on the relative performance of alternative strategies (Moreno-Garcia et al., 2017).

The unit operations involved in the microalgal biomass supply chain entail different inputs, and hence, their corresponding outputs to the environment are different. The major challenges in the industrial-scale production of microalgal biomass lie in the high demand for energy, nutrients (including

CO_2) and water. Moreover, the spatiotemporal variability in the availability of photosynthetically active radiation (PAR), the low atmospheric concentration of CO_2 and the relatively low biomass density in the open cultivation systems necessitate advanced technological interventions and thus add to the production cost of microalgal biomass (Gifuni et al., 2019). The lifecycle impacts of microalgal biomass/biofuel production are often characterized in terms of the energy return on energy investment (EROEI), the use of non-renewable energy, greenhouse gas (GHG) balance, ozone depletion potential, exergy, acidification potential, water footprint, etc. (Sills et al., 2020).

An overwhelming majority of published studies on microalgal biorefineries deal with the production of biofuels/bioenergy as the primary product along with the valorization and/or processing of residual biomass (Singh and Olsen, 2011). Despite the paucity of published literature on microalgal biorefineries wherein the primary product(s) are non-biofuel entities (such as pigments and nutraceuticals), the information on biofuel/bioenergy-based biorefineries can provide valuable insights regarding their potential impacts as well. The cultivation, harvesting and drying (except wet biomass processing routes) stages are common to all microalgal biorefineries and have been recognized as the most influential contributor to the environmental impacts of a microalgal biorefinery (Collet et al., 2015).

In this chapter, we highlight the attractiveness of different biomass to biofuels processing strategies integrated into a microalgal biorefinery. Characterizing the relative impacts of alternative microalgal biomass processing routes, identification of hotspots in the biomass supply chain exerting a disproportionately large influence on the overall attractiveness of a biorefinery, and the most attractive alternatives to such hotspots are central to the idea of this chapter.

Microalgal biorefinery

The inexhaustible demand for food, animal feed and chemicals, along with the pursuit of energy security and sustainable development, is central to the idea of a biorefinery. The non-renewable supplies of fuels and chemicals are becoming increasingly vulnerable due to rapidly changing geopolitical scenarios and their extraction unfavorable. The gap between the availability and the demand for biomass has been identified to be the single most significant deterrent in the realization of a bioeconomy (D'Amato et al., 2017). Although the lignocellulosic biomass is abundant, slow rate of growth, recalcitrant nature, widespread variation in the relative abundance of different biomass fractions, and limited valorization strategies have somewhat restricted its utility in a biorefinery (Cherubini and Strømman, 2011).

Microalgae offer several important advantages over their terrestrial counterparts. These include high photosynthetic efficiency, high rate of growth, high lipid accumulation, avoidance of arable land, ability to uptake nutrients from waste streams (including CO_2 from industrial discharges), short production cycle, significant protein content, availability of multiple metabolic routes of carbon assimilation (autotrophic/heterotrophic/mixotrophic) and biomass processing strategies, and the possibility of controlled manipulation of growth conditions for optimizing the biomass composition, among several others (Rizwan et al., 2015; Chew et al., 2017). Various studies have assessed the utility of alternative products, processes and system configurations for the realization of a microalgal biorefinery. A schematic representation of potential products and opportunities offered by a microalgal biorefinery is provided in Figure 12.1.

Depending on the scope, LCA studies can take different forms. The most comprehensive is the well-to-wheel (WTW) approach, wherein all the stages of the biomass supply chain (cultivation to end-use) are taken into account. The choice of impact assessment categories for any LCA study is governed by the desired product(s). For instance, in the case of bioenergy, the primary impacts of concern are the EROEI and GHG balance, but these (particularly EROEI) may not hold much significance in cases where other products such as pigments or antioxidants are major deliverable. In such cases, a different set of impact assessment categories for different products can be opted for.

The typical processes involved in a microalgal biorefinery are the cultivation of microalgae, harvesting of biomass, dewatering and drying of biomass, extraction of valuables and/or processing of biomass/biomass fractions and valorization of residual biomass/coproducts. The following section highlights the challenges and opportunities in the realization of a sustainable microalgal biorefinery.

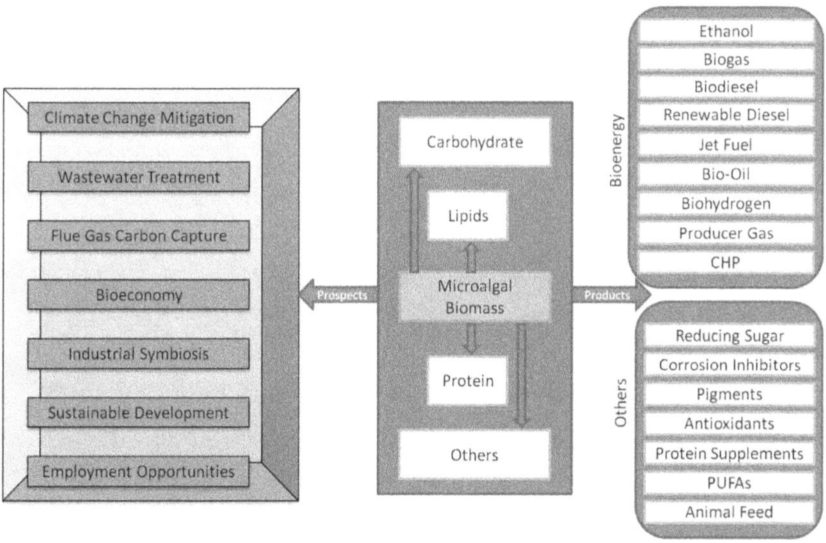

Figure 12.1 Schematic representation of potential products and opportunities offered by a microalgal biorefinery.

Lifecycle assessment of microalgal biorefinery

Various combinations of cultivation, harvesting, extraction, processing and coproduct/residual biomass allocation strategies have been analyzed, yielding a different set of results with varying degrees of uncertainty (Brownbridge et al., 2014). The availability of multiple alternative routes at various stages of the biomass supply chain and the unavailability of industrial data often give rise to low-confidence results (Pérez-López et al., 2018). The problem is further compounded by the definition of an ideal strain of microalgae, which is greatly influenced by a host of other strategic decisions including the preferred choice of primary product(s), cultivation strategy and infrastructure and location specificity.

Figure 12.2 depicts the potential alternatives at various stages of the biomass supply chain in a microalgal biorefinery. Microalgae-to-bioenergy production pathways involve several thermal, thermochemical and biochemical routes, including (but not limited to) pyrolysis, gasification, hydrothermal liquefaction, transesterification, anaerobic digestion, fermentation, biophotolysis and cogeneration (Adeniyi et al., 2018). A majority of studies on microalgal biorefinery (biofuel based) have investigated the lifecycle impacts of biomass/lipids/lipid-based biofuel (including biodiesel and renewable diesel) production (Soh et al., 2014). These studies have also assessed several alternative residual biomass (lipid-extracted microalgal biomass) processing/valorization routes, including the production of bioenergy (biofuels, heat and electricity), nutraceuticals, protein meal for livestock and organic manure, or as a source of nutrients in aquaculture (Chandra et al., 2019). Such studies usually assess the lifecycle impact of microalgal biorefinery in terms of EROEI, GHG balance and ozone depletion potential, and many studies have also included the

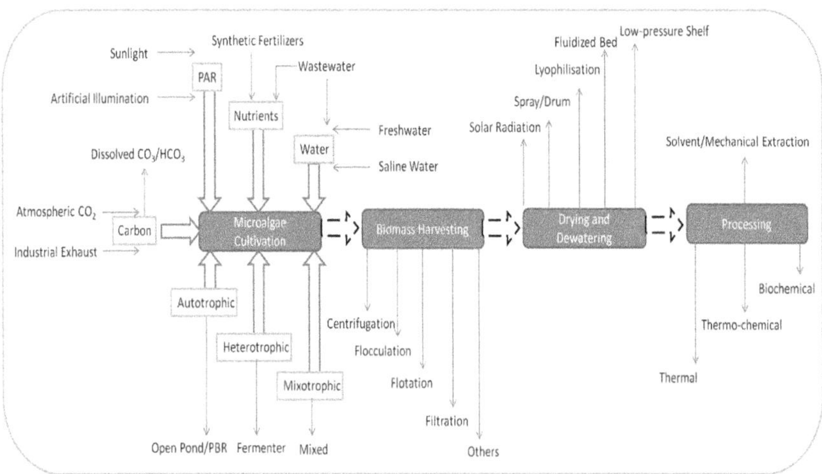

Figure 12.2 Potential alternatives at different stages of a biomass supply chain in a microalgal biorefinery.

calculation of exergy (Peralta-Ruiz et al., 2013). However, a comparative analysis of such studies is rather tricky, as there are inherent differences in the choice of a functional unit, assumptions regarding the biomass/lipid productivity and process efficiency, system boundary, cutoff criteria for inclusion of impacts, etc. (Sills et al., 2013; Yuan et al., 2015).

In general, attributes that favor high biomass and/or lipid productivity and low energy, transportation and nutrient requirements exert a favorable impact on the economic and environmental credentials of a biorefinery (Moreno-Garcia et al., 2017). The selection of an ideal microalgal strain is a critical requirement as it dictates the overall feasibility and attractiveness of a biorefinery. A strain with high photosynthetic efficiency and biomass and/or lipid productivity rates favors the economic and environmental performance of a biorefinery. Besides, strains tolerant to wide ranges of environmental stress and those that keep the unwanted species in check exhibit self-flocculation tendency, and the ability to grow in wastewater is desirable (Soh et al., 2014). The uncertainties in the overall impact of a microalgal biorefinery are primarily attributed to the different sets of assumptions regarding the biomass and/or lipid productivity and lipid content of the microalgae.

The studies on different modes of microalgae growth (auto-/hetero-/mixotrophic) have provided valuable insights in terms of the tunability of biomass composition in a microalgal strain. The search for sustainable sources of growth substrates for supporting the hetero-/mixotrophic growth is part of an ongoing process, and an overwhelming majority of published literature deals with autotrophic cultivation of microalgae (Perez-Garcia and Bashan, 2015). The challenges in low-cost open-autotrophic cultivation systems lie in the low biomass density and growth rate, the possibility of contamination and exposure to environmental stresses. In contrast, the closed cultivation systems (photobioreactors, PBRs) offer better quality control, high cell density and productivity, reduced likelihood of contamination and better area-to-production ratio. However, the high energy requirements somewhat negate the attractive attributes of PBRs, and unless the biomass productivity is significantly higher, the PBRs are unlikely to compete favorably with the conventional open cultivation arrangements (e.g., raceway ponds) (Narala et al., 2016). This is despite the fact that the cell density is higher (supports easy harvest and biomass drying) in the case of PBRs.

Often artificial PAR is required to support the optimal growth of microalgae in PBRs. Microalgae can source their nutrient demand from waste streams, including industrial flue gas rich in CO_2 and municipal/industrial wastewater. Utilization of concentrated sources of waste CO_2, including flue gases from coal-fired thermal power plants and the cement industry, can support high biomass productivity as the low atmospheric concentration of CO_2 is often a limiting factor in microalgal photosynthesis (Cheah et al., 2015). Moreover, the utilization of waste CO_2 opens the prospects for carbon trading/credit and can significantly affect the GHG balance of a microalgal biorefinery. Likewise, municipal and industrial effluents are reservoirs of cheap and abundant supplies of macro- and micronutrients.

Microalgae can uptake the nutrients from wastewater with simultaneous remediation of pollution load, and the effectiveness of such a synergy has been established in several pilot-scale studies (Abdel-Raouf et al., 2012). The synergy solves two major issues in one go, including (i) avoidance of expenditure and emissions involved in the synthesis and use of synthetic fertilizers and (ii) partial/complete avoidance of tertiary water treatment (Christenson and Sims, 2011).

The current supplies of synthetic fertilizers cannot absorb any significant scale demand from microalgal biorefineries, and the synthesis and use of fertilizers (especially nitrogenous) have been identified to be a significant source of GHGs (Chisti, 2007). Thus, the diversion of nutrient-laden waste streams toward a microalgae cultivation system offers novel opportunities in augmenting the commercial appeal and environmental credentials of the biorefinery. In addition, microalgae while growing can modify their local environment in a manner that affects the saturation levels with respect to Ca^{++} and CO_3^{2-}, and under favorable conditions, these ions can precipitate out of the solution (microbially induced calcite precipitation). While settling, the $CaCO_3$ precipitate can entrap and support the removal of pollutants of similar nature and size (Santomauro et al., 2012).

EROEI (also called the net energy ratio, NER) is one of the most significant indicators of sustainability for biomass-to-bioenergy conversion pathways which helps visualization of whether a bioenergy carrier is a source (EROEI>1) or sink of energy (EROEI<1) (Hall et al., 2014). Many LCA studies on microalgal biomass/bioenergy production pathways have identified the process of biomass harvesting and dewatering to be a significant drain of energy (Zaimes and Khanna, 2013). Low biomass/cell density, as in the case of open cultivation systems, further compounds the problem.

For non-energy biomass conversion pathways, EROEI may not prove to be an ideal measure, but energy demands can provide estimates for other indicators of sustainability, including GHG balance, production cost and ozone depletion potential. Wet biomass processing routes including gasification, hydrothermal liquefaction, wet extraction and in situ transesterification have significantly improved energy returns over biomass processing routes that require prior drying (Xu et al., 2011). The utility of solar irradiance in the drying of harvested biomass is marred by spatiotemporal variability and inherent process inefficiencies. The idea of industrial symbiosis wherein the utilization of waste heat and materials from one industry/unit operation is diverted toward other points of use can significantly improve the EROEI and values of other sustainability indicators for biomass-to-bioenergy pathways (Trivedi et al., 2015). Depending on the type of biomass pretreatment, materials and energy requirements, process efficiency and the environmental footprints of output streams, the aggregate impacts of alternative biomass conversion pathways can be significantly different (Xu et al., 2011).

In energetic terms, fuels with higher energy density are expected to fare better. A vast majority of studies suggest that the stand-alone production of biofuels (from microalgal biomass) is economically unviable and coproduction of high-value products such as nutraceuticals, animal feed and pigments is warranted for economic gains (Banu et al., 2020). For any significant scale displacement of fossil fuels, biofuels, and processes that are compatible with the existing infrastructure of internal combustion engines and petroleum refineries are particularly attractive.

Drop-in biofuels such as renewable diesel closely resemble petroleum diesel, and the utilization of such fuels is expected to outperform biofuel–fossil fuel blends (Knothe, 2010). Likewise, higher blends of fuels such as ethanol and biodiesel offer better environmental performance than their lower blends with fossil fuels. Unless energy demand reduction strategies, especially those involved in biomass harvesting, concentration and drying, are integrated into a microalgal biorefinery, the system is not expected to perform desirably on energetic balance. The energy ratio has a direct bearing on the GHG balance as the EROEI calculation often accounts only for non-renewable energy demand (Zaimes and Khanna, 2013). Thus, the coproduction of heat and/or energy from biofuels or a portion of the residual biomass can significantly enhance the energy ratio. Careful metabolic control during cultivation can further help the cause of sustainability. For instance, since N-starvation has been identified to be a gateway for enhanced lipid accumulation in microalgae, it can favor the GHG balance by keeping the requirements of external fertilizers in check (Ho et al., 2012). Utilization of wastewater along with a high degree of nutrient and water recycling appears to be the most promising prospect for reducing the nutrient demand and water footprint of a microalgal biorefinery.

The current supplies of biodiesel are majorly sourced from edible vegetable oil, as in the case of soybean in the USA, rapeseed in Europe and oil palm in South-East Asia (Boutesteijn et al., 2017). The lifecycle impacts of edible oil-based biodiesel usually exhibit poor energy ratio and GHG balance (Carneiro et al., 2017). The case of South-East Asia, where massive oil palm plantations have been developed at the expense of tropical rainforests, has been highlighted in several studies (Afriyanti et al., 2016; Szulczyk and Khan, 2018). The non-edible feedstocks grown on wasteland/marginal land are expected to fare better on GHG balance, but a detailed carbon budgeting of the landscape before and after plantations is often poorly characterized (Jingura and Kamusoko, 2016). The utilization of waste cooking oil (WCO) offers advantages on both economic and environmental fronts as it can be sourced cheaply, and the impacts associated with the production of vegetable oil are avoided (Caldeira et al., 2016). Microalgae cultivation is expected not to cause any significant scale land use/land cover alterations, and unlike WCO, the traveling and collection requirements for microalgal oil are also expected to be low. However, any comparative study on the relative impacts of these feedstocks is long overdue.

Given the immense flexibility on offer, a microalgal biorefinery can be tuned for more targeted applications, outputs and performance. Advancements in biotechnology, genetic, process and metabolic engineering are also expected to improve the economics of scale (Rodionova et al., 2017). To overcome the spatiotemporal variability in the availability of sunlight, several artificial sources of PAR have been investigated, including incandescent bulbs, halogen lamps, fluorescent lamps, Gro-Lux fluorescent lamps, light-emitting diodes and laser diodes (Schulze et al., 2014). Among these, the later four alternatives offer the highest efficiency with minimum loss of energy in the form of heat.

Careful integration of other efficient technologies at various stages of the biomass supply chain can significantly improve the energy return and GHG balance of microalgal biorefinery. Since the processes upstream of biomass processing are more or less identical, the published literature on LCA of microalgal biofuels can be used as a proxy for microalgal biorefinery. This can be facilitated by using biomass production as a functional unit. It can then be integrated with the LCA of biomass processing in a biorefinery. LCA studies on microalgae-based products have primarily examined the credentials of biofuels/bioenergy, and there is a dearth of data on other value-added products. Table 12.1 provides an assessment of different stages and the alternative approaches at each of the stages in a microalgal biorefinery.

Batan et al. (2010) performed a comparative analysis of lifecycle impacts emanating from the production of petroleum diesel, soybean biodiesel and microalgal biodiesel using an industrial-scale engineering model for *Nannochloropsis* sp. grown in a PBR. The energy requirement for supporting the growth of microalgae was 2.1 times higher than the requirement for the growth stage of soybean plantations, and despite an ambitious assumption for 50% lipid content (18% in soybean), the EROEI for microalgal biodiesel was merely 0.93. The corresponding values for soybean biodiesel and conventional diesel were estimated to be 1.64 and 0.19, respectively. The findings suggest that microalgal biodiesel offers significantly greater energy returns than conventional diesel, but further energy demand reductions are required for it to be sustainable. In terms of avoided emission of GHGs, microalgal biodiesel (75.3 g $CO_{2\text{-eq}}$/MJ) performed better than both soybean biodiesel (71.7 g $CO_{2\text{-eq}}$/MJ) and conventional diesel (17.3 g $CO_{2\text{-eq}}$/MJ).

The impact of biorefinery design choices on the overall appeal (EROEI, freshwater savings and other sustainability indicators), including ten specific cases for two locations (Texas and Hawaii), was analyzed (Beal et al., 2015). The findings of the study suggest that an EROEI value of 3.24–8.35 can be attained under optimistic conditions for growth, lipid content and energy use efficiency. An LCA study on biodiesel production from freshwater alga *Chlorella vulgaris* using a hybrid cultivation system (airlift tubular PBR and raceway pond in two stages for improved biomass and lipid accumulation) reported a 42% reduction in global warming potential and a corresponding reduction of 38% in fossil energy requirement (Adesanya et al., 2014). The most significant reduction in global warming potential (76%) and fossil energy use (75%) was observed under mixotrophic conditions. The findings of similar studies on EROEI and GHG balance are presented in Tables 12.2 and 12.3, respectively.

Table 12.1 Comparative assessment of different stages and the alternative approaches at each of the stages in a microalgal biorefinery

Process/ parameter	*Alternative approaches*	*Advantages*	*Limitations*
Microalgae	Single species	• Easy control of growth • More targeted production of selected products	• Limitations of the chosen species/strain
	Microalgae consortia	• Simultaneous production of species-specific products • One species can compensate for the limitations of the other	• Biomass separation for targeted products can be tricky • Interaction is often poorly characterized
Growth mode	Autotrophic	• Simple and easily controllable	• Limited by photosynthetic efficiency of microalgae in photosynthesis
	Heterotrophic	• Independent of sunlight • High lipid accumulation • Biomass/lipid loss due to respiration in the night can be compensated	• The search for cost-effective growth substrates is ongoing • High chance of contamination
	Mixotrophic	• Combines the advantages of autotrophic and heterotrophic modes	• Complicated control • Combines the limitations of autotrophic and heterotrophic modes
Cultivation system	Open pond	• Cost-effective • Low energy demand • Easy maintenance	• Poor quality control • High chances of contamination • Low productivity • High space requirement • High water footprint
	Photobioreactor	• High productivity • Better quality control • Low chance of contamination • Low space requirement • Low water footprint	• High cost and energy demand • Difficult maintenance

(*Continued*)

Process/ parameter	*Alternative approaches*	*Advantages*	*Limitations*
Cultivation medium	Freshwater	• Production of various freshwater species has been optimized	• Highest water footprint
	Saline water	• Low water footprint • Low nutrient requirement • Abundant resource	• Evaporative loss has to be compensated by freshwater
	Wastewater	• Often contains most of the nutrients • Low water footprint • Simultaneous water treatment • Avoided eutrophication/contamination of receiving water body	• May contain toxic chemicals and elements • Detailed water characterization is required • The efficiency of removal for different pollutants varies
Nutrient requirement	Synthetic fertilizer	• Fixed concentration and strength of nutrients • Easy to use and control	• Synthesis is an energy-intensive process • A significant contribution to the release of GHGs
	Wastewater	• Often contains all/most of the required nutrients for microalgae • Low water footprint • Cost-effective • Significant energy and cost savings	• May contain toxic chemicals and elements • Detailed water characterization is required • The efficiency of removal for different pollutants varies
Light requirement	Solar radiation	• Freely available • Eco-friendly	• Spatiotemporal variability • Low conversion efficiency
	Artificial PAR	• High conversion efficiency • Can be switched on/off as per the requirement	• Adds to overall energy usage and high efficiency comes at a high cost
Carbon requirement	Flue gas	• Concentrated source of CO_2 • Prospects of carbon credit and trading • Can contribute toward carbon neutrality	• Often a very hot mixture of gases may contain other undesirable/toxic gases • Transportation can be challenging
	Dissolved carbon	• Can contribute to infrastructure-related savings • Concentrated source	• Can add to the overall cost • Can significantly alter the chemistry of water
	Atmospheric CO_2	• Freely available	• Low concentration may prove to be a limiting factor in photosynthesis

Biomass harvesting	Centrifugation	• Efficient process • Easily controllable	• Very high energy demand
	Self-flocculating species	• Avoided use of chemicals/energy • High residual moisture	• Often inefficient • Few species with promising prospects • Biomass slurry contains high moisture
	Chemical flocculants	• Established process for water treatment • Few natural flocculants have also exhibited promising results	• Can affect the chemistry of water • Recovery and reuse of chemicals is rather challenging • Added cost • Many flocculants only work in the presence of alkali
	Floatation	• Avoided use of chemicals	• Limited efficiency
	Filtration	• High efficiency	–
Drying and dewatering	Freeze-drying	• Effective process	–
	Solar radiation	• Freely available	• Spatiotemporal variability • Low efficiency • May require large areas of land
Extraction of lipid or metabolites	Solvent extraction	• Can easily be tuned for different metabolites	• Involves hazardous solvents • Solvent recovery is energy-intensive
	Ultrasonication	• Can be easily coupled with other extraction techniques • Ensures decomposition and degradation of temperature-sensitive metabolites • Operates faster	• Particle size is often a critical factor
	Mechanical extraction	• Straightforward • Avoids use of chemicals/solvents	• Often less efficient than solvent/mechanical extraction

(*Continued*)

Process/ parameter	*Alternative approaches*	*Advantages*	*Limitations*
Biomass processing to bioenergy	Fermentation	• Established process	• The low energy density of the fuel
	Renewable diesel	• Drop-in-fuel • Superior fuel properties	• Techno-economic viability is uncertain
	Pyrolysis	• Simultaneous production of bio-oil, biochar and gases	• Bio-oil may need to be upgraded • The product stream is a complex mixture
	Gasification	• Produces syngas and biochar	• Requires product purification
	Combined heat and power	• Reduces grid dependence and non-renewable energy usage	• –
	Transesterification	• Established and efficient process • Produces glycerol as a by-product	• Primarily involves non-renewable and hazardous solvent (methanol) • Conventional catalysts (homogeneous acid/alkali) cannot be reused • Water wash generates a large quantity of effluents
	Hydrothermal liquefaction	• Processes wet biomass	• High temperature and pressure
Residual biomass valorization	Manure	• Readily available nutrients	• Nutrient recycling
	Animal feed	• Compares favorably with other alternatives such as soybean	–
	Antioxidants	–	–
	Pigments	–	–
	Nutraceuticals	• High demand and commercial value	–

Table 12.2 Greenhouse gas balance of bioenergy/bioenergy carriers derived from microalgal biomass

Biomass/bioenergy	*EROEI*	*Process parameters and assumptions*	*References*
Microalgal biodiesel	0.93	• *Nannochloropsis* spp. grown in a PBR • Well-to-pump modeling of impacts using GREET	Batan et al. (2010)
Bio-crude, bioethanol and animal feed	8.35	• *Desmodesmus* sp. • Seawater from a saline aquifer within the facility premises • Lipid content of 38% • Biomass productivity of 23 $g/m^2/day$	Beal et al. (2015)
Microalgal biodiesel (lab-scale data on production of 1 kg biodiesel were extrapolated)	0.975	• *Chlorella vulgaris* grown in open raceways • Biomass harvested by flocculation and dried using belt drier • Solvent extraction of lipids • Transesterification to biodiesel and combustion in a diesel engine	Lardon et al. (2009)
	3.545	• *Chlorella vulgaris* grown in open raceways • Biomass harvested by flocculation • Lipid extraction from wet biomass slurry • Transesterification to biodiesel and combustion in a diesel engine	
	1.274	• *Chlorella vulgaris* grown in open raceways under nitrogen starvation • Biomass harvested by flocculation and dried using belt drier • Solvent extraction of lipids • Transesterification to biodiesel and combustion in a diesel engine	
	4.343	• *Chlorella vulgaris* grown in open raceways under nitrogen starvation • Biomass harvested by flocculation • Lipid extraction from wet biomass slurry • Transesterification to biodiesel and combustion in a diesel engine	

(*Continued*)

Biomass/bioenergy	*EROEI*	*Process parameters and assumptions*	*References*
Microalgal oil	3.05	• Open raceway pond • 29.6% lipid content	Jorquera et al. (2010)
	1.65	• Flat-plate photobioreactor (PBR)	
	0.07	• Tubular PBR	
Microalgal biomass	8.34	• Cultivation in open raceway pond • Lipid content of 29.6% • *Nannochloropsis* spp.	
	4.51	• Flat-plate PBR • *Nannochloropsis* spp.	
	0.20	• Tubular PBR • *Nannochloropsis* spp.	
Microalgal bio-oil	6.67	• Hydrothermal liquefaction of *Botryococcus braunii* at 300°C • Higher heating value of bio-crude: 45.9 MJ/kg	Dote et al. (1994)
	2.94	• Hydrothermal liquefaction of *Dunaliella tertiolecta* • Higher heating value of bio-crude: 34.9 MJ/kg	Minowa et al. (1995)
Microalgal biomass	2.56	• *Nannochloropsis* spp. grown in an open pond • Total energy input of 450 GJ/year	Jorquera et al. (2010)
	>1	• *Nannochloropsis* spp. grown in a PBR • Total energy input of 729 GJ/year	
Microalgal oil	7.01	• *Nannochloropsis* spp. grown in an open pond	
	>1	• *Nannochloropsis* spp. grown in a PBR	

Table 12.3 Greenhouse gas balance of bioenergy/bioenergy carriers derived from microalgal biomass

Energy carrier	*GHG balance ($g\ CO_{2\text{-}eq}/MJ$)*	*Process parameters*	*References*
Electricity	0.061	• Flue gas CO_2 from the coal-fired thermal power plant was used as a source of carbon • Algae and biomass were used to produce electricity (WTW)	Kadam (2002)
Electricity	48.7	• Direct combustion of biomass for bioelectricity • Production (WTW)	Clarens et al. (2010)
Biodiesel	59.9	• Open-pond cultivation under nitrogen starvation • Dry extraction (WTW)	Lardon et al. (2009)
Biodiesel	18.5	• Cultivation in PBR maintained under a greenhouse • Waste heat used for heating requirements (WTP)	Baliga and Powers (2010)
Biodiesel	75.3	• Cultivation in PBR • GREET modeling (WTP) without combustion	Batan et al. (2010)
	1.31	• Cultivation in PBR • GREET modeling (WTW)	
Biomass	56.8	• Cultivation in open raceway pond and synthetic fertilizer (WTG)	Clarens et al. (2010)
Biodiesel	−18.0	• Two-stage cultivation • GREET modeling (WTP)	Sander and Murthy (2010)
Biodiesel	13.6	• Anaerobic digestion of lipid-extracted biomass • Digestates used as a nutrient source (WTW)	Stephenson et al. (2010)
Biodiesel	−0.729	• Cultivation in open ponds • CO_2 from the fertilizer industry as a carbon source (WTW)	Campbell et al. (2011)
Biomethane	61.02	• Cultivation in open raceways • Anaerobic digestion of biomass • Digestate used as a nutrient source (WTW)	Collet et al. (2011)
Biodiesel	15.0	• Cultivation in open raceways • Seawater (WTW)	Hou et al. (2011)
Biodiesel	310	• Cultivation in two phases: initially in PBRs and then in open raceway (WTP)	Khoo et al. (2011)
Biodiesel	534	• Cultivation in open raceways • Solvent extraction of lipid from dried biomass • Methanolic transesterification • Oilcake as waste	Brentner et al. (2011)
	80.5	• Cultivation in PBR • In situ extraction and esterification by supercritical methanol • Anaerobic digestion of oilcake • Digestate used as a nutrient source (WTP)	

Table 12.4 List of a few companies engaged in the business of microalgal biorefinery

Company	*Production capacity*	*Products*	*Process*	*Revenue*
Algae.Tec	100,000 tons of biomass/year	• Biofuel • Animal feed • Aquaculture feed	PBR	$93,000
Algatech	2–3 million tons/year	• Astaxanthin • Fish feed • Fucoxanthin	PBR	$7.7 million
Algenol Biotech LLC	8,000 gallons of biofuel/acre/year	• Bioethanol • Biodiesel • Gasoline • Jet fuel	PBR	$3.1 million
Cellana Inc.		• Pigments • Omega-3-rich oils • Biofuels • Animal feed	PBR with open pond	$4.8 million
Seambiotic Ltd.	100–200 gallons of biofuels	• Bioethanol	Open pond	$10 million
Solix Biofuels	3,000 gallons of biofuels/year	• Jet fuel • Biodiesel • Green diesel • Omega-3 fatty acids • β-Carotene • Astaxanthin • Animal feed	PBR	$4 million
Synthetic Genomics Inc.	1,600 gallons/acre/year	• Biodiesel • Ethanol • Jet fuel	PBR	$16.8 million
TerraVia Inc.	100,000–300,000 metric tons of biomass/year	• Biofuel • Algal oil rich in omega-3 fatty acids	Dark fermenters, heterotrophic algae	$12.1 million

Note: Adapted from Banu et al. (2020).

An overwhelming majority of studies have identified the biomass production, harvesting and drying stages to be the major hotspots as they typically entail high energy demand. It holds for all the biorefinery designs irrespective of the preferred choice of products and coproducts. Considering the growing demand for organic products and their commercial appeal, many companies have entered into the business of microalgal products. Table 12.4 lists a few of the major companies involved in the business of algal products and microalgal biorefineries. In addition to those listed in Table 12.4, companies such as AFS BioOil Co. and AFS Biofarm™ (San Francisco, USA), Aeon Biogroup (Chile) and IGV Biotech (Nuthetal, Germany) are also engaged in the business of microalgal biorefineries (Jacob-Lopes et al., 2015).

Conclusions

Despite the significant commercial appeal, the realization of a biorefinery is hampered by several strategic decisions that affect the overall sustainability of the biomass supply chain. LCA studies have identified the biomass production, harvest and drying stages to be a significant contributor to total energy demand, and these hotspots deserve the most attention. Unprecedented GHG savings, energy return and appealing values for other sustainability indicators have been reported in a few optimistic cases, and improvement in process engineering, system configurations, energy use efficiency, better metabolic control, coproduction of high-value commodities, integration of waste streams, and genetic/metabolic engineering are all expected to play a significant role.

References

Abdel-Raouf, N., Al-Homaidan, A.A., and Ibraheem, I.B.M. (2012) Microalgae and wastewater treatment. *Saudi Journal of Biological Sciences* 19:257–275.

Adeniyi, O.M., Azimov, U., and Burluka, A. (2018) Algae biofuel: current status and future applications. *Renewable and Sustainable Energy Reviews* 90:316–335.

Adesanya, V.O., Cadena, E., Scott, S.A., and Smith, A.G. (2014) Life cycle assessment on microalgal biodiesel production using a hybrid cultivation system. *Bioresource Technology* 163:343–355.

Afriyanti, D., Kroeze, C., and Saad, A. (2016) Indonesia palm oil production without deforestation and peat conversion by 2050. *Science of the Total Environment* 557:562–570.

Baliga, R., and Powers, S.E. (2010) Sustainable algae biodiesel production in cold climates. *International Journal of Chemical Engineering* 2010:102179.

Banu, J.R., Kavitha, S., Gunasekaran, M., Kumar, G., and others (2020) Microalgae based biorefinery promoting circular bioeconomy-techno economic and life-cycle analysis. *Bioresource Technology* 302:122822.

Batan, L., Quinn, J., Willson, B., and Bradley, T. (2010) Net energy and greenhouse gas emission evaluation of biodiesel derived from microalgae. *Environmental Science & Technology* 44:7975–7980.

Beal, C.M., Gerber, L.N., Sills, D.L., Huntley, M.E., Machesky, S.C., Walsh, M.J., Tester, J.W., Archibald, I., Granados, J., and Greene, C.H. (2015) Algal biofuel production for fuels and feed in a 100-ha facility: A comprehensive techno-economic analysis and life cycle assessment. *Algal Research* 10:266–279.

Boutesteijn, C., Drabik, D., and Venus, T.J. (2017) The interaction between EU biofuel policy and first-and second-generation biodiesel production. *Industrial Crops and Products* 106:124–129.

Brentner, L.B., Eckelman, M.J., and Zimmerman, J.B. (2011) Combinatorial life cycle assessment to inform process design of industrial production of algal biodiesel. *Environmental Science & Technology* 45:7060–7067.

Brownbridge, G., Azadi, P., Smallbone, A., Bhave, A., Taylor, B., and Kraft, M. (2014) The future viability of algae-derived biodiesel under economic and technical uncertainties. *Bioresource Technology* 151:166–173.

Caldeira, C., Queirós, J., Noshadravan, A., and Freire, F. (2016) Incorporating uncertainty in the life cycle assessment of biodiesel from waste cooking oil addressing different collection systems. *Resources, Conservation and Recycling* 112:83–92.

Campbell, P.K., Beer, T., and Batten, D. (2011) Life cycle assessment of biodiesel production from microalgae in ponds. *Bioresource Technology* 102:50–56.

Carneiro, M.L.N.M., Pradelle, F., Braga, S.L., Gomes, M.S.P., Martins, A.R.F.A., Turkovics, F., and Pradelle, R.N.C. (2017) Potential of biofuels from algae: Comparison with fossil fuels, ethanol and biodiesel in Europe and Brazil through life cycle assessment (LCA). *Renewable and Sustainable Energy Reviews* 73:632–653.

Chandra, R., Iqbal, H.M.N., Vishal, G., Lee, H.-S., and Nagra, S. (2019) Algal biorefinery: A sustainable approach to valorize algal-based biomass towards multiple product recovery. *Bioresource Technology* 278:346–359.

Cheah, W.Y., Show, P.L., Chang, J.-S., Ling, T.C., and Juan, J.C. (2015) Biosequestration of atmospheric CO_2 and flue gas-containing CO_2 by microalgae. *Bioresource Technology* 184:190–201.

Cherubini, F., and Strømman, A.H. (2011) Chemicals from lignocellulosic biomass: Opportunities, perspectives, and potential of biorefinery systems. *Biofuels, Bioproducts and Biorefining* 5:548–561.

Chew, K.W., Yap, J.Y., Show, P.L., Suan, N.H., Juan, J.C., Ling, T.C., Lee, D.-J., and Chang, J.-S. (2017) Microalgae biorefinery: High value products perspectives. *Bioresource Technology* 229:53–62.

Chisti, Y. (2007) Biodiesel from microalgae. *Biotechnology Advances* 25:294–306.

Christenson, L., and Sims, R. (2011) Production and harvesting of microalgae for wastewater treatment, biofuels, and bioproducts. *Biotechnology Advances* 29:686–702.

Clarens, A.F., Resurreccion, E.P., White, M.A., and Colosi, L.M. (2010) Environmental life cycle comparison of algae to other bioenergy feedstocks. *Environmental Science & Technology* 44:1813–1819.

Collet, P., Hélias, A., Lardon, L., Ras, M., Goy, R.-A., and Steyer, J.-P. (2011) Life-cycle assessment of microalgae culture coupled to biogas production. *Bioresource Technology* 102:207–214.

Collet, P., Hélias, A., Lardon, L., Steyer, J.-P., and Bernard, O. (2015) Recommendations for Life Cycle Assessment of algal fuels. *Applied Energy* 154:1089–1102.

D'Amato, D., Droste, N., Allen, B., Kettunen, M., Lähtinen, K., Korhonen, J., Leskinen, P., Matthies, B.D., and Toppinen, A. (2017) Green, circular, bio economy: A comparative analysis of sustainability avenues. *Journal of Cleaner Production* 168:716–734.

Dote, Y., Sawayama, S., Inoue, S., Minowa, T., and Yokoyama, S. (1994) Recovery of liquid fuel from hydrocarbon-rich microalgae by thermochemical liquefaction. *Fuel* 73:1855–1857.

Gifuni, I., Pollio, A., Safi, C., Marzocchella, A., and Olivieri, G. (2019) Current bottlenecks and challenges of the microalgal biorefinery. *Trends in Biotechnology* 37:242–252.

Hall, C.A.S., Lambert, J.G., and Balogh, S.B. (2014) EROI of different fuels and the implications for society. *Energy Policy* 64:141–152.

Ho, S.-H., Chen, C.-Y., and Chang, J.-S. (2012) Effect of light intensity and nitrogen starvation on CO_2 fixation and lipid/carbohydrate production of an indigenous microalga *Scenedesmus obliquus* CNW-N. *Bioresource Technology* 113:244–252.

Hou, J., Zhang, P., Yuan, X., and Zheng, Y. (2011) Life cycle assessment of biodiesel from soybean, jatropha and microalgae in China conditions. *Renewable and Sustainable Energy Reviews* 15:5081–5091.

Jacob-Lopes, E., Mérida, L.G.R., Queiroz, M.I., and Zepka, L.Q. (2015) Microalgal biorefineries. In: Jacob-Lopes, E., and Zepka, L.Q. (eds.) *Biomass Production and Uses*. Rijeka: InTech, pp. 81–106.

Javed, F., Aslam, M., Rashid, N., Shamair, Z., Khan, A.L., Yasin, M., Fazal, T., Hafeez, A., Rehman, F., Rehman, M.S.U., and others (2019) Microalgae-based biofuels, resource recovery and wastewater treatment: A pathway towards sustainable biorefinery. *Fuel* 255:115826.

Jingura, R.M., and Kamusoko, R. (2016) Evaluation of life-cycle assessment of Jatropha biodiesel. *Energy Sources, Part B: Economics, Planning, and Policy* 11:396–403.

Jorquera, O., Kiperstok, A., Sales, E.A., Embiruçu, M., and Ghirardi, M.L. (2010) Comparative energy life-cycle analyses of microalgal biomass production in open ponds and photobioreactors. *Bioresource Technology* 101:1406–1413.

Kadam, K.L. (2002) Environmental implications of power generation via coal-microalgae cofiring. *Energy* 27:905–922.

Khoo, H.H., Sharratt, P.N., Das, P., Balasubramanian, R.K., Naraharisetti, P.K., and Shaik, S. (2011) Life cycle energy and CO_2 analysis of microalgae-to-biodiesel: Preliminary results and comparisons. *Bioresource Technology* 102:5800–5807.

Knothe, G. (2010) Biodiesel and renewable diesel: A comparison. *Progress in Energy and Combustion Science* 36:364–373.

Lardon, L., Hélias, A., Sialve, B., Steyer, J.-P., and Bernard, O. (2009) Life-cycle assessment of biodiesel production from microalgae. *Environmental Science & Technology* 43:6475–6481.

Minowa, T., Yokoyama, S., Kishimoto, M., and Okakura, T. (1995) Oil production from algal cells of *Dunaliella tertiolecta* by direct thermochemical liquefaction. *Fuel* 74:1735–1738.

Moreno-Garcia, L., Adjallé, K., Barnabé, S., and Raghavan, G.S.V. (2017) Microalgae biomass production for a biorefinery system: Recent advances and the way towards sustainability. *Renewable and Sustainable Energy Reviews* 76:493–506.

Narala, R.R., Garg, S., Sharma, K.K., Thomas-Hall, S.R., Deme, M., Li, Y., and Schenk, P.M. (2016) Comparison of microalgae cultivation in photobioreactor, open raceway pond, and a two-stage hybrid system. *Frontiers in Energy Research* 4:29.

Peralta-Ruiz, Y., González-Delgado, A.-D., and Kafarov, V. (2013) Evaluation of alternatives for microalgae oil extraction based on exergy analysis. *Applied Energy* 101:226–236.

Perez-Garcia, O., and Bashan, Y. (2015) Microalgal heterotrophic and mixotrophic culturing for bio-refining: From metabolic routes to techno-economics. In: *Algal Biorefineries*. Cham: Springer, pp. 61–131.

Pérez-López, P., Montazeri, M., Feijoo, G., Moreira, M.T., and Eckelman, M.J. (2018) Integrating uncertainties to the combined environmental and economic assessment of algal biorefineries: A Monte Carlo approach. *Science of the Total Environment* 626:762–775.

Rizwan, M., Lee, J.H., and Gani, R. (2015) Optimal design of microalgae-based biorefinery: Economics, opportunities and challenges. *Applied Energy* 150:69–79.

Rodionova, M.V., Poudyal, R.S., Tiwari, I., Voloshin, R.A., Zharmukhamedov, S.K., Nam, H.G., Zayadan, B.K., Bruce, B.D., Hou, H.J.M., and Allakhverdiev, S.I. (2017) Biofuel production: Challenges and opportunities. *International Journal of Hydrogen Energy* 42:8450–8461.

Sadhukhan, J., Martinez-Hernandez, E., Murphy, R.J., Ng, D.K.S., Hassim, M.H., Ng, K.S., Kin, W.Y., Jaye, I.F.M., Hang, M.Y.L.P., and Andiappan, V. (2018) Role of bioenergy, biorefinery and bioeconomy in sustainable development: Strategic pathways for Malaysia. *Renewable and Sustainable Energy Reviews* 81:1966–1987.

Sander, K., and Murthy, G.S. (2010) Life cycle analysis of algae biodiesel. *The International Journal of Life Cycle Assessment* 15:704–714.

Santomauro, G., Baier, J., Huang, W., Pezold, S., Bill, J., and others (2012) Formation of calcium carbonate polymorphs induced by living microalgae. *Journal of Biomaterials and Nanobiotechnology* 3:413.

Schulze, P.S.C., Barreira, L.A., Pereira, H.G.C., Perales, J.A., and Varela, J.C.S. (2014) Light emitting diodes (LEDs) applied to microalgal production. *Trends in Biotechnology* 32:422–430.

Seghetta, M., Hou, X., Bastianoni, S., Bjerre, A.-B., and Thomsen, M. (2016) Life cycle assessment of macroalgal biorefinery for the production of ethanol, proteins and fertilizers—A step towards a regenerative bioeconomy. *Journal of Cleaner Production* 137:1158–1169.

Sills, D.L., Paramita, V., Franke, M.J., Johnson, M.C., Akabas, T.M., Greene, C.H., and Tester, J.W. (2013) Quantitative uncertainty analysis of life cycle assessment for algal biofuel production. *Environmental Science & Technology* 47:687–694.

Sills, D.L., Van Doren, L.G., Beal, C., and Raynor, E. (2020) The effect of functional unit and co-product handling methods on life cycle assessment of an algal biorefinery. *Algal Research* 46:101770.

Singh, A., and Olsen, S.I. (2011) A critical review of biochemical conversion, sustainability and life cycle assessment of algal biofuels. *Applied Energy* 88:3548–3555.

Soh, L., Montazeri, M., Haznedaroglu, B.Z., Kelly, C., Peccia, J., Eckelman, M.J., and Zimmerman, J.B. (2014) Evaluating microalgal integrated biorefinery schemes: Empirical controlled growth studies and life cycle assessment. *Bioresource Technology* 151:19–27.

Stephenson, A.L., Kazamia, E., Dennis, J.S., Howe, C.J., Scott, S.A., and Smith, A.G. (2010) Life-cycle assessment of potential algal biodiesel production in the United Kingdom: A comparison of raceways and air-lift tubular bioreactors. *Energy & Fuels* 24:4062–4077.

Szulczyk, K.R., and Khan, M.A.R. (2018) The potential and environmental ramifications of palm biodiesel: Evidence from Malaysia. *Journal of Cleaner Production* 203:260–272.

Trivedi, J., Aila, M., Bangwal, D.P., Kaul, S., and Garg, M.O. (2015) Algae based biorefinery—how to make sense? *Renewable and Sustainable Energy Reviews* 47:295–307.

Ubando, A.T., Felix, C.B., and Chen, W.-H. (2020) Biorefineries in circular bioeconomy: A comprehensive review. *Bioresource Technology* 299:122585.

Xu, L., Brilman, D.W.F.W., Withag, J.A.M., Brem, G., and Kersten, S. (2011) Assessment of a dry and a wet route for the production of biofuels from microalgae: Energy balance analysis. *Bioresource Technology* 102:5113–5122.

Yuan, J., Kendall, A., and Zhang, Y. (2015) Mass balance and life cycle assessment of biodiesel from microalgae incorporated with nutrient recycling options and technology uncertainties. *GCB Bioenergy* 7:1245–1259.

Zaimes, G.G., and Khanna, V. (2013) Environmental sustainability of emerging algal biofuels: A comparative life cycle evaluation of algal biodiesel and renewable diesel. *Environmental Progress & Sustainable Energy* 32:926–936.

Index

www.ingramcontent.com/pod-product-compliance
Lightning Source LLC
LaVergne TN
LVHW021118300926
849160LV00010B/770
* 9 7 8 0 3 6 7 5 7 0 0 6 4 *